Sixth Workshop on Computational Linguistics and Clinical Psychology: From Keyboard to Clinic (CLPsych 2019)

Minneapolis, Minnesota, USA
6 June 2019

ISBN: 978-1-5108-8784-8

NAACL HLT 2019

Computational Linguistics and Clinical Psychology: From Keyboard to Clinic

Proceedings of the Sixth Workshop

June 6, 2019
Minneapolis, MN

Gold Sponsor

CHIB | **Center for Health-related Informatics and Bioimaging**
at the University of Maryland

Silver Sponsor

Introduction

Mental health continues to be one of the most significant global health problems we face, affecting approximately 450 million people worldwide (World Health Organization, 2017). Mental disorders have a significant detrimental effect on quality of life, accounting for 13% of disability-adjusted life years and 32.4% of years lived with disability globally (Vigo, Thornicroft, & Atun, 2016). Additionally, mental illness can have substantial economic consequences. Mental disorders cost US$2.5 trillion globally, and economic output loss due to mental disorders is anticipated to be US$16.3 trillion worldwide between 2011 and 2030 (Trautmann, Rehm, & Wittchen, 2016). Effective treatments exist for mental illness, however many of those affected do not have access.

According to Professor Shekhar Saxena of Harvard T.H. Chan School of Public Health, no countries are developed when it comes to mental health (Davies, 2018). Approximately 35-50% of those affected by mental disorders do not receive treatment in high-income countries (Saxena, Thornicroft, Knapp, & Whiteford, 2007). Worse still, in middle and low-income countries, 76-85% of affected individuals do not receive treatment (Saxena, Thornicroft, Knapp, & Whiteford, 2007). Key barriers to accessing effective treatment include a shortage in supply of trained mental health workers relative to demand for services, and low funding for treatment and prevention (Rathod et al., 2017). One way to increase the supply of mental healthcare is through technology.

Language technology may be particularly well-suited to improve supply of mental health services. Conversations are a fundamental part of the diagnostic and therapeutic process for mental health. This is because language provides crucial insights into a patient's symptoms, thoughts, feelings, and functioning (Pennebaker, Mehl, & Niederhoffer, 2003). Given the advent of the internet and personal electronic devices, linguistic data is readily available, and can be found in and outside of treatment contexts in text and oral form. Applying language technology to mental healthcare can open the door to creating scalable, inexpensive screening measures or risk assessments that may be administered by a wider variety of healthcare professionals in a broad range of contexts. Additionally, conversational agents can assist with the provision of therapy exercises or emotional support beyond treatment settings (Fitzpatrick, Darcy, & Vierhile, 2017). Public social media posts have been used to infer a community's mental health following crisis events (Kumar, Dredze, Coppersmith, & De Choudhury, 2015), and triage tools have been used to present messages to online support workers by order of crisis severity (Milne, Pink, Hachey, & Calvo, 2016). Evidently, language technology shows incredible promise for increasing the supply of quality mental health support services, and further research and development efforts are needed. While at the same time, remaining cognizant of ethical issues that may arise in the process (Benton et al., 2017; Chancellor et al., 2019).

The Computational Linguistics and Clinical Psychology (CLPsych) workshop series aims to support and accelerate the development of language technology for mental healthcare. CLPsych brings together computational linguists and mental health clinicians to discuss and develop tools and data that can support clinicians, service organizations, and/or individuals with lived experience of mental disorders. Given its multidisciplinary community, CLPsych values clear communication of relevant computational methods and results, and all presentations are followed by clinical commentary.

CLPsych has been held annually at the meeting of the Association of Computational Linguistics (ACL) or the North American Association of Computational Linguistics (NAACL) since 2014. During this time, CLPsych has helped to define the start of the art in language technology for mental health, introduced a clinically-oriented workshop structure to the ACL community, and established a shared task tradition in which participants work on common datasets and tasks to develop systems or techniques that aid in the detection of mental disorders. Prior shared tasks have involved working with data from ReachOut.com and the UK Data Service.

The Sixth Workshop on Computational Linguistics and Clinical Psychology (CLPsych 2019) was held at the North American Association for Computational Linguistics and Human Language Technology's (NAACL-HLT) annual meeting in Minneapolis, MN on June 6th. The focus of 2019's workshop was reconciling outcomes, with the goal of fostering discussions on the outcomes that are most important to pursue as a community. Continuing CLPsych's traditional interdisciplinary approach, practicing clinicians and clinical researchers were included as part of our program committee, and were invited to submit papers and serve as discussants of presented work.

The workshop also included a keynote talk by Becky Inkster, a UK neuroscientist active in digital innovation for mental health, as well as a technologist/clinician panel discussion including Nick Allen (University of Oregon), Glen Coppersmith (Qntfy), Nazli Goharian (Georgetown University), and Michelle Kuchuk (National Suicide Prevention Lifeline).

2019's workshop had two submission formats: full papers and position papers. Overall, 17 submissions were received. Accepted submissions included 11 full papers and 2 position papers, which were presented as 6 talks and 7 posters.

A shared task was held that focused on predicting individuals' suicide risk from de-identified, public Reddit data. Teams could participate in three tasks. Task A involved predicting level of risk for users posting to the r/SuicideWatch subreddit based on their SuicideWatch posts. Task B involved the same risk assessment, but with additional access to all the users' posts elsewhere on Reddit. Task C involved a screening/monitoring scenario in which user risk was assessed based only on their Reddit posts excluding SuicideWatch or other mental health forums. A total of 83 entries were provided by 15 teams who participated in at least one task each. Accepted shared task paper submissions were presented as an additional 9 posters and 2 full talks at the workshop. 2019's shared task was organized by Ayah Zirikly, Philip Resnik, Özlem Uzuner, and Kristy Hollingshead.

The organizers wish to thank all who contributed to the success of CLPsych 2019. This includes authors and shared task participants for their insightful contributions, Program Committee members for their thoughtful reviews, our keynote speaker, panelists, and clinical discussants for their valuable insights, and shared task organizers for putting together a series of challenging exercises with important applications. The organizers also wish to thank the generous workshop sponsors, Amazon and the University of Maryland Center for Health-Related Informatics and Bioimaging (CHIB), as well as the North American chapter of the Association for Computational Linguistics, for making this workshop possible.

Kate Niederhoffer, Kristy Hollingshead, Philip Resnik, Rebecca Resnik, & Kate Loveys

Organizing Committee

    Kate Niederhoffer, 7 Cups
    Kristy Hollingshead, IHMC
    Philip Resnik, University of Maryland
    Rebecca Resnik, Rebecca Resnik and Associates, LLC
    Kate Loveys, University of Auckland School of Medicine

Shared Task Organizers:

    Ayah Zirikly, NIH
    Philip Resnik, University of Maryland
    Özlem Uzuner, George Mason University
    Kristy Hollingshead, IHMC

Keynote Speaker:

    Becky Inkster, University of Cambridge

Panelists:

    Nick Allen, University of Oregon
    Glen Coppersmith, Qntfy
    Nazli Goharian, Georgetown University
    Michelle Kuchuk, National Suicide Prevention Lifeline

Program Committee:

    Nazli Goharian, Georgetown University
    H. Andrew Schwartz, Stony Brook University
    Glen Coppersmith, Qntfy
    Ayah Zirikly, NIH
    Cindy Chung, 7 Cups
    Adam Miner, Stanford University
    Frank Rudzicz, University of Toronto
    Nick Allen, University of Oregon
    Steven Bedrick, Oregon Health & Science University
    Mike Conway, University of Utah
    Danielle Mowery, University of Pennsylvania
    Eric Morley, Google
    Sean Murphy, Small Data Consulting, LLC
    Daniel Preotiuc-Pietro, Bloomberg
    Mark Rosenstein, Pearson Knowledge Technologies
    Masoud Rouhizadeh, Johns Hopkins University
    Maarten Sap, University of Washington
    J. Ignacio Serrano, Spanish National Research Council (CSIC)
    Matthew Purver, Queen Mary University of London
    Shervin Malmasi, Harvard Medical School
    Molly Ireland, Texas Tech University
    Craig Pfeifer, Mitre

Dimitrios Kokkinakis, University of Gothenburg
Loring Ingraham, George Washington University
Archna Bhatia, IHMC
Hiroki Tanaka, Naist
Kathleen C. Fraser, National Research Council Canada
Craig Harman, Johns Hopkins University
Tong Liu, RIT
William Jarrold, Nuance
Paul Thompson, Dartmouth
Christopher Homan, RIT
Micah Iserman, Texas Tech University
Jill Dolata, Oregon Health & Science University
Patrick Crutchley, Qntfy
Jonathan Schler, Bar Ilan University
Raymond Tucker, Louisiana State University
Joseph Costello, Western Michigan University School of Medicine
James Sexton, George Washington University

Table of Contents

Towards augmenting crisis counselor training by improving message retrieval
Orianna Demasi, Marti A. Hearst and Benjamin Recht 1

Identifying therapist conversational actions across diverse psychotherapeutic approaches
Fei-Tzin Lee, Derrick Hull, Jacob Levine, Bonnie Ray and Kathy McKeown 12

CLPsych 2019 Shared Task: Predicting the Degree of Suicide Risk in Reddit Posts
Ayah Zirikly, Philip Resnik, Ozlem Uzuner and Kristy Hollingshead 24

CLaC at CLPsych 2019: Fusion of Neural Features and Predicted Class Probabilities for Suicide Risk Assessment Based on Online Posts
Elham Mohammadi, Hessam Amini and Leila Kosseim 34

Suicide Risk Assessment with Multi-level Dual-Context Language and BERT
Matthew Matero, Akash Idnani, Youngseo Son, Sal Giorgi, Huy Vu, Mohammad Zamani, Parth Limbachiya, Sharath Chandra Guntuku and H. Andrew Schwartz 39

Using natural conversations to classify autism with limited data: Age matters
Michael Hauser, Evangelos Sariyanidi, Birkan Tunc, Casey Zampella, Edward Brodkin, Robert Schultz and Julia Parish-Morris 45

The importance of sharing patient-generated clinical speech and language data
Kathleen C. Fraser, Nicklas Linz, Hali Lindsay and Alexandra Konig 55

Depressed Individuals Use Negative Self-Focused Language When Recalling Recent Interactions with Close Romantic Partners but Not Family or Friends
Taleen Nalabandian and Molly Ireland 62

Linguistic Analysis of Schizophrenia in Reddit Posts
Jonathan Zomick, Sarah Ita Levitan and Mark Serper 74

Semantic Characteristics of Schizophrenic Speech
Kfir Bar, Vered Zilberstein, Ido Ziv, Heli Baram, Nachum Dershowitz, Samuel Itzikowitz and Eiran Vadim Harel 84

Computational Linguistics for Enhancing Scientific Reproducibility and Reducing Healthcare Inequities
Julia Parish-Morris 94

Temporal Analysis of the Semantic Verbal Fluency Task in Persons with Subjective and Mild Cognitive Impairment
Nicklas Linz, Kristina Lundholm Fors, Hali Lindsay, Marie Eckerström, Jan Alexandersson and Dimitrios Kokkinakis 103

Mental Health Surveillance over Social Media with Digital Cohorts
Silvio Amir, Mark Dredze and John W. Ayers 114

Reviving a psychometric measure: Classification and prediction of the Operant Motive Test
Dirk Johannßen, Chris Biemann and David Scheffer 121

Coherence models in schizophrenia
Sandra Just, Erik Haegert, Nora Kořánová, Anna-Lena Bröcker, Ivan Nenchev, Jakob Funcke, Christiane Montag and Manfred Stede 126

Overcoming the bottleneck in traditional assessments of verbal memory: Modeling human ratings and classifying clinical group membership
Chelsea Chandler, Peter W. Foltz, Jian Cheng, Jared C. Bernstein, Elizabeth P. Rosenfeld, Alex S. Cohen, Terje B. Holmlund and Brita Elvevag 137

Analyzing the use of existing systems for the CLPsych 2019 Shared Task
Alejandro González Hevia, Rebeca Cerezo Menéndez and Daniel Gayo-Avello 148

Similar Minds Post Alike: Assessment of Suicide Risk Using a Hybrid Model
Lushi Chen, Abeer Aldayel, Nikolay Bogoychev and Tao Gong 152

Predicting Suicide Risk from Online Postings in Reddit The UGent-IDLab submission to the CLPysch 2019 Shared Task A
Semere Kiros Bitew, Giannis Bekoulis, Johannes Deleu, Lucas Sterckx, Klim Zaporojets, Thomas Demeester and Chris Develder 158

CLPsych2019 Shared Task: Predicting Suicide Risk Level from Reddit Posts on Multiple Forums
Victor Ruiz, Lingyun Shi, Wei Quan, Neal Ryan, Candice Biernesser, David Brent and Rich Tsui 162

Suicide Risk Assessment on Social Media: USI-UPF at the CLPsych 2019 Shared Task
Esteban Rissola, Diana Ramírez-Cifuentes, Ana Freire and Fabio Crestani 167

Using Contextual Representations for Suicide Risk Assessment from Internet Forums
Ashwin Karthik Ambalavanan, Pranjali Dileep Jagtap, Soumya Adhya and Murthy Devarakonda 172

An Investigation of Deep Learning Systems for Suicide Risk Assessment
Michelle Morales, Prajjalita Dey, Thomas Theisen, Daniel Belitz and Natalia Chernova 177

ConvSent at CLPsych 2019 Task A: Using Post-level Sentiment Features for Suicide Risk Prediction on Reddit
Kristen Allen, Shrey Bagroy, Alex Davis and Tamar Krishnamurti 182

Dictionaries and Decision Trees for the 2019 CLPsych Shared Task
Micah Iserman, Taleen Nalabandian and Molly Ireland 188

Conference Program

Thursday June 6, 2019

9:00–9:15   *Opening Remarks*

9:15–10:30   *Workshop Session I: Presentations with Discussant Commentary*

Towards augmenting crisis counselor training by improving message retrieval
Orianna Demasi, Marti A. Hearst and Benjamin Recht

Identifying therapist conversational actions across diverse psychotherapeutic approaches
Fei-Tzin Lee, Derrick Hull, Jacob Levine, Bonnie Ray and Kathy McKeown

10:30–10:45   *Break*

10:45–11:45   *Keynote Speaker and Discussion: Becky Inkster*

11:45–12:45   *Workshop Session II: Shared Task Presentations with Discussant Commentary*

CLPsych 2019 Shared Task: Predicting the Degree of Suicide Risk in Reddit Posts
Ayah Zirikly, Philip Resnik, Ozlem Uzuner and Kristy Hollingshead

CLaC at CLPsych 2019: Fusion of Neural Features and Predicted Class Probabilities for Suicide Risk Assessment Based on Online Posts
Elham Mohammadi, Hessam Amini and Leila Kosseim

Suicide Risk Assessment with Multi-level Dual-Context Language and BERT
Matthew Matero, Akash Idnani, Youngseo Son, Sal Giorgi, Huy Vu, Mohammad Zamani, Parth Limbachiya, Sharath Chandra Guntuku and H. Andrew Schwartz

12:45–1:45   *Lunch and Poster Session*

1:45–2:45   *Workshop Session III: Presentations with Discussant Commentary*

Using natural conversations to classify autism with limited data: Age matters
Michael Hauser, Evangelos Sariyanidi, Birkan Tunc, Casey Zampella, Edward Brodkin, Robert Schultz and Julia Parish-Morris

The importance of sharing patient-generated clinical speech and language data
Kathleen C. Fraser, Nicklas Linz, Hali Lindsay and Alexandra Konig

2:45–3:45   ***Workshop Session IV: Presentations with Discussant Commentary***

Depressed Individuals Use Negative Self-Focused Language When Recalling Recent Interactions with Close Romantic Partners but Not Family or Friends
Taleen Nalabandian and Molly Ireland

Linguistic Analysis of Schizophrenia in Reddit Posts
Jonathan Zomick, Sarah Ita Levitan and Mark Serper

3:45–4:00   ***Break***

4:00–5:00   ***Panel***

5:00–6:00   ***Happy Hour and Posters***

Semantic Characteristics of Schizophrenic Speech
Kfir Bar, Vered Zilberstein, Ido Ziv, Heli Baram, Nachum Dershowitz, Samuel Itzikowitz and Eiran Vadim Harel

Computational Linguistics for Enhancing Scientific Reproducibility and Reducing Healthcare Inequities
Julia Parish-Morris

Temporal Analysis of the Semantic Verbal Fluency Task in Persons with Subjective and Mild Cognitive Impairment
Nicklas Linz, Kristina Lundholm Fors, Hali Lindsay, Marie Eckerström, Jan Alexandersson and Dimitrios Kokkinakis

Mental Health Surveillance over Social Media with Digital Cohorts
Silvio Amir, Mark Dredze and John W. Ayers

Thursday June 6, 2019 (continued)

Reviving a psychometric measure: Classification and prediction of the Operant Motive Test
Dirk Johannßen, Chris Biemann and David Scheffer

Coherence models in schizophrenia
Sandra Just, Erik Haegert, Nora Kořánová, Anna-Lena Bröcker, Ivan Nenchev, Jakob Funcke, Christiane Montag and Manfred Stede

Overcoming the bottleneck in traditional assessments of verbal memory: Modeling human ratings and classifying clinical group membership
Chelsea Chandler, Peter W. Foltz, Jian Cheng, Jared C. Bernstein, Elizabeth P. Rosenfeld, Alex S. Cohen, Terje B. Holmlund and Brita Elvevag

Analyzing the use of existing systems for the CLPsych 2019 Shared Task
Alejandro González Hevia, Rebeca Cerezo Menéndez and Daniel Gayo-Avello

Similar Minds Post Alike: Assessment of Suicide Risk Using a Hybrid Model
Lushi Chen, Abeer Aldayel, Nikolay Bogoychev and Tao Gong

Predicting Suicide Risk from Online Postings in Reddit The UGent-IDLab submission to the CLPysch 2019 Shared Task A
Semere Kiros Bitew, Giannis Bekoulis, Johannes Deleu, Lucas Sterckx, Klim Zaporojets, Thomas Demeester and Chris Develder

CLPsych2019 Shared Task: Predicting Suicide Risk Level from Reddit Posts on Multiple Forums
Victor Ruiz, Lingyun Shi, Wei Quan, Neal Ryan, Candice Biernesser, David Brent and Rich Tsui

Suicide Risk Assessment on Social Media: USI-UPF at the CLPsych 2019 Shared Task
Esteban Rissola, Diana Ramírez-Cifuentes, Ana Freire and Fabio Crestani

Using Contextual Representations for Suicide Risk Assessment from Internet Forums
Ashwin Karthik Ambalavanan, Pranjali Dileep Jagtap, Soumya Adhya and Murthy Devarakonda

An Investigation of Deep Learning Systems for Suicide Risk Assessment
Michelle Morales, Prajjalita Dey, Thomas Theisen, Daniel Belitz and Natalia Chernova

ConvSent at CLPsych 2019 Task A: Using Post-level Sentiment Features for Suicide Risk Prediction on Reddit
Kristen Allen, Shrey Bagroy, Alex Davis and Tamar Krishnamurti

Dictionaries and Decision Trees for the 2019 CLPsych Shared Task
Micah Iserman, Taleen Nalabandian and Molly Ireland

Towards Augmenting Crisis Counselor Training by Improving Message Retrieval

Orianna DeMasi
University of California
Berkeley

Marti A. Hearst
University of California
Berkeley

Benjamin Recht
University of California
Berkeley

Abstract

A fundamental challenge when training counselors is presenting novices with the opportunity to practice counseling distressed individuals without exacerbating a situation. Rather than replacing human empathy with an automated counselor, we propose simulating an individual in crisis so that human counselors in training can practice crisis counseling in a low-risk environment. Towards this end, we collect a dataset of suicide prevention counselor role-play transcripts and make initial steps towards constructing a CRISISbot for humans to counsel while in training. In this data-constrained setting, we evaluate the potential for message retrieval to construct a coherent chat agent in light of recent advances with text embedding methods. Our results show that embeddings can considerably improve retrieval approaches to make them competitive with generative models. By coherently retrieving messages, we can help counselors practice chatting in a low-risk environment.

1 Introduction

Suicide prevention hotlines can provide immediate care in critical times of need (Gould et al., 2012, 2013; Ramchand et al., 2016). These hotlines are expanding services to text to meet growing demands and adapt to shifts in communication trends (Smith and Page, 2015). Crisis helplines rely on counselors who are trained in a variety of skills, such as empathy, active listening, assessing risk of suicide, de-escalation, and connecting individuals to longer term solutions (Gould et al., 2013; Paukert et al., 2004).

Properly training counselors is critical yet difficult as, resource costs aside, counselors need to practice and develop expertise in realistic environments that are low-risk, i.e., they do not put distressed individuals in danger. Because novice counselors are unable to assume full responsibility for a crisis situation until they have some experience, training often includes human-to-human role-playing (American Association of Suicidology, 2012; Suicide Prevention Resource Center, 2007). Role-playing has been shown to improve crisis intervention training (Cross et al., 2011). However, such training takes a lot of human time, which centers struggle to provide.

Instead of attempting to scale services by replacing human counselors and trying to automate the generation of empathetic responses, we seek to build a training tool that can augment hotline training and empower more counselors. As a first component, we develop a chat interface where novices can practice formulating responses by interacting with a simulated distressed individual.

To build such a system, we collect synthetic role-play transcripts that provide example scenarios and example messages.Because real transcripts may contain scenarios that cannot be fully de-identified, we hope that synthetic transcripts will enable the development of a training system without violating the confidentiality of anyone contacting a real hotline. Here, we consider the one-sided case of simulating the individual in distress with the intention of eventually providing a training environment for novice counselors to practice counseling without putting anyone in danger.

In the application we consider, and in many similarly data-constrained applications, language generation methods may be challenged by the limited data that can initially be collected. To surmount this issue, we explore the extent to which retrieval methods can be improved to provide an engaging chat experience. More specifically, we consider whether improved embedding methods, which enable better representation of text, improve retrieval models through better comparisons of text similarity. Briefly stated, we ask two research questions:

Proceedings of the Sixth Workshop on Computational Linguistics and Clinical Psychology, pages 1–11
Minneapolis, Minnesota, June 6, 2019. ©2019 Association for Computational Linguistics

RQ1 Do improved embedding methods retrieve coherent responses to a single turn of context more often than commonly-used TF-IDF or generative models?

RQ2 Can we extend retrieval baseline models to consider more than one turn of context when selecting a response?

Our results show that recent developments in embedding methods have considerably improved dialogue retrieval, which is promising for the use of these methods in data-limited applications. We also find that extending retrieval to consider additional messages of context does improve baselines. This indicates the potential for retrieval methods to benefit data-limited dialogue systems and the need to re-evaluate baselines for generative models. Within the setting that we study, our results provide promise for building a chat module that can enable crisis counselors to practice before interacting with individuals in need.

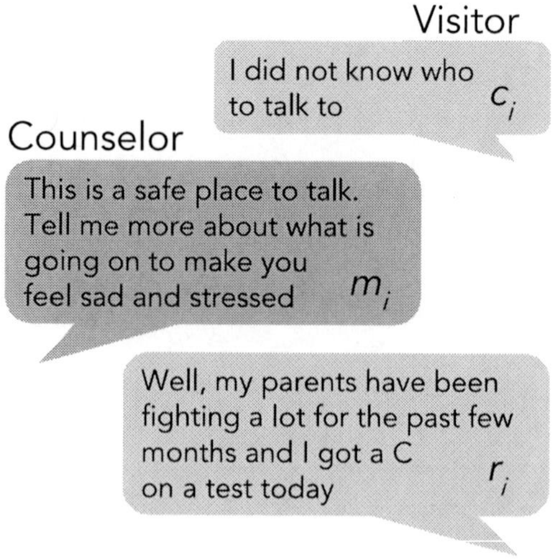

2 Related Work

Considerable potential for automating a counselor was shown with the initial rule-based Eliza system (Weizenbaum, 1966) and recent developments have sought to target systems for delivering cognitive behavioral therapy (Fitzpatrick et al., 2017). Other studies have looked at the effect of suicide prevention counselor training (Gould et al., 2013), identifying patterns of successful crisis hotline counselors (Althoff et al., 2016), automating counselor evaluation (Pérez-Rosas et al., 2017), and building a dashboard for crisis counselors (Dinakar et al., 2015). There is additional work to identify supportive and distressed behaviors and language in online forums (Balani and De Choudhury, 2015; De Choudhury and De, 2014; Wang and Jurgens, 2018) and support forum moderators (Hussain et al., 2015). Most similar to our study, was one study that showed the potential for an avatar system to help train medical doctors to deliver news to patients (Andrade et al., 2010). However, this study did not target counselors or train conversation strategies. To our knowledge, there has been no work on automating the individual seeking help to improve counselor training.

2.1 Text Retrieval for Dialogue Systems

Previous systems have explored the use of retrieving messages from related contexts for continuing

Figure 1: A conversation snippet showing a visitor's response r_i to a counselor's message m_i with preceding context, i.e., a visitor's message c_i.

dialogue. Some studies have looked at defining or learning scoring functions over IDF weights to construct retrieval scores (Krause et al., 2017; Ritter et al., 2011). Most similar to our work is a system that considered similarities of full histories of dialogues in addition to a previous turn of context (Banchs and Li, 2012) and another study that hand-tuned weights in a scoring function on IDF weights to include additional messages of context (Sordoni et al., 2015). However, these works used similarities calculated over TF-IDF (Baeza-Yates et al., 2011) and bag-of-words of representations, instead of more recent embedding methods (Bojanowski et al., 2016; Conneau et al., 2017; Pennington et al., 2014; Peters et al., 2018; Subramanian et al., 2018), which we explore.

3 Dataset

We collected a dataset of synthetic chat transcripts between suicide prevention counselors and hotline visitors. An example of such a conversation is shown in Figure 1 and additional examples are discussed in the Results section. Artificial or role-play transcripts were generated by trained counselors in order to protect the identity of any individuals who may contact crisis hotlines. We chose this approach because retrieval should not be used on datasets consisting of real conversations. Such datasets have been explored in prior work to understand effective hotline conversations (Althoff et al., 2016).

Role-playing between experienced and novice counselors is a common tool for crisis counselor training, and is a task counselors are often exposed to before being approved to work on a hotline (American Association of Suicidology, 2012; Kalafat et al., 2007). In addition to expecting role-playing to be a natural task for hotline counselors, prior work on short, unstructured social dialogues between peers found that self-dialogues, i.e., where an individual would produce both sides of a two-person dialogue, generated high quality and creative example conversations (Krause et al., 2017). We followed this work and asked experienced counselors to self-role-play scenarios of a counselor working with a hotline visitor. We collected transcripts in three phases: full role-plays, visitor-only role-plays, and counselor-paraphrase role-plays.

3.1 Collection

After consenting to participate in the study, counselors were invited to the first of three phases. In the first phase, counselors were asked to role-play both sides of a potential crisis text conversation. To be representative of common demographic of individuals who contact a helpline over text, counselors were prompted to role-play a youth experiencing trouble in school and with their parents. This persona was chosen to represent a common scenario that a counselor may encounter in a text-based conversation. The counselors were able to decide if the fictional youth was experiencing suicidal thoughts, specific issues they were having, and if they felt better by the end of the conversation. Transcripts were required to be 20 turns for each counselor and visitor (40 turns total). However, participants were able to extend the conversation to at most 60 turns total, if they chose. Messages were unconstrained in length, but it was suggested that they resemble SMS messages.

Counselors who participated in a second phase of the study were given the counselor's side of a transcript generated in the first phase of the study and asked to role-play only the youth experiencing trouble in a way that fit with the counselor's messages. Participants in the third phase of the study were given a full transcript generated in the first phase and asked to generate counselor paraphrases that reworded and possibly improved the original counselor messages. The second and third phases were designed to increase the variety of responses

	Phase	Count
Unique conversations	1	254
Visitor-only role-plays	2	182
Counselor-only role-plays	3	118
Visitor messages	1-2	9062
Counselor messages	2	5320
Counselor paraphrases	3	2999

Table 1: Statistics on role-play transcripts. Phase indicates the study phase during which each set of data was collected. Each counselor paraphrase reworded a single counselor message.

that might be made.

Additional data were collected for evaluating models, as will be discussed below. All study methods were approved by the university's Internal Review Board.

3.2 Dataset Statistics

In total, 32 crisis counselors participated in the study and wrote example messages. In general, the transcripts represent a broad range of scenarios. Statistics on the resulting dataset are in Table 1. In the following results, we do not include messages generated in the second phase of the study.

4 Methods

After preprocessing, we consider two tasks: how to return a visitor response to a single input counselor message and how to return a visitor response when considering a counselor input message and preceding conversation context. For responding to a single counselor input message, we consider two approaches: one based on cosine similarity of vector representations and the other based on likelihood. For responding to a counselor message when considering additional conversation context, we extend retrieval to consider additional messages of context, i.e., an additional message preceding the counselor's last message. For generating responses, we consider a popular Seq2Seq model (Sutskever et al., 2014; Vinyals and Le, 2015) and a hierarchical neural model (Park et al., 2018).

4.1 Data Preprocessing

Names were standardized to be popular American male or female baby names from the last 5 decades. Entire messages were tokenized with appropriate tokenizers for each embedding method and converted to lowercase, as appropriate.

4.2 Response Retrieval Considering a Single Message

For the first retrieval approach we consider, let a message input to the system be m_i. Let M_N and R_N be all the N messages and responses, respectively, in the training set and m_j and r_j indicate individual messages and responses in the training set. The first method considers all the messages in the training set and returns the response $r_{j'}$ to the message $m_{j'}$ that shares the highest cosine similarity with the input message, i.e., $j' = \arg\max_j \text{sim}(m_i, m_j)$ where j indexes over the messages in the training set.

Similarity is commonly calculated as cosine similarity between TF-IDF vector representations of the input (i.e., counselor) message m_i and messages in the training set. We compare the TF-IDF representation with additional vector representations of the counselor input. Exhaustive comparison of embedding methods is not feasible, so we chose popular, successful, and diverse embeddings: GloVe (Pennington et al., 2014), FastText (Bojanowski et al., 2016), Attract-Repel (Vulić et al., 2017), and ELMo (Peters et al., 2018; Gardner et al., 2018). We also consider two sentence embeddings: InferSent (Conneau et al., 2017) and GenSen (Subramanian et al., 2018). Messages are embedded by summing the embeddings of their elements, e.g., across words or sentences for appropriate embeddings.

For the second retrieval approach, we select the response from the training data that is most probable, i.e, $j' = \arg\max_j P(r_j|m_i)$ where m_i is again the input message and j indexes over training examples. With this approach, which we will refer to as *S2S-retrieve*, the probability of a response is calculated by a Seq2Seq model trained on counselor-visitor message-response pairs. All Seq2Seq models were trained in the OpenNMT framework (Klein et al., 2017).

4.2.1 Response Retrieval Considering More than One Message of Context

When multiple messages of context are present, we propose including the additional context in the retrieval methods in three ways. For this work, we consider only one message in the conversation that precedes the counselor's input message to be additional context, as indicated in Figure 1.

First, we consider the response from the training data $r_{j'}$ that has the highest similarity calculated over the sum of the previous messages embeddings, i.e., considering contexts c_i and c_j that precede a test message m_i and a training message m_j respectively, we choose $r_{j'}$ such that $j' = \arg\max_j \text{sim}(m_i + c_i, m_j + c_j)$.

As a second approach, we measure context similarity as the weighted sum of context and message similarities: $j' = \arg\max_j \text{sim}(m_i, m_j) + \lambda\text{sim}(c_i, c_j)$. The weight parameter λ is found via cross-validation to optimize the similarity of embedded responses returned with true responses on a development set.

Third, for the likelihood based model, we again consider the response from the training set that returns the highest likelihood, as calculated by a Seq2Seq model. To include an additional context message, we concatenate preceding messages before encoding and decoding.

4.3 Response Generation

For generating a response to a single counselor message, we consider a Seq2Seq model (Sutskever et al., 2014).

When considering an additional message of context, we first use the Seq2Seq model with the preceding messages concatenated into a single input. Second, we use a Variational Hierarchical Conversation RNN (VHCR) that explicitly models prior conversation state with a hierarchical structure of latent variables (Park et al., 2018). This model has been shown to improve on other models that adjust for context when there is more than one preceding utterance (Park et al., 2018). Seq2Seq and VHCR model embeddings are initialized with GloVe vectors (Pennington et al., 2014).

5 Experiments

For the two response selection tasks, we randomly separated transcripts into training, development, and test sets, with the training set accounting for 80% of the conversations and the rest evenly distributed between development and test sets. Counselor paraphrases were assigned to the set that their original message was assigned to. Messages were not randomly shuffled, but separated by conversation, to avoid training on data related to the test data. For both research questions, a response was either generated from a model trained on the training set or retrieved from the bank of training examples for every counselor message or paraphrased counselor message in the test set.

	Method	Embedding unit	Selection metric	Percent that made sense	Avg. tokens in response	Avg. tokens in MS
	Random	–	–	25.30	15.1	12.6
retrieval	TF-IDF	word	cos-sim	60.34	13.1	12.4
	Attract-Repel	word	cos-sim	58.50	18.3	**16.2**
	ELMo	word	cos-sim	65.88	14.5	14.0
	FastText	word	cos-sim	62.71	**16.2**	15.5
	GloVe	word	cos-sim	58.63	15.9	15.1
	GenSen	sentence	cos-sim	64.16	14.5	14.2
	InferSent	sentence	cos-sim	61.79	14.9	14.0
	S2S-retrieve	–	likelihood	**67.46**	8.8	8.2
gen.	S2S-generate	–	–	64.16	11.7	10.8
	Ground truth	–	–	**89.33**	**14.6**	**14.6**

Table 2: Performance of methods used to return a response to a single input message. MS indicates the set of responses that crowdworkers judged as making sense in context, rather than all the responses that the method returned. Both the best performing method and ground truth results are in bold.

5.1 Evaluation

To evaluate the overall quality of responses that methods returned, we follow prior work that indicated there is currently no automatic equivalent and used human judges (Liu et al., 2016). These judges were crowdworkers on Amazon Mechanical Turk[1] who had been granted Masters status and were located in the United States. Crowdworkers were presented with instructions, labeled examples, and batches of 10 cases where they were asked to judge responses to messages.

To evaluate methods for the first research question, crowdworkers were given a single message and a response and asked to judge the response. For the second research question, crowdworkers were given two messages of context and a highlighted response and asked to judge the response.

In contrast to studies that rank on scales (Lowe et al., 2017), we directly asked the workers to decide if a response made sense or not. In addition to indicating that a response did or did not make sense, we allowed a third class for workers to indicate if they were unsure without additional context. We found these classes to be sufficiently descriptive to consistently label messages between researchers. In preliminary trials with crowdworkers, there was insufficient agreement on labels. This instability of labels could stem from a variety of causes, including uncertainty about whether a change of topic should be considered a coherent response. To surmount this ambiguity, we asked two crowdworkers to label each response and a third crowdworker to break any ties. All cases where crowdworkers indicated that they were unsure were considered to be labeled as not coherent. With this voting approach, on a trial set of message and response pairs, crowdworker labels corresponded with researcher determined labels with a Cohen's Kappa of 0.69 (Cohen, 1968), indicating considerable agreement.

5.2 Performance Metrics

To assess the quality of a method at returning responses, we take messages from a held-out test set and return a response to it by either selecting a message from the training set or generating a response with a model trained on the message and response pairs in the training set. The split into training, development, and test sets is held constant across methods. We ask crowdworkers to judge whether each response makes sense as a possible response to the given message and aggregate multiple crowdworker decisions into a single label for each returned response. We then use the percent of responses returned by a method that were labeled as making sense as an indicator of method performance. The higher percent of messages that made sense as responses, the better the method is at responding coherently. We also consider the number of tokens in each response returned by a method and average the number across all the responses returned as a surrogate for how interesting the responses are. Presumably, longer messages are more interesting than short responses.

[1] https://www.mturk.com/

Decision	Subcategory	Count
Makes sense	Answers the counselor's question(s)	17
	Logical response, fits the conversation	15
	Not perfect, but conceivable someone could respond this way	7
	Agrees/disagrees with counselor's statement	2
Mismatched	Doesn't answer or respond to the question	11
	Messages are unrelated	9
	Doesn't fit, seem right, or make sense	4
	Responses answers a different question	3
	Response is a bad, incoherent message	3
	Message is from a different part of the conversation	2
Unclear	Response is vague or confusing	4
	Worker just didn't know	3
	Can't tell without more context	2
	Explanation of why worker is unsure	1
Other	Researchers were unsure what rationale meant	13
	Description of message content	4

Table 3: Themes in crowdworker rationales for why a response made sense or not. The count is the number of rationales out of a subset of 100 pairs that shared the theme.

5.3 Random and Ground Truth Baselines

For the first research question, we included a method that randomly selected responses from the training set to messages in the test set. This method is intended as a baseline for how easy the task was for a method to guess responses.

For both the first and second research questions, we included a method that returned ground truth visitor responses from the test set as an indicator of how hard the task was for humans to determine response quality without additional context.

5.4 Assessing Why Responses Are Coherent

To understand how crowdworkers decided if a response was coherent, we asked crowdworkers to evaluate responses on a set of 100 message-response pairs and additionally provide a rationale for their decision. For each of 50 test messages, we made two pairs: one with a response randomly selected from the training messages and the other with the ground truth response from the test set. These two methods where chosen to generate pairs that were not likely and likely to be coherent. We directly asked whether the response was coherent and "Why did you choose that option?" with an open text box for crowdworkers to enter a rationale. We read and grouped the rationales into themes of why responses did or did not make sense.

6 Results

We present results on two tasks corresponding to our two research questions: retrieving a response to a counselor's message and extending retrieval to consider an additional message of context. We also consider rationales for why responses do or do not make sense.

6.1 Comparing Retrieval Methods for a Single Message of Context

Retrieval methods showed a clear benefit over randomly selecting responses, i.e., retrieval methods returned a higher percent of coherent messages, as judged by crowdworkers (Table 2). ELMo embeddings and three other embeddings (FastText, InferSent, and GenSen) improved on the commonly used TF-IDF retrieval baseline. This suggests that retrieval methods with recent embeddings provide stronger baselines for generative methods and may continue to improve alongside better embedding methods. However, only the ELMo embeddings surpassed the Seq2Seq generative model in percent of coherent messages returned, indicating the potential of generative models even in this data-limited setting.

Retrieving responses with a Seq2Seq likelihood score returned the highest percent of coherent responses, but the shortest messages on average. Short messages may indicate that, while coherent, the messages may be less engaging or interesting for an interactive environment.

When assessing crowdworker rationales for deciding if a response was coherent, we found 16 themes or reasons why crowdworkers would decide whether a response made sense (Table 3). Crowdworkers cited more ways that responses did not make sense than ways a response would be coherent. Examples of messages in the test set and the corresponding responses that each method re-

	Method	Example 1	Example 2	Example 3
Message		Wow, not even 18 but already paying the bills. How old are you?	Sounds like a great plan, take care Jessica... we're here 24/7 in case you need to talk to us again	Hi, I'm Michael! What is your name? Location?
Responses	Random	I took a small amount. (N)	Sometimes I have to. They don't really like hanging out with him. (N)	Yes, please. (N)
	Tf-idf	I'm 17. (Y)	Thanks, it's good to talk about this stuff. I have to go eat dinner now (Y)	Hi Michael, this is Christopher. I'm in Daly City and I need help. (Y)
	Attract-Repel	I feel lonely (N)	Maybe. Maybe next time. (N)	I'm Jennifer (Y)
	ELMo	14 (Y)	ok thx (Y)	i'd rather not say my name. is this space confidential (Y)
	FastText	14 (Y)	ok thx (Y)	i'd rather not say my name. is this space confidential (Y)
	GloVe	Im 15, i have a twin but hes a boy. Im a girl. They let him get away with evvvverything at home and school (Y)	Maybe. Maybe next time. (N)	i'd rather not say my name. is this space confidential (Y)
	GenSen	14 (Y)	Okay. I got it. Thank you. You were really helpful. I was feeling really awful but I'm feeling better now. (Y)	i'd rather not say my name. is this space confidential (Y)
	InferSent	14 (Y)	Yes please. I'll be right back. (N)	i'd rather not say my name. is this space confidential (Y)
	S2S-retrieve	I'm 15 and I'm pretty sure. Life is lame and I can't go on like this anymore. (Y)	Maybe (Y)	i'd rather not say my name. is this space confidential (Y)
	S2S-generate	I'm 15 and I'm pretty sure . Life is lame and I can't go on like this anymore . (Y)	Okay, I will message you tomorrow. Thank you (Y)	i'd rather not say my name. is this space confidential (Y)
	Ground truth	Yea it's awkward. Im 17, be 18 in 4mo (Y)	You too (Y)	My name is Christopher and I'm in Golden Gate Park. (Y)

Table 4: Examples of three counselor messages and the corresponding visitor response output from each method. These examples are from the first research question, where only one preceding counselor message is considered. Whether crowdworkers thought a response made sense or not is indicated parentheses as "Y" and "N", respectively.

turned for them are shown in Table 4.

6.2 Extending Retrieval to Include Additional Messages of Context

Providing crowdworkers with an additional message of context appeared to impact their impression of whether responses made sense in context. When presented with an additional message of context, i.e, one visitor message and one counselor message, crowdworkers found a larger percent of the ground truth responses from the test set to make sense (Table 5). In contrast, when provided with an additional message of context to evaluate a response, crowdworkers judged a lower percent of responses returned by the ELMo-based retrieval method to be coherent (61.40%, Table 5) than when they were only presented with a single message of context (65.88%, Table 2). Incorporating a previous message of context into a similarity score increased the percent of coherent messages returned, but by less than 1%. We only consid-

ered the ELMo embeddings, as they were found to perform best in the first research question. Three out of four retrieval methods returned a higher percent of coherent messages than both generative models, indicating that including more context for generative models is challenging. Again using the Seq2Seq likelihood to retrieve responses returned the highest percent of messages that made sense. However, these responses also had the fewest tokens, implying generic, short messages that might score low on a qualitative scale of how engaging an interactive system is.

7 Discussion

In contrast to many popular dialogue datasets (Serban et al., 2015), the transcripts we collected have a relatively high number of turns (minimum 40 total turns per conversation), implying rich conversations. These conversations are also interesting for their unique position of having distinct roles for participants, a counselor and a distressed

	Method	Incorporation of additional context	Percent that made sense	Avg. tokens in response	Avg. tokens in MS
retrieval	ELMo	–	61.40	14.6	13.6
	ELMo-sum	Measure similarity of sum of embedded messages	51.78	**15.6**	**15.2**
	ELMo-weight	Weight similarities of previous messages	61.66	14.9	13.9
	S2S-retrieve	Concatenate context	**65.48**	5.5	4.6
gen.	S2S-generate	Concatenate context	58.89	8.3	7.3
	VHCR-generate	Models conversation	55.07	10.8	8.4
	Ground truth	–	**91.30**	**14.6**	**14.7**

Table 5: Performance of methods used to retrieve or generate responses when an additional message of context is considered, i.e., two total messages. MS denotes only responses that were considered to make sense in context. Both the best performing method and ground truth results are in bold.

youth, and related themes. We find retrieval to be a competitive approach with generative models and return responses that make sense for more than 60% of input messages. We also find themes for how responses can seem to be coherent.

Giving crowdworkers an additional message of context to judge whether a response was coherent or not affected their decisions. It appeared that ground truth responses were easier to distinguish as coherent and fewer retrieved messages were judged as coherent if an additional message of context was presented. This indicates the importance of context, especially during evaluation.

The results we present are on a specific, data-limited setting, but the implications of our results may be broader both for other important applications, which commonly have data limitations, and for retrieval baselines that are used to assess generative models. As embeddings have improved, so too have retrieval baselines, which need to be updated for appropriate evaluation of generative models in any language generation setting.

Our results are not without limitations. The data-limited setting presented a challenge to training generative models, and perhaps extensive hyper-parameter tuning could influence results. However, limited data and non-exhaustive parameter tuning are common limitations. Further, as datasets increase in size, so does the potential for relevant, related contexts to be present and thus the potential for successful retrieval increases as well. Thus, even on larger datasets, competitive retrieval models, such as those we have presented, should be considered for baseline comparisons.

Another limitation of our approach is the extent to which we have considered context so far. Because the conversations we collected are long relative to some other datasets it is likely more context will be necessary to produce a coherent simulation. We have begun to methodically look at the effects of incrementally including more context and extending retrieval models beyond a single message. These initial steps indicate the impact context has and provide important baselines for comparing future, more general models.

8 Conclusion

Our work shows promise that data-limited applications may build initial systems with retrieval methods powered by recently developed embeddings. By collecting role-play transcripts and showing results in a data-limited context, we have demonstrated the potential to develop a successful simulation of a hotline visitor that novice counselors can practice with during training. We found that retrieval methods became more competitive with improved embedding methods and surpassed generative methods when more context was considered. We also found that context had impact on how difficult it was for crowdworkers to evaluate responses.

As a next step, we plan to explore better leveraging rich structure in the conversations, with a focus on the protocol that the counselors are trained to follow. There has been increased interest in blending retrieval and generation approaches by modifying prototypes retrieved from training data (Li et al., 2018; Weston et al., 2018). It is possible that such an approach would enable modifying and thus tailoring responses to similar contexts.

References

Tim Althoff, Kevin Clark, and Jure Leskovec. 2016. Large-scale Analysis of Counseling Conversations: An Application of Natural Language Processing to Mental Health. *Transactions of the Association for Computational Linguistics*.

American Association of Suicidology. 2012. *Organization Accreditation Standards Manual*.

Allen D Andrade, Anita Bagri, Khin Zaw, Bernard A Roos, and Jorge G Ruiz. 2010. Avatar-mediated training in the delivery of bad news in a virtual world. *Journal of palliative medicine*, 13(12):1415–1419.

Ricardo Baeza-Yates, Berthier de Araújo Neto Ribeiro, et al. 2011. *Modern information retrieval*. New York: ACM Press; Harlow, England: Addison-Wesley,.

Sairam Balani and Munmun De Choudhury. 2015. Detecting and characterizing mental health related self-disclosure in social media. In *Proceedings of the 33rd Annual ACM Conference Extended Abstracts on Human Factors in Computing Systems*, pages 1373–1378. ACM.

Rafael E Banchs and Haizhou Li. 2012. Iris: a chat-oriented dialogue system based on the vector space model. In *Proceedings of the ACL 2012 System Demonstrations*, pages 37–42. Association for Computational Linguistics.

Piotr Bojanowski, Edouard Grave, Armand Joulin, and Tomas Mikolov. 2016. Enriching word vectors with subword information. *arXiv preprint arXiv:1607.04606*.

Jacob Cohen. 1968. Weighted kappa: Nominal scale agreement provision for scaled disagreement or partial credit. *Psychological bulletin*, 70(4):213.

Alexis Conneau, Douwe Kiela, Holger Schwenk, Loïc Barrault, and Antoine Bordes. 2017. Supervised learning of universal sentence representations from natural language inference data. In *Proceedings of the 2017 Conference on Empirical Methods in Natural Language Processing*.

Wendi F Cross, David Seaburn, Danette Gibbs, Karen Schmeelk-Cone, Ann Marie White, and Eric D Caine. 2011. Does practice make perfect? a randomized control trial of behavioral rehearsal on suicide prevention gatekeeper skills. *The journal of primary prevention*, 32(3-4):195.

Munmun De Choudhury and Sushovan De. 2014. Mental health discourse on reddit: Self-disclosure, social support, and anonymity. In *ICWSM*.

Karthik Dinakar, Jackie Chen, Henry Lieberman, Rosalind Picard, and Robert Filbin. 2015. Mixed-initiative real-time topic modeling & visualization for crisis counseling. In *Proceedings of the 20th international conference on intelligent user interfaces*, pages 417–426. ACM.

Kathleen Kara Fitzpatrick, Alison Darcy, and Molly Vierhile. 2017. Delivering cognitive behavior therapy to young adults with symptoms of depression and anxiety using a fully automated conversational agent (woebot): a randomized controlled trial. *JMIR mental health*, 4(2).

Matt Gardner, Joel Grus, Mark Neumann, Oyvind Tafjord, Pradeep Dasigi, Nelson F. Liu, Matthew Peters, Michael Schmitz, and Luke S. Zettlemoyer. 2018. AllenNLP: A deep semantic natural language processing platform. In *ACL workshop for NLP Open Source Software*.

Madelyn S Gould, Wendi Cross, Anthony R Pisani, Jimmie Lou Munfakh, and Marjorie Kleinman. 2013. Impact of applied suicide intervention skills training on the national suicide prevention lifeline. *Suicide and Life-Threatening Behavior*, 43(6):676–691.

Madelyn S Gould, Jimmie LH Munfakh, Marjorie Kleinman, and Alison M Lake. 2012. National suicide prevention lifeline: enhancing mental health care for suicidal individuals and other people in crisis. *Suicide and Life-Threatening Behavior*, 42(1):22–35.

M Sazzad Hussain, Juchen Li, Louise A Ellis, Laura Ospina-Pinillos, Tracey A Davenport, Rafael A Calvo, and Ian B Hickie. 2015. Moderator assistant: A natural language generation-based intervention to support mental health via social media. *Journal of Technology in Human Services*, 33(4):304–329.

John Kalafat, Madelyn S Gould, Jimmie Lou Harris Munfakh, and Marjorie Kleinman. 2007. An evaluation of crisis hotline outcomes. part 1: Nonsuicidal crisis callers. *Suicide and Life-threatening behavior*, 37(3):322–337.

Guillaume Klein, Yoon Kim, Yuntian Deng, Jean Senellart, and Alexander M. Rush. 2017. OpenNMT: Open-source toolkit for neural machine translation. In *Proceedings of ACL*.

Ben Krause, Marco Damonte, Mihai Dobre, Daniel Duma, Joachim Fainberg, Federico Fancellu, Emmanuel Kahembwe, Jianpeng Cheng, and Bonnie Webber. 2017. Edina: Building an open domain socialbot with self-dialogues. *arXiv preprint arXiv:1709.09816*.

Juncen Li, Robin Jia, He He, and Percy Liang. 2018. Delete, retrieve, generate: a simple approach to sentiment and style transfer. In *Proceedings of the 2018 Conference of the North American Chapter of the Association for Computational Linguistics: Human Language Technologies*, pages 1865–1874.

Chia-Wei Liu, Ryan Lowe, Iulian Serban, Mike Noseworthy, Laurent Charlin, and Joelle Pineau. 2016. How not to evaluate your dialogue system: An empirical study of unsupervised evaluation metrics for dialogue response generation. In *Proceedings of the 2016 Conference on Empirical Methods in Natural Language Processing*, pages 2122–2132. Association for Computational Linguistics.

Ryan Lowe, Michael Noseworthy, Iulian Vlad Serban, Nicolas Angelard-Gontier, Yoshua Bengio, and Joelle Pineau. 2017. Towards an automatic turing test: Learning to evaluate dialogue responses. In *Proceedings of the 55th Annual Meeting of the Association for Computational Linguistics*, pages 1116–1126.

Yookoon Park, Jaemin Cho, and Gunhee Kim. 2018. A hierarchical latent structure for variational conversation modeling. In *Proceedings of the 2018 Conference of the North American Chapter of the Association for Computational Linguistics: Human Language Technologies*, pages 1792–1801.

Amber Paukert, Brian Stagner, and Kerry Hope. 2004. The assessment of active listening skills in helpline volunteers. *Stress, Trauma, and Crisis*, 7(1):61–76.

Jeffrey Pennington, Richard Socher, and Christopher D. Manning. 2014. Glove: Global vectors for word representation. In *Empirical Methods in Natural Language Processing (EMNLP)*, pages 1532–1543.

Verónica Pérez-Rosas, Rada Mihalcea, Kenneth Resnicow, Satinder Singh, Lawrence Ann, Kathy J Goggin, and Delwyn Catley. 2017. Predicting counselor behaviors in motivational interviewing encounters. In *Proceedings of the 15th Conference of the European Chapter of the Association for Computational Linguistics: Volume 1, Long Papers*, volume 1, pages 1128–1137.

Matthew E. Peters, Mark Neumann, Mohit Iyyer, Matt Gardner, Christopher Clark, Kenton Lee, and Luke Zettlemoyer. 2018. Deep contextualized word representations. In *Proc. of NAACL*.

Rajeev Ramchand, Lisa Jaycox, Pat Ebener, Mary Lou Gilbert, Dionne Barnes-Proby, and Prodyumna Goutam. 2016. Characteristics and proximal outcomes of calls made to suicide crisis hotlines in california. *Crisis*.

Alan Ritter, Colin Cherry, and William B Dolan. 2011. Data-driven response generation in social media. In *Proceedings of the conference on empirical methods in natural language processing*, pages 583–593. Association for Computational Linguistics.

Iulian Vlad Serban, Ryan Lowe, Peter Henderson, Laurent Charlin, and Joelle Pineau. 2015. A survey of available corpora for building data-driven dialogue systems. *arXiv preprint arXiv:1512.05742*.

Aaron Smith and Dana Page. 2015. Us smartphone use in 2015. *Pew Research Center*, 1.

Alessandro Sordoni, Michel Galley, Michael Auli, Chris Brockett, Yangfeng Ji, Margaret Mitchell, Jian-Yun Nie, Jianfeng Gao, and Bill Dolan. 2015. A neural network approach to context-sensitive generation of conversational responses. In *Proceedings of the 2015 Conference of the North American Chapter of the Association for Computational Linguistics: Human Language Technologies*, pages 196–205.

Sandeep Subramanian, Adam Trischler, Yoshua Bengio, and Christopher J Pal. 2018. Learning general purpose distributed sentence representations via large scale multi-task learning. In *ICLR*.

Suicide Prevention Resource Center. 2007. Applied suicide intervention skills training (ASIST).

Ilya Sutskever, Oriol Vinyals, and Quoc V Le. 2014. Sequence to sequence learning with neural networks. In *Advances in neural information processing systems*, pages 3104–3112.

Oriol Vinyals and Quoc Le. 2015. A neural conversational model. *arXiv preprint arXiv:1506.05869*.

Ivan Vulić, Nikola Mrkšić, Roi Reichart, Diarmuid Ó Séaghdha, Steve Young, and Anna Korhonen. 2017. Morph-fitting: Fine-tuning word vector spaces with simple language-specific rules. In *Proceedings of ACL*, pages 56–68.

Zijian Wang and David Jurgens. 2018. Its going to be okay: Measuring access to support in online communities. In *Proceedings of the 2018 Conference on Empirical Methods in Natural Language Processing*, pages 33–45.

Joseph Weizenbaum. 1966. Elizaa computer program for the study of natural language communication between man and machine. *Communications of the ACM*, 9(1):36–45.

Jason Weston, Emily Dinan, and Alexander H Miller. 2018. Retrieve and refine: Improved sequence generation models for dialogue. *arXiv preprint arXiv:1808.04776*.

A   Appendices

Modified model parameters are shared below for reproducibility.

A.1   Seq2Seq Model Parameters

More information on model parameters can be found in the OpenNMT-py online documentation[2].

-dynamic_dict on

---

[2] `http://opennmt.net/OpenNMT-py/index.html`

-share_vocab on

-src_seq_length $= 200$

-tgt_seq_length $= 200$

-rnn_size $= 500$

-src_word_vec_size $= 300$

-tgt_word_vec_size $= 300$

-share_embeddings on

-encoder_type $=$ brnn

-decoder_type $=$ rnn

-rnn_type $=$ LSTM

-layers $= 2$

-global_attention $=$ general

-optim $=$ adam

-learning_rate $= 0.001$

-batch_size $= 4$

pre-trained embedding glove.840B.300d.txt

A.2 VHCR Model Parameters

More info can be found about model parameters in the online repository[3].

-model $=$ VHCR

-batch_size $= 4$

-embedding_size $= 300$

-encoder_hidden_size $= 500$

-decoder_hidden_size $= 500$

-context_size $= 500$

-z_sent_size $= 50$

-z_conv_size $= 50$

pre-trained embedding glove.840B.300d.txt

-max_sentence_length $= 60$

-max_conversation_length $= 5$

-min_vocab_frequency $= 3$

---

[3] https://github.com/ctr4si/
A-Hierarchical-Latent-Structure\
-for-Variational-Conversation-Modeling

Identifying therapist conversational actions across diverse psychotherapeutic approaches

Fei-Tzin Lee[*], **Derrick Hull**[†], **Jacob Levine**[†], **Bonnie Ray**[†], **Kathleen McKeown**[*]
[*]Columbia University, Department of Computer Science
[†]Talkspace
{feitzin, kathy}@cs.columbia.edu
{derrick, bonnie.ray, jacob.levine}@talkspace.com

Abstract

While conversation in therapy sessions can vary widely in both topic and style, an understanding of the underlying techniques used by therapists can provide valuable insights into how therapists best help clients of different types. Dialogue act classification aims to identify the conversational "action" each speaker takes at each utterance, such as sympathizing, problem-solving or assumption checking. We propose to apply dialogue act classification to therapy transcripts, using a therapy-specific labeling scheme, in order to gain a high-level understanding of the flow of conversation in therapy sessions. We present a novel annotation scheme that spans multiple psychotherapeutic approaches, apply it to a large and diverse corpus of psychotherapy transcripts, and present and discuss classification results obtained using both SVM and neural network-based models. The results indicate that identifying the structure and flow of therapeutic actions is an obtainable goal, opening up the opportunity in the future to provide therapeutic recommendations tailored to specific client situations.

1 Introduction

Dialogue act classification is a task in which utterances in a conversation (or dialogue) are labeled with the *action* that utterance performs in the context of the dialogue - essentially, the intention of the speaker at that point in the conversation. In the general case, this might be something like a question, an agreement, or a backchannel, though specific acts of interest depend on the application. This type of classification generally lends itself to a more thorough understanding of the flow of a conversation. For our application, psychotherapy, it can be particularly helpful in clarifying the specific patterns of behavior exhibited by the therapist in response to different client statements.

Mental health treatment is unique in that, unlike other specialties, intervention can take place directly through the interaction between a patient and the care provider or therapist (Gaut et al., 2017; Hull, 2014). This places critical emphasis on research to understand the dynamics and mechanisms of change within the interaction itself, just as medical investigators would perform for a newly advanced drug or surgical procedure. Historically, however, it has been too labor intensive to manually summarize sessions and therapist notes for record keeping, or to implement a process for reliably quantifying the flow and quality of the conversation, especially for large numbers of sessions or among large, heterogeneous samples. An automated avenue for labeling clinically relevant dialogue acts would allow us to learn patterns of discourse associated with differing clinical outcomes, potentially even uncovering patterns and effects that had previously remained hidden. The results could be used to inform the development of automated clinical assistants, conversational agents, and recommender or supervisory systems for therapists delivering care through technology.

In this paper we provide preliminary results towards this end on a dataset of therapy transcripts labeled with a novel set of high-level therapy-specific acts at the sentence level. While we are not at liberty to make the annotated corpus available publicly, we do include a description of the annotation scheme, and will release examples of our annotations. Our analyses result in two key findings: firstly, the *context* of the sentence provides the clearest and most stable signal of the act; and secondly, on our limited dataset, simple methods can achieve performance as good as or better than that of more complex approaches (i.e., our simple SVM classifier significantly outperformed more complex neural methods). We present a detailed error analysis of our models' performance on the development set

Proceedings of the Sixth Workshop on Computational Linguistics and Clinical Psychology, pages 12–23
Minneapolis, Minnesota, June 6, 2019. ©2019 Association for Computational Linguistics

to better understand where the approach works well and where it encounters the most challenges, and discuss future avenues of research to potentially address these challenges.

Our contributions include (1) a simple therapy-specific dialogue act classification scheme for therapist utterances relevant across a broad range of therapeutic approaches; (2) a sample of annotated utterances for a large corpus of diverse therapy transcripts; and (3) initial classification results on this dataset, with analysis.

2 Related Work

Several papers in recent years have developed machine learning approaches for the coding of dialogue in a psychotherapy context. Early work (Can et al., 2015) leveraged n-grams, dictionary-based features constructed based on psycho-linguistic norms such as LIWC (Pennebaker et al., 2015), and features used in more general dialog act classification modeling, such as that of (Jurafsky et al., 1997), to automate coding of therapist skill usage. More recent work has leveraged the methods of deep learning to incorporate the sequential aspects of client-therapist interactions, using variations on recurrent neural network models to improve the ability of the model to accurately classify therapist behaviors. See, for example, (Xiao et al., 2016; Gibson et al., 2016, 2017). This body of work has focused primarily on identifying therapist skills in Motivational Interviewing, a highly structured psychotherapy approach used for resolving ambivalence related to the treatment of conditions such as substance or alcohol use, or to engaging with treatment in general (Miller and Rollnick, 2012). Independently, Flemotomos et al. 2018 and Rojas Barahona et al. 2018 applied machine learning approaches to code behaviors common in the context of Cognitive Behavior Therapy (CBT). Rojas Barahona et al. developed neural network models for classification of various types of client 'thinking errors' identified as part of cognitive behavioral treatment, while Flemotomos et al. built SVM models to classify the overall quality of a CBT treatment session, looking at the distribution of different types of therapist behaviors used within the session, both process and content-oriented (e.g. homework assignments). CBT, while widely used, is again a fairly structured and goal-oriented approach to psychotherapy, making it more amenable to machine learning of underlying linguistic patterns. Other recent work (Gibson and Narayanan, 2018) has applied multi-task learning to transcripts representing both Motivational Interviewing and CBT-based approaches, an important advance due to the difficulty of obtaining large corpora of annotated transcripts for any single psychotherapy approach. Multi-label learning for concurrently classifying individual therapist utterances as well as the overall 'quality' of a session was also explored in the same paper.

Our work differs from these previous works in that our corpus of psychotherapy transcripts includes therapists using a variety of therapeutic approaches, including second- and third-wave CBT, psychodynamic, motivational interviewing, supportive/Rogerian, and an integrative or eclectic approach blending aspects of several approaches, thus providing less consistency in the language and behaviors exhibited by the therapists and making the automated coding task more difficult. To handle the greater heterogeneity of therapist speech, we have developed a broader annotation scheme that captures a wide variety of therapist behaviors common to the general therapeutic process, combining these with a small range of labels specific to particular approaches.

3 Data

3.1 Corpus

Our dataset consists of an annotated selection of transcripts from a corpus maintained by the publisher Alexander Street Press[1], available through library subscription; the full collection consists of approximately four thousand transcripts, 340 of which we labeled. In the base corpus, transcript lengths ranged from approximately 200 to 900 sentences. The client tended to speak more than the therapist, with client sentences ranging from 162 to 614 per transcript, while therapists spoke between 54 and 473 sentences (the entire dataset contained around 126,000 client sentences, and only 53,000 therapist sentences).

Transcripts were labeled with dialogue acts at the sentence level; some sentences were judged to contain no dialogue act in the annotation set and thus were left unlabeled. This left us with 8,420 labeled sentences from clients, and 9,056 labeled sentences from therapists. We focus on therapist act classification in this work, as it has proven easier

---

[1] https://alexanderstreet.com/products/counseling-and-psychotherapy-transcripts-series

Code	Description
Simple Reflection	Repeats client statement with minimal alteration.
Makes Needs Explicit	Identifies an implied or background need for the client.
Makes Emotions Explicit	Identifies an implied or background emotion for the client.
Makes Values Explicit	Identifies an implied or background value or set of values for the client.
Makes Relational Patterns Explicit	Identifies an implied or background relational pattern for the client.
Makes Consequences Explicit	Identifies an implied or background consequence of a client's action.
Makes Conflict Explicit	Identifies an implied or background emotional or situational conflict for the client.
Problem-Solving	Therapist offers possible solutions to a client problem.
Evokes Concrete Elaboration	More information about a specific event or statement is sought.
Evokes Perspective Elaboration	Client is asked to consider an experience from a different perspective or vantage point.
Narrowing	Therapist guides client to focus on a specific area of concern.
Planning	Therapist works with client to construct a specific plan of action.
Assumption Checking	Helps client determine if a thought or assumption is accurate or helpful.
Metaprocessing	Asks client to express how they are feeling in the immediate present about something that just happened in the therapy.
Makes Strengths/Resources Explicit	Identifies an implied or background strength or resource that the client exhibits.
Normalization	Client's experience is classified as "normal" or expectable by the therapist.
Sympathizing	Brief statements expressing regret for the challenges the client is having.
Reassuring	Therapist attempts to convince client that painful experiences are in fact okay or will get better.
Counterprojection	Makes assumptions the client might be making about the therapist or therapy explicit.
Teaching/Psychoeducation	Therapeutically relevant information about psychological principles is provided.
Self-Disclosure of Therapist Affect	Therapist expresses how they feel about what the client has said.

Table 1: Clinical codes for therapist. Sections indicate clinical codes in the categories Reflection, Question, Normalization/Misc, and Meta, in order.

both to define useful act categories and to practically classify acts for the therapist. Even though we are capturing several therapeutic approaches, therapists tend to deploy a limited range of dialogue acts and expressions, likely owing to the common elements among different psychotherapies and to shared aspects of clinical training and the clinical setting. Clients, on the other hand, are not operating from a handful of theoretical frameworks. They exhibit behavior that is less easy to organize and categorize, especially when drawing primarily on language.

3.2 Annotation scheme

To define the general section of the annotation scheme we drew from the dialogue acts identified in (Jurafsky et al., 1997) and selected those most pertinent to psychotherapy dialogue. The acts chosen were *Agreement, Disagreement, Apology, Thanking, Hedge, Opinion, Yes-No Question, Opening, Closing,* and *Signal Non-understanding.* These codes were used for both therapist and client. Clinical codes were identified for both therapist and client as relevant to psychotherapy and were derived from Emotion Focused Therapy (Pascual-Leone, 2018; Pascual-Leone and Greenberg, 2005), Cognitive Behavioral Therapy (Beck and Beck,

2011), Motivational Interviewing (Miller and Rollnick, 2012), and Accelerated Experiential Dynamic Psychotherapy (Fosha et al., 2009). There were 17 codes for client statements derived from the frameworks above and 21 therapist codes (see Table 1); when combined with the general codes, this resulted in 27 codes for the client and 31 for the therapist. As the client codes are not the focus of this work, we omit them from this paper. Therapist Statement codes are organized around whether the therapist is offering a statement to the client, making an observation, or emphasizing something in what the client said. Therapist Question codes cover the various kinds of questions or requests for more information that a therapist might invoke. Therapist codes were chosen that are determined by theory or previous research to be helpful, as well as those determined to be unhelpful. It is likely useful to identify both kinds of therapist behaviors for other clinical and analytic tasks.

3.3 Annotation process

A random sample of the total Alexander Street Corpus was annotated by 30 Masters level counseling and clinical psychology trainees using a spreadsheet annotation tool we adapted from Microsoft Excel functions. Annotators were trained by a clinical psychology researcher and could confer with others and the researcher when unsure about a particular annotation. The implementation allowed annotators to see each statement within the context of the overall therapy session and to annotate each statement with an individual general code and/or a clinical code when applicable. Each statement could receive both a general or clinical code, but only one of each. Codes were designed to minimize conceptual overlap at the sentence level.

3.4 Category selection

As the act classes were extremely unbalanced (see section 3.5) and due to annotator reliability concerns (see section 3.6), we merged our act codes into higher level categories (see Table 2) that would be more stable and easier to classify, while still clinically meaningful. We ended up with five classes: agreement (consisting of only the general code Agreement); reflection (consisting of the first section of Table 1); question (the second section of Table 1, and the general codes Yes-No Question and Signal Non-Understanding); Normalization/Misc (the third section, as well as Disagreement, Apology, Hedge, Opinion, and Opening from the gen-

Category	Sentences
Agreement	1277
Reflection	4016
Question	3164
Normalization/Misc	1715
Meta	790

Table 2: Class sizes for categories.

eral codes); and Meta (the final section, and the general code Closing).

3.5 Data imbalance

Due to the already limited quantity of annotated data, we did not subsample classes to produce a balanced dataset. This resulted in a notable imbalance in our data, even at the category level, though much more so at the act level. Class sizes for categories are provided in table 2. Due to space constraints, we have left the class sizes at the act level for the appendix, but the largest act class for therapist was agreement, with 1277 samples, while there were nine classes with under a hundred samples.

3.6 Inter-annotator agreement

Agreement on low-level codes was fairly low for the client, though relatively high for the therapist: on the subset of sentences which were coded by two annotators, Cohen's kappa was 0.3164 for client sentences, and 0.7900 for therapist. Agreement on categories was higher: 0.6303 for client, and 0.8577 for therapist. Category agreement was computed by aggregating the total number of low-level acts that received a label within the category. The greater category-level agreement than act-level agreement indicates that most disagreements at the act level nevertheless fell within the same category - that is, for the same sentence, different annotators were more likely to mark two different act codes in the same category than they were to mark two different act codes corresponding to different categories altogether. Whether due to the complex and compound structure of certain sentences where multiple codes were possible or to the similar psychological function of different codes, the high-level categories appear to be more stable.

3.7 Data handling and preprocessing

Sentences were tokenized using the NLTK Tweet-Tokenizer[2], with automatic lowercasing. In cases

---

[2] https://www.nltk.org/api/nltk.tokenize.html

where sentences had both a general and a clinical label, the clinical label was given precedence (i.e. the clinical label was used as the single "true" label). We used a 70/15/15% data split, yielding 6335, 1357, and 1359 sentences for our train, development and test sets, respectively.

4 Methods

4.1 Models

Our primary models include an SVM based on discrete features (n-grams, dialogue information, context features, and length) as well as two different neural network models - a feedforward neural net on the discrete features alone, and a convolutional neural network (CNN) over the text as well as the discrete features. For baselines, we used an SVM over n-grams only and a CNN over text only. In our initial experiments we also investigated recurrent models (RNNs), but found that convolutional models strongly outperformed these, and so we did not include an RNN in our final set of models.

4.2 Discrete features

We experimented with a number of different features, using n-grams from the sentence as our baseline. As features about the sentence itself, we included the length of the sentence (in tokens), as well as position information including the index of the sentence within the conversation (sentence position); as dialog features, we included the index of the speaker turn (turn position), and the index of the sentence within the current speaker turn (utterance position). As context-related information, we used labels from the immediate history of the sample sentence, with varying window sizes, as well as n-grams from those previous sentences. We also experimented with sentiment features for the sentence itself (minimum, maximum, and average word scores using SentiWordNet (Baccianella et al., 2010)); counts of words from two different psychologically meaningful dictionaries, LIWC (Pennebaker et al., 2015) and DAAP (Bucci and Maskit, 2005); part-of-speech tags; word embeddings; and metadata for the transcript. Of these, position and length information, context labels, and context n-grams provided a boost to performance over the baseline, and so we omitted the others from our final model. Thus, our final sets of features included sentence features (sentence position, length, and n-grams), context features (labels and n-grams), and dialogue features (speaker change, turn index,

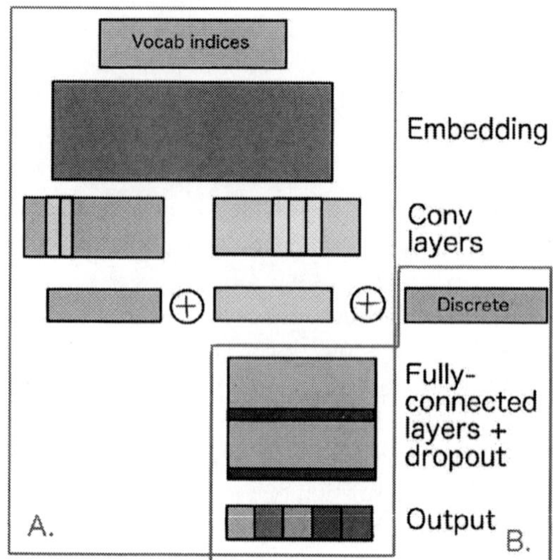

Figure 1: Architecture diagram for the full CNN model.

and sentence index within current turn). Interestingly, we found that using category-level labels as context labels provided better performance for category classification than using the more fine-grained act labels, perhaps due to therapists focusing on a particular approach, e.g. reflection, for multiple utterances in sequence before moving to a different type of intervention.

5 Experiments

5.1 Convolutional baseline

As our baseline model we use a convolutional neural network that takes as input only the text of the sentence and outputs a prediction in the form of a distribution over the category classes. We followed previous work (Liu et al., 2017) in the design of our architecture. The text is originally represented as a series of vocabulary indices; thus, the input to our model is initially a matrix whose dimensions are batch size (number of sentences) and sequence length (predefined number of words), where each element is a vocabulary index (see section A. of Figure 1). Sentences longer than the fixed maximum sequence length are clipped to that length, and shorter sentences are zero-padded. This array is passed through a 64-dimensional embedding layer with 0.5 dropout, followed by two parallel convolutional layers, one with window size 2 and one with window size 3. The representations produced by these two layers are concatenated and fed into a series of two fully-connected dense layers with 0.5 dropout after each; our final layer performs softmax to produce the classification prediction. Intermediate layers use ReLU activation.

5.2 Other neural models

We experimented with two neural network architectures beyond the baseline. The first was a simple feedforward network running on the discrete features only (i.e. without word embeddings - see section B. of Figure 1), identical to the final component of the full architecture, consisting of two fully-connected layers with 128 nodes each, with dropout of 0.5 after each layer, and finally softmax over the classes.

The second was a convolutional net over the text combined with a feedforward component on the discrete features (see Figure 1). We used the same setup as the baseline, but concatenated the discrete features to the intermediate representations produced by the convolutional layers; the concatenated output was then processed by the fully-connected layers, mimicking the feedforward setup. Of our neural models, this latter model performed best.

5.3 Parameters and tuning

We performed gridsearch to find the optimal SVM parameters on different combinations of features. We found that a linear-kernel SVM performed best, with balanced class weights, l2 penalty, regularization parameter C = 0.01, and tolerance 0.3.

For the neural models, we used a batch size of 256, embedding dimension of 64, and maximum sequence length of 128 tokens; we trained for 16 epochs using Nadam optimization with .0002 learning rate, and crossentropy loss. We experimentally determined these parameters to be the best on the development set.

For the embedding layers in both convolutional nets we used random normal initialization and did not fix the weights, training the embedding weights along with the model parameters. Of the embedding initialization settings we tried (uniform random, random normal, and pretrained) this performed the best.

6 Results and Discussion

6.1 Category classification

Our evaluation task involved classifying individual sentences with one of the five act categories. Because of the high imbalance in class size, we used

Classifier	Acc.	Pr.	Rc.	F1
Baseline SVM	70.20	61.77	60.75	60.27
Baseline CNN	49.99	28.26	36.39	29.60
Feedforward*†	74.07	**71.58**	65.96	67.66
CNN + features*†	74.52	70.61	66.68	68.00
SVM*†	**74.98**	70.91	**69.71**	**69.94**

Table 3: Classifier performance on test categories: accuracy, precision, recall and f-measure. Neural network results are reported as an average over five runs to account for variation in random initialization. (*) indicates significance over the SVM baseline, and (†) over the CNN baseline. More detailed results are presented in the appendix.

macro-F1 score as our primary statistic.

For all models, we experimented with feature selection, using Scikit-learn's SelectKBest feature selector, but found that reducing the number of features in this manner had a negative impact on development set performance. Thus, all final models equipped with discrete features used the full number of features. Although it seems likely that there would have been some uninformative features present in the large number (approximately 144,000) we ended up with, the lack of success of feature selection may be due to the small size of the training and validation sets, so that the features most informative on one may not have been the most informative on the other.

All final models performed significantly better than the baselines. Accuracy did not vary greatly between non-baseline classifiers (see Table 3). This is somewhat as expected - the majority classes were the easiest to classify, and classifiers performed well on them, while minority-class performance varied more but had less weight in the accuracy score. The other metrics (particularly recall and f-measure) showed more evident differences in performance, as they were weighted equally between classes. In overall performance, measured by macro-F1, the SVM was clearly the best. Interestingly, this was mostly due to a markedly higher recall than the neural methods, while its precision was between that of the feedforward net and the CNN. We used the Approximate Randomization Test (Riezler and Maxwell, 2005) to measure significance; oddly, the SVM achieved significance over every other method except the feedforward net. Considering that the SVM and feedforward net were the only two methods to receive exactly the same set of input features, this is perhaps due

	Text	SVM	True
1	Tell me your thoughts at that moment.	Meta	Question
2	So you've sort of ceased to mean all that much to him either?	Question	Reflection
3	Your mind really is just refusing to do it ... cause it doesn't want to and it's going to (inaudible).	Reflection	Reflection
4	Well, it's time for us to end but I guess I'm thinking ahead to the anniversary of your sister's death and I'm hoping that you get what you want.	Meta	Reflection

Table 4: Example classified sentences.

Field	Value	F1	
Therapy style	Client-centered therapy	71.29	1050
	Brief dynamic-relational therapy	48.96	201
	Experiential psychotherapy	58.78	65
	Cognitive behavioral therapy	84.17	41
Symptoms	Anger	65.86	430
	Anxiety	69.46	361
	Depression	71.13	322
	Low self-esteem	72.96	145
	Fearfulness	76.46	92
Therapist gender	Male	66.99	852
	Female	73.44	505

Table 5: Performance breakdown by metadata information on the development set. The final column contains the number of sentences present for the particular value of the specified field.

to some similarity in their outputs - possibly the feedforward net essentially performed as a slightly worse SVM, whereas the convolutional net had markedly different predictions, though with slightly better performance than the feedforward net.

6.2 Error analysis

In this section we analyze the performance of our best-performing model, the SVM with full feature sets. Agreement seemed easiest to classify, as one might expect; there were relatively few errors in that category. Unsurprisingly, the SVM tended to have difficulty with sentences that were requests for information not explicitly phrased as a question (e.g. example 1 in Table 4), as well as sentences phrased as questions that were not, in fact, questions - for instance, reflection-type rephrasings of the client's previous statement (example 2). Another major source of error was misclassification of normalization/misc statements as reflections. Both are similar in grammatical form and speak to the client's emotional experience. However, the intended psychological effect is different (reflections move to clarify and specify, normalizations act to reframe feelings in order to bring them down), and this difference was easy to miss or confuse. There was also a slight tendency to classify very short sentences as agreement, even if they were not - as agreement sentences are on average under four words per sentence, as opposed to most classes' 10-20, sentence length was a very strong signal for this class. On the other hand, the SVM was occasionally able to recover the labels of even sentences containing transcription artifacts such as (inaudible) or (ph) (see example 3).

One other quite interesting phenomenon we observed was that, upon close inspection, a number of the sentences that the SVM 'misclassified' in fact seemed to have been annotated incorrectly in the first place - for instance, example 4, which had been annotated as a reflection, but in fact should fall into the meta category, as the SVM predicted. This suggests the possibility of using a similar model as an annotation-checker of sorts, calling attention to sentences which coders might want to take a second or closer look at.

We also analyzed results across different therapy styles and other information about the transcript using the metadata available for the corpus (Table 5). One of the goals of the project was to develop a coding system capable of capturing important elements of several different therapies. The therapy style results suggest some progress in that direction. Interestingly, there was larger variation across therapy style than the other types of metadata. For

true/pred.	agr.	ref.	q.	misc	meta
agreement	153	7	1	3	2
reflection	20	444	62	22	25
question	4	50	302	9	14
norm/misc	3	53	3	55	9
meta	3	38	6	9	60

Table 6: Confusion matrix for SVM on development set categories.

Feature(s) removed	p	r	f
None	70	69.16	69.40
Sentence n-grams	64.64	66.80	65.50
Length	70.11	69.09	69.27
Sentence position	69.82	68.55	68.87
Context unigrams	69.67	68.43	68.43
Context labels	61.95	60.92	60.86
Speaker-change	69.85	68.95	69.2
Turn and intra-turn position	69.45	68.69	68.82

Table 7: Feature ablation for the SVM: precision, recall, and f-measure after removing features.

example, accuracy for sentences taken from Brief Dynamic-Relational Therapy achieved an f-score of only 48.96 with the SVM, while Client-Centered Therapy had an f-score of 71.29. The SVM also did quite well with Cognitive Behavioral Therapy, but this class had only 41 samples. An examination of the annotated sentences for each therapy style themselves revealed two possible explanations for differences in accuracy. The first is that the sentences for Brief Dynamic-Relational and Experiential therapies tended to be nearly twice as long as those for Cognitive Behavioral therapies. They also tended to contain more comma splices and center embedding of clauses suggestive of more complex sentence structure. Secondly, the therapy styles with lower f-scores tended to have a smaller proportion of Agreement sentences (14% for Experiential and just 5% for Brief Dynamic-Relational compared to 46% for Cognitive Behavioral). The greater consistency in category distribution in these transcripts may have contributed to it being easier to guess the categories of their component sentences. Nevertheless, as there was generally very little data for each style, we presume that increasing the annotated data set for each style would help to diminish these differences and bring the therapy style f-scores closer together.

6.3 Ablation studies

From the final configuration of the SVM, we also performed ablation studies to determine which features had the most impact (Table 7). Context labels seemed to be by far the most important, with sentence n-grams second.

6.4 Negative results

In addition to the methods discussed here, we attempted a number of other techniques that were not successful (details presented in the appendix). To address the data scarcity issue, we pretrained on the Switchboard corpus; we tried a few different ways of distantly labeling the unlabeled data; we trained word embeddings on the unlabeled transcripts; we attempted to augment our dataset by "noising" sentences; and we attempted self-training with the unlabeled data. To address the discrepancy between reliability on act-level and category-level codes, we trained a cascading setup for the SVM, where a high-level classifier would first predict the category, and then the corresponding low-level classifier for that category would predict the act within that category. Finally, we attempted a basic weighted-average ensemble of our three non-baseline classifiers (SVM, feedforward net, and CNN with discrete features), as well as a more conservative ensemble that returned the SVM's prediction except when the SVM had low confidence, in which case it backed off to a weighted average.

7   Conclusions and Future Work

We have created a new annotated corpus for therapy dialog act classification with labels at two levels of granularity, and analyzed classification results at each level. Our results indicate that context was very important, followed by sentence information, and that an SVM classifier is sufficient to make use of this information - our SVM model had significantly better performance than both the baselines and the neural methods we tried, aside from a feedforward net on exactly the same features.

One of the major challenges for this task was the limited size of the dataset. To address this, possible future directions include additional work on semisupervised learning, as well as an investigation into active learning for more efficient labeling. More broadly, future work might also focus more closely on the client's statements rather than only the therapist's, in order to glean a more comprehensive picture of the conversation.

References

Stefano Baccianella, Andrea Esuli, and Fabrizio Sebastiani. 2010. Sentiwordnet 3.0: An enhanced lexical resource for sentiment analysis and opinion mining. In *Proceedings of the Seventh conference on International Language Resources and Evaluation (LREC'10)*. European Languages Resources Association (ELRA).

Judith S Beck and Aaron T Beck. 2011. *Cognitive Behavior Therapy*. Guilford Press, New York.

Wilma Bucci and Bernard Maskit. 2005. Building a weighted dictionary for referential activity. *Computing attitude and affect in text*, pages 49–60.

Doğan Can, David C Atkins, and Shrikanth S Narayanan. 2015. A dialog act tagging approach to behavioral coding: A case study of addiction counseling conversations. In *Sixteenth Annual Conference of the International Speech Communication Association*.

Nikolaos Flemotomos, Victor Martinez, James Gibson, David Atkins, Torrey Creed, and Shrikanth Narayanan. 2018. Language features for automated evaluation of cognitive behavior psychotherapy sessions. *Proc. Interspeech 2018*, pages 1908–1912.

Diana Fosha, Daniel J Siegel, and Marion Solomon. 2009. *The healing power of emotion: Affective neuroscience, development & clinical practice*. WW Norton & Company.

Garren Gaut, Mark Steyvers, Zac E Imel, David C Atkins, and Padhraic Smyth. 2017. Content coding of psychotherapy transcripts using labeled topic models. *IEEE journal of biomedical and health informatics*, 21(2):476–487.

James Gibson, Doğan Can, Panayiotis Georgiou, David C Atkins, and Shrikanth S Narayanan. 2017. Attention networks for modeling behaviors in addiction counseling. *Proc. Interspeech 2017*, pages 3251–3255.

James Gibson, Doğan Can, Bo Xiao, Zac E Imel, David C Atkins, Panayiotis Georgiou, and Shrikanth S Narayanan. 2016. A deep learning approach to modeling empathy in addiction counseling. *Interspeech 2016*, pages 1447–1451.

James Gibson and Shrikanth Narayanan. 2018. Multi-label multi-task deep learning for behavioral coding. *arXiv preprint arXiv:1810.12349*.

Thomas D Hull. 2014. Neuropsychiatric mhealth: Design strategies from emotion research. *mHealth Multidisciplinary Verticals*, page 199.

Daniel Jurafsky, Rebecca Bates, Noah Coccaro, Rachel Martin, Marie Meteer, Klaus Ries, Elizabeth Shriberg, Andreas Stolcke, Paul Taylor, and Carol Van Ess-Dykema. 1997. Automatic detection of discourse structure for speech recognition and understanding. In *1997 IEEE Workshop on Automatic Speech Recognition and Understanding Proceedings*, pages 88–95. IEEE.

Yang Liu, Kun Han, Zhao Tan, and Yun Lei. 2017. Using context information for dialog act classification in dnn framework. In *Proceedings of the 2017 Conference on Empirical Methods in Natural Language Processing*, pages 2170–2178. Association for Computational Linguistics.

William R Miller and Stephen Rollnick. 2012. *Motivational interviewing: Helping people change*. Guilford Press, New York.

A Pascual-Leone and LS Greenberg. 2005. Classification of affective-meaning states. *A. Pascual-Leone, Emotional processing in the therapeutic hour: Why the only way out is through*, pages 289–367.

Antonio Pascual-Leone. 2018. How clients change emotion with emotion: A programme of research on emotional processing. *Psychotherapy Research*, 28(2):165–182.

James W Pennebaker, Ryan L Boyd, Kayla Jordan, and Kate Blackburn. 2015. The development and psychometric properties of liwc2015. Technical report.

Stefan Riezler and John T Maxwell. 2005. On some pitfalls in automatic evaluation and significance testing for mt. In *Proceedings of the ACL workshop on intrinsic and extrinsic evaluation measures for machine translation and/or summarization*, pages 57–64.

Lina M. Rojas Barahona, Bo-Hsiang Tseng, Yinpei Dai, Clare Mansfield, Osman Ramadan, Stefan Ultes, Michael Crawford, and Milica Gasic. 2018. Deep learning for language understanding of mental health concepts derived from cognitive behavioural therapy. In *Proceedings of the Ninth International Workshop on Health Text Mining and Information Analysis*, pages 44–54. Association for Computational Linguistics.

Anand Venkataraman, Andreas Stolcke, and Elizabeth Shriberg. 2002. Automatic dialog act labeling with minimal supervision.

Bo Xiao, Doğan Can, James Gibson, Zac E Imel, David C Atkins, Panayiotis Georgiou, and Shrikanth S Narayanan. 2016. Behavioral coding of therapist language in addiction counseling using recurrent neural networks. *Interspeech 2016*, pages 908–912.

Code	Samples	Wps
Agreement	1277	3.01
Disagreement	87	6.77
Apology	18	12.36
Thanking	7	7.4
Hedge	526	12.57
Opinion	676	14.19
Yes-no question	875	9.17
Signal non-understanding	215	8.52
Opening	63	8.30
Closing	90	6.74

Table 8: Distribution over general therapist act classes. "Wps" indicates the average number of words per sentence for that code.

A   Code details

In this section we include more detailed statistics on the distribution of act-level classes in our data. Tables 8 and 9 include the number of sentences as well as the average number of words per sentence for each therapist act. The imbalance at the act level is far greater than that at the category level; the largest category is agreement, with 1277 sentences, while the smallest is thanking, with 7.

B   Annotation process

A screenshot of the annotation spreadsheet is presented in Figure 2. Annotators were presented with a list of sentences and asked to choose an act or "u" (unlabeled) for each one.

Additionally, a confusion matrix for annotators' category labels is presented in Table 10. While the first annotator to give a label for each sentence was treated universally as "Annotator 1" and the second as "Annotator 2", not every sentence with two annotations was labeled by the same two annotators, and so this distinction is somewhat arbitrary. Nevertheless, this matrix still provides some notion of where disagreements occurred.

C   Details of results

Further details of results are presented here. Table 11 contains performance broken down by category for the SVM classifier.

D   Negative results

D.0.1   Distant labeling and data augmentation

As the most evident challenge with this dataset is the relatively small size - especially in the case of

Code	Samples	Wps
Simple reflection	638	9.10
Makes needs explicit	696	15.86
Makes emotions explicit	999	15.63
Makes values explicit	248	14.98
Makes relational patterns explicit	680	18.92
Makes consequences explicit	373	18.54
Makes conflict explicit	382	22.31
Makes strengths/resources explicit	122	18.01
Counterprojection	115	17.12
Teaching/psychoeducation	212	18.82
Problem-solving	166	16.93
Evokes concrete elaboration	1029	10.37
Evokes perspective flexibility	182	14.52
Narrowing	121	14.25
Planning	39	16.46
Assumption checking	426	14.77
Check in/metaprocessing	111	13.46
Self-disclosure	373	18.20
Normalization	77	17.15
Sympathizing	81	13.83
Reassuring	65	15.22

Table 9: Distribution over clinical therapist act classes. "Wps" indicates the average number of words per sentence for that code.

	agr.	refl.	q.	misc.	meta
agr.	46	0	0	5	0
refl.	0	136	24	7	6
q.	0	49	121	2	1
misc.	2	5	5	26	1
meta	1	3	0	0	11

Table 10: Annotator confusion matrix. Rows correspond to labels from the annotator 1, columns to labels from annotator 2.

category	precision	recall	F1
agreement	80.20	95.18	87.05
reflection	75.42	78.57	76.96
question	79.46	77.37	78.40
norm/misc	54.17	42.28	47.49
meta	65.31	55.17	59.81

Table 11: SVM performance by category.

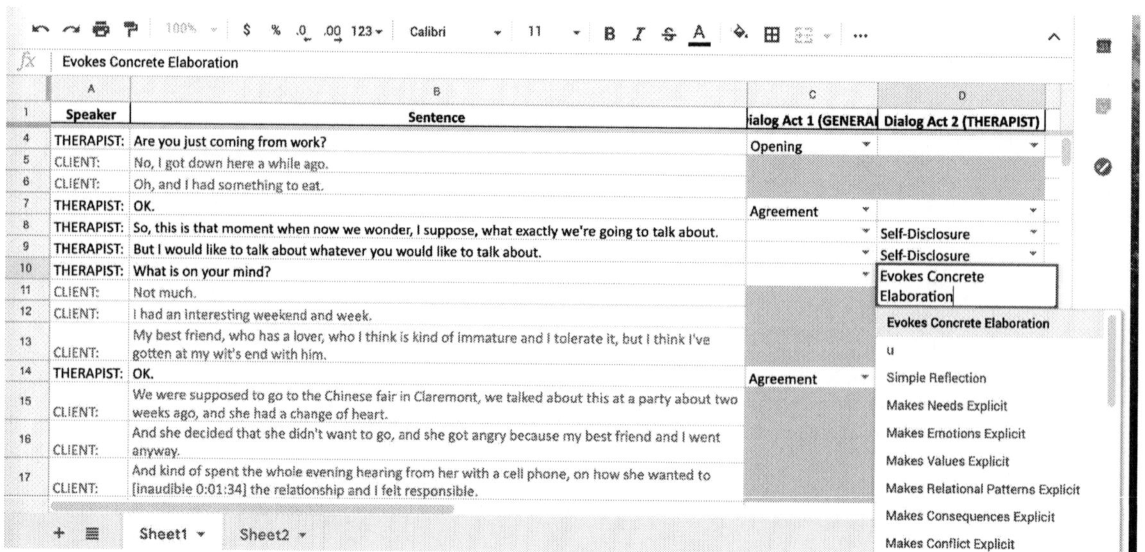

Figure 2: The interface that annotators used.

classification at the act level, in which the category classes are further subdivided - a natural course of inquiry was whether we could find additional data for transfer learning, produce noisy labels by some method on our much larger set of available unlabeled data, or leverage the unlabeled data in some other way.

Our first attempt in this direction was simply to add to our dataset the subset of labeled data from the Switchboard corpus corresponding to the labels that we had selected for our own annotation scheme. Surprisingly, this improved performance neither on the clinical labels nor even on the corresponding general labels. The fact that the Switchboard data was relatively uninformative for our own classification task suggests that the content of general-topic conversation (as in Switchboard) markedly differs from that found in therapy, as in our own corpus.

We next turned our attention to the remaining transcripts in the Alexander Street corpus that had not been labeled. We trained word embeddings on this data (using Word2Vec, with varying dimensionalities, and a window size of 7 and minimum count of 4); however, random initialization proved superior to both these and the publicly available pretrained embeddings trained on the Google News corpus.

As our SVM model had found success with relatively simple features, we also attempted to augment our dataset with distant labels generated by a few simple heuristic rules - if a sentence ends with '?', label it as a question; if it has relatively many agreement words, label it as agreement; return counterprojection if it has many "I" words (I, me, my, etc.); return normalization/misc if it has a high sentiment score; return reflection if it has many "you" words; and guess nothing otherwise.

Finally, observing the typical suite of tactics employed to boost the size and robustness of image datasets, we attempted to develop a similar technique for data augmentation in text. In essence, we drop or replace words randomly (with uniform probability, or with probability proportional to their smoothed unigram frequency). With a high base rate, this should produce highly noisy sentences that nevertheless contain some amount of signal approximating the original training data, hopefully improving classifier robustness. Unfortunately, this did not in fact improve performance.

D.0.2 Semisupervised learning

Partially inspired by the work of (Venkataraman et al., 2002), we explored self-training the SVM on sentences from the unlabeled transcripts. We experimented with a number of different learning schedules - adding all data labeled above a fixed confidence threshold to the training set in the next iteration; progressively increasing the confidence threshold by a fixed step at each iteration; halving the distance from the threshold to 100% confidence at each iteration; and scaling the base threshold by the ratio of current average confidence to original confidence over all unlabeled sentences at each iteration. Very small improvements were found under some settings in preliminary work, but we did not explore this direction thoroughly as it yielded a dramatic increase in training time but only very minor gains in performance. Nevertheless, this might be worth revisiting in a more principled fashion in future work.

D.0.3 Ensembling

We attempted a couple simple methods of ensembling, in the hopes that our classifiers were different enough that this would yield useful information. The most basic of these was a simple weighted average of the prediction scores in each of the classes, with the highest averaged score being the final prediction. We also tried an ensemble-based method where we used the SVM's prediction unless its confidence was beneath a certain threshold, in which case we backed off to a weighted ensemble. Neither of these produced a performance improvement over the SVM; only the best weight assignment for classifiers that we found in the former case even approached the SVM's performance. This may be due to the high agreement between classifiers (agreement percentages between 86-92% for all three pairs of classifiers), meaning that none of them contributes new information relative to the others.

E   Metadata analysis

We include breakdowns of performance by other metadata fields on the following page.

Psychological subject	F1	Samples
Emotional states	67.65	1458
Relationships	66.80	1244
Personality traits	70.10	516
Frustration	66.40	463
Spousal relationships	67.57	302
Behavior	74.20	285
Guilt	75.78	277
Family	64.35	267
Diagnosis	76.05	252
Sexual behavior	72.72	235
Communication	64.40	230
Client-counselor relations	64.36	193
Parent-child relationships	64.14	187
Personality factors	71.68	143
Ability	66.87	129
Self-confidence	67.19	97
Family relations	59.63	76

Table 12: Performance breakdown by psychological subject.

Experience	F1	Samples
Under 10 years	71.35	1102
11-20 years	75.35	118

Table 13: Performance breakdown by therapist experience.

Client age	F1	Samples
21-30 years	70.69	1200
31-40 years	77.13	40
41-50 years	42.94	109
51-60 years	54.17	8

Table 14: Performance breakdown by client age.

Client gender	F1	Samples
Male	69.22	744
Female	69.21	613

Table 15: Performance breakdown by client gender.

CLPsych 2019 Shared Task: Predicting the Degree of Suicide Risk in Reddit Posts

Ayah Zirikly[1], **Philip Resnik**[2], **Özlem Uzuner**[3,4], and **Kristy Hollingshead**[5]

[1]Rehabilitation Medicine Department, National Institutes of Health, Bethesda, MD, USA
[2]Linguistics and UMIACS CLIP Laboratory, University of Maryland, College Park, MD, USA
[3]George Mason University, Fairfax, VA, USA
[4]Massachusetts Institute of Technology, Cambridge, MA, USA
[5]Florida Institute for Human and Machine Cognition, Pensacola, FL, USA
`ayah.zirikly@nih.gov, resnik@umd.edu, ouzuner@gmu.edu, kseitz@ihmc.us`

Abstract

The shared task for the 2019 Workshop on Computational Linguistics and Clinical Psychology (CLPsych'19) introduced an assessment of suicide risk based on social media postings, using data from Reddit to identify users at no, low, moderate, or severe risk. Two variations of the task focused on users whose posts to the r/SuicideWatch subreddit indicated they might be at risk; a third task looked at screening users based only on their more everyday (non-SuicideWatch) posts. We received submissions from 15 different teams, and the results provide progress and insight into the value of language signal in helping to predict risk level.

1 Introduction

Predicting risk of suicide is hard. McHugh et al. (2019), reviewing 70 studies, conclude that suicidality cannot be predicted effectively using the standard practice of clinicians asking people in person about suicidal thoughts: 80% of patients who were not already undergoing psychiatric treatment and who died of suicide denied having suicidal thoughts when asked by a general practitioner. They conclude that their study, along with with other recent meta-analyses, "highlight a high degree of uncertainty about the statistical strength of commonly used approaches to suicide risk assessment."

On a similar theme, after carefully reviewing more than three hundred studies, Franklin et al. (2016) conclude that predictive ability for suicidal thoughts and behaviors (STBs) has not improved across 50 years of research. Nock et al. (2019) observe that, in contrast to other fatal problems like flu or tuberculosis, deaths by suicide are as prevalent now as they were a hundred years ago, a lack of progress resulting in large part because "we lack a firm understanding of the fundamental properties of STBs, and when, why, and among whom they unfold" — not least because suicidal thoughts and behaviors rarely occur in a research laboratory.

Coppersmith et al. (2018) offer a powerful example of the information that is available beyond the research laboratory. They observe that for many people the "clinical whitespace" — long intervals between healthcare encounters — is occupied by frequent use of social media, an opportunity for obtaining data "in situ" (Nock et al., 2019), and they demonstrate that this information can be tapped effectively in order to build create automated binary classifiers for screening.

This progress raises two new problems, though. First, when binary screening systems are deployed, the number of people flagged as at risk will far exceed clinical capacity for intervention. So, rather than a binary classification, a finer grained assessment for degree of risk is needed, in order to support decisions about intervention priority. Second, obtaining relevant data for developing, improving, and validating classifiers is extremely difficult. Coppersmith and colleagues, for example, went to considerable effort to obtain donations of private social media data for research on suicide, and these sensitive materials are not easy to share with the broader research community.[1]

With these considerations in mind, we have formulated a new shared task for research community participation, based on a dataset introduced by Shing et al. (2018). In order to address the limits of binary classification, we formulate tasks based on a multi-level assessment of suicide risk

---

[1]In particular, Coppersmith et al. (2018) have introduced the OurDataHelps.org platform, which permits donors to authorize research access to their data from numerous social media sources, as well as information from wearables and other technologies. The platform has been adapted by their collaborators for research on other mental health topics, as well; for example, UMD.OurDataHelps.org collects data donations for a project focused on schizophrenia.

Proceedings of the Sixth Workshop on Computational Linguistics and Clinical Psychology, pages 24–33
Minneapolis, Minnesota, June 6, 2019. ©2019 Association for Computational Linguistics

designed for social media, similar in spirit to previous CLPsych shared tasks on four-way assessment of crisis risk in a peer support forum (Milne et al., 2016; Milne, 2017). In order to address ethical access to and sharability of data, we focus on materials collected from Reddit, where posts are public and anonymous, and further de-identified by us; see Section 2. A limitation of the tasks is that we lack information about actual outcomes (suicide attempts or competions); we instead use human annotations of risk level as a starting point. In that regard this year's exercise can be viewed at minimum as establishing face validity for the idea of extracting meaningful signal related to suicidality from Reddit posts, and more optimistically as a step along the path to clinically meaningful predictions.

2 Data

2.1 Source dataset

We derived our shared task data from the dataset introduced by Shing et al. (2018). Shing et al. began with a collection intended to contain essentially every publicly available Reddit posting from its beginning in 2005 into summer 2015, and identified a subset of users potentially at risk by extracting all users who had posted to the *r/SuicideWatch* subreddit.[2] The process was analogous to the data collection method pioneered by Coppersmith et al. (2014) for a variety of mental health conditions, where an explicit signal for candidate (potentially relevant) Twitter users was defined by specifying a self-report pattern, e.g. *I have been diagnosed with [condition]*, and then matching posts were reviewed manually to identify candidates where the signal does not appear genuine, such as sarcastic or joking references. For the suicidality dataset, posting on SuicideWatch constituted the signal, and Shing et al. (2018) collected 11,129 candidate users on SuicideWatch, accounting for a total of 1,556,194 posts across Reddit, along with a comparable number of control users who did not post on SuicideWatch.[3]

2.2 User-level annotation

As discussed in more detail by Shing et al. (2018), annotation involved the assessment of risk for a randomly selected subset of 621 users on a four-level scale, based on their SuicideWatch posts. A detailed set of annotation instructions drawing on prior literature (Joiner et al., 1999; Corbitt-Hall et al., 2016), created in consultation with suicide prevention experts, identified four families of risk factors, described as follows:

- *Thoughts* includes not only explicit ideation but also, e.g., feeling they are a burden to others or having a "f*** it" (screw it, game over, farewell) thought pattern;

- *Feelings* includes, e.g., a lack of hope for things to get better, or a sense of agitation or impulsivity (mixed depressive state, Popovic et al. (2015));

- *Logistics* includes, e.g., talking about methods of attempting suicide (even if not planning), or having access to lethal means like firearms;

- *Context* includes, e.g. previous attempts, a significant life change, or isolation from friends and family.

Using this assessment scheme, Shing et al. obtained annotations both from experts and from crowdsource workers for a randomly selected subset of users based on their SuicideWatch postings, assigning one of the following risk levels (a to d):

(a) No Risk (or "None"): I don't see evidence that this person is at risk for suicide;

(b) Low Risk: There may be some factors here that could suggest risk, but I don't really think this person is at much of a risk of suicide;

(c) Moderate Risk: I see indications that there could be a genuine risk of this person making a suicide attempt;

---

[2]The r/SuicideWatch subreddit, `https://www.reddit.com/r/SuicideWatch/`, is a forum providing "peer support for anyone struggling with suicidal thoughts, or worried about someone who may be at risk". Henceforth we refer to it simply as SuicideWatch.

[3]It is worth noting that, subsequent to Shing et al.'s collection and annotation, Gaffney and Matias (2018) reported on an analysis showing that the widely used Baumgartner Reddit collection, which Shing et al. had used as their start-ing point, has a number of gaps and limitations. However, Gaffney and Matias identify the greater risks as pertaining to user history analyses, network analysis, or comparison of participation across communities. They posit lower risk from coverage gaps for machine learning work on predictive modeling, commenting, "since the purpose of this kind of machine learning research is to make inferences about out-of-sample observations rather than to test hypotheses about a population, such research may be less sensitive to variation due to missing data."

(d) Severe Risk: I believe this person is at high risk of attempting suicide in the near future.

It is important to note that this process produced risk assessments at the level of individual users, not of individual posts. Inter-rater reliability was achieved for experts (Krippendorff's $\alpha = 0.81$) (to our knowledge the first published demonstration of reliability for clinical assessment of suicidality based on social media), along with fair agreement among crowdsourcers (Krippendorff's $\alpha = 0.55$). Analysis of the results also showed that when crowdsource workers make mistakes relative to experts' judgments, they tend to err on the side of caution — a good thing in a setting where false positives are a better kind of error than false negatives.

In the absence of data about outcomes (see discussion in Section 6), we expect the expert annotations to represent "truth" more accurately than crowdsourced judgments. However, for the shared task we elected to create both training and test data using the crowdsourced annotations, rather than using expert judgments as test data. We made this choice for two reasons. First, at least this first time creating a shared task on Reddit suicidality assessment, we wished to avoid the extra difficulties encountered in machine learning when there are mismatches between the training set and the test set. Second, we anticipate the possibility of repeating this shared task, and would like to lay the groundwork for a head-to-head comparison of results; obtaining crowdsourced judgments to create fresh test data will be considerably more practical than obtaining more expert judgments.

2.3 Reddit posts and metadata

For our tasks, the evidence we have about users' mental state comes from their Reddit posts. Information provided to participant teams included post_id (a unique identifier for the post), user_id (a unique numeric identifier for the user who authored the post), timestamp (time the post was created, encoded as a Unix epoch), subreddit (the name of the subreddit where the post appeared), post_title (title of the post) and post_body (text contents of the post).[4]

As discussed further in Section 7, although Reddit data are publicly available and the site was

created specifically for anonymous posting, discussions on the platform nonetheless need to be viewed as sensitive and subject to careful ethical consideration (Benton et al., 2017; Chancellor et al., 2019). For that reason, a number of steps were taken to further remove identifying information from the dataset for the shared task.

First, although Reddit is a site for anonymous discussion, it is possible for users to put identifying information in their self-selected user names; although most select names like *awesomeprogrammer*, in principle nothing on the site would prevent someone from naming herself *mary-smith-UMDsophomore-born7July2002*. Therefore the dataset replaces the self-selected user names with arbitrary numeric identifiers for the user_id.

Second, automatic processing was performed on post titles and bodies, to replace IP addresses, email addresses, URLs, and person entities with special tokens.[5] For example, a processed post body might resemble this made-up example: *Taking a great class from _PERSON_ _PERSON_. If you want to learn more about it drop me a line at _EMAIL_ or check it out at _URL_.*

In addition, we filtered out all posts containing Arabic using the langdetect library.[6] We also performed data-cleaning steps to remove encoding issues or special string sequences that tokenizers such as spaCy's would fail to handle.

3 Tasks

Teams participated in one or more of the following three tasks.

- Task A is about risk assessment: the task simulates a scenario in which there is already online evidence that a person might be in need of help (e.g., because they have posted to a relevant online forum or discussion, in this case r/SuicideWatch), and the goal is to assess their level of risk from what they posted. This task uses the smallest amount of data, with each user typically having no more than a few SuicideWatch posts.

- Task B is the same risk assessment problem as task A, but in addition to the SuicideWatch posts (which identify that they may need help), teams can also use the users posts

---

[4]Unix epochs are a widely used standard for encoding time. Any timestamp is represented as the number of seconds that have passed since 00:00:00 Thursday, 1 January 1970, Coordinated Universal Time (UTC), minus leap seconds.

[5]We used spaCy for named entity recognition.
[6]https://pypi.org/project/langdetect/

elsewhere on Reddit (which might tell you more about them or their mental state). On average each user we collected data for has more than 130 posts on Reddit, and the subreddit categories are wildly diverse, from *Accounting* to *mylittlepony* to *SkincareAddiction* to *zombies*.

- Task C is about screening. This task simulates a scenario in which someone has opted in to having their social media monitored (e.g., a new mother at risk for postpartum depression, a veteran returning from a deployment, a patient whose therapist has suggested it) and the goal is to identify whether they are at risk even if they have not explicitly presented with a problem. Here predictions are made only from users posts that are not on SuicideWatch.

For all tasks, we provided participating teams with training and test data using an 80-20 split. In order to keep the original labels' distribution in the split, we applied the proportional training/test split separately for each label. The statistics of the data are shown in Tables 1 and 2. Note the large number of posts in tasks B and C, which makes these two tasks more challenging given the extra information and noise the participants have about each user.

	train	test	total
a	127	32	159
b	50	13	63
c	113	28	141
d	206	52	258
control	497	124	621
total	993	249	1242

Table 1: Number of users in training and test data

	Task A	Task B	Task C
train	919	57015	56096
test	186	9610	14231

Table 2: Number of posts for each task per split

4 Shared task submissions

Fifteen teams participated in at least one task, with 12 participating in task A, 11 in task B, and 8 in task C. Each team was permitted to submit up to 3 runs per task, and each identified a primary system that would be used in the official results and rankings. The full number of submissions we received for tasks A, B, and C were 33, 28, and 22, respectively. Teams were given in total (training and testing) about four weeks to develop their systems, generating predictions on test data during a roughly week-long interval at the end. Table 3 shows the participating teams and the tasks they submitted to, with per-task rankings (see Section 5).

In this section, we list the common preprocessing steps that the teams used prior to training and testing. Additionally, we descibe the approaches followed (machine learning models and features if applicable) in Sections 4.2 and 4.3. In section 6, we provide more details about the top systems per task.

Team	A	B	C
Affective_Computing	7	7	
ASU (Ambalavanan et al., 2019)	2		5
CAMH †	5	2	2
Chen et al. (2019)		4	
CLaC (Mohammadi et al., 2019)	1	5	1
CMU (Allen et al., 2019)	8		
IBM data science (Morales et al., 2019)	12	10	4
IDLab (Bitew et al., 2019)	4		
JXUFE †	9	8	
SBU-HLAB (Matero et al., 2019)	3	1	3
TsuiLab (Ruiz et al., 2019)		3	
TTU (Iserman et al., 2019)	6		8
UniOvi-WESO (Hevia et al., 2019)	10	11	
uOttawa †		9	7
USI-UPF (Ríssola et al., 2019)	11	6	6

Table 3: CLPsych 2019 participating teams and rankings (no paper is available for teams indicated with †)

4.1 Data preprocessing

The most common preprocessing steps that teams followed was removing stop words and punctuation, in addition to lowercasing. Some teams opted to remove the special deidentification tokens (e.g., _PERSON_, _URL_), and to apply number normalization or removal. Some filtered out posts that contain more than thirty _PERSON_ special tokens. An interesting preprocessing step suggested ordering the posts by timestamp, following the intuition that recent posts have more impact on the risk assessment. Additionally, some teams aggregated the name of the subreddit to the post for task B and C (Ambalavanan et al., 2019). Most teams employed commonly used tokenizers such as spaCy (Honnibal and Montani, 2017); an exception is Ríssola et al. (2019), who used Ekphrasis (Baziotis et al., 2017), a tool set that is tailored for social media data. Iserman et al. (2019) applied two-stage error spelling correction using edit distance from augmented dictionary entries.

4.2 Approaches

4.3 Model inputs

The submitted systems used a wide range of input representations on the post and user level. We can distinguish several main categories:

- Embeddings on the word, sentence or document (post/posts) level. In addition to GloVe and word2vec, we mostly see the more recently introduced contextualized embedding techniques such as ELMo (Peters et al., 2018) and BERT (Devlin et al., 2018).

- Lexicon-based features. Teams used dictionaries mainly to capture emotions represented in the user's posts. Examples of dictionaries used are NRC (Mohammad, 2017) and LIWC (Tausczik and Pennebaker, 2010). These features were generally represented as the normalized count of post per category. Other lexicons were employed to capture user-level features including age and gender (Sap et al., 2014), and assessment of the Big-5 personality traits (Schwartz et al., 2013).

- N-gram features, mainly in the form of unigrams with TF-IDF weighting.

- Meta-features such as the time when the post was made available (i.e. timestamp) or the post's subreddit (Tasks B and C).

- Topic models such as LDA (Blei et al., 2003) and Empath (Fast et al., 2016).

We also see keywords to identify certain behaviors such as motivations linked to suicidality using a set of keywords; clinical findings in terms of UMLS (Bodenreider, 2004) keywords in the posts, flagging the suicide-related unique identifiers (CUIs); and language style similarity between posts in the same subreddit.

4.4 Models

The submissions for the shared task range from conventional machine learning approaches to deep neural network models. Support vector machines (SVM) and logistic regression are frequently used, in addition to the occasional decision tree and random forests approach. These approaches often involve feature engineering, where we see a wide variety and extensive combinations of the features mentioned above (Section 4.3).

The neural network models, on the other hand, depend mainly on embeddings, though teams opted to use the embeddings output from the language models in different ways. Many teams fine-tune the embeddings on either the full training data, the SuicideWatch subset, or on each of the title and body of the posts to create separate language models. Some teams used models that were pre-trained on Wikipedia and some other large corpora as-is in their system.

The most commonly used neural architecture is convolutional neural networks (CNN) on the user or post level, where in the latter case an aggregation step is needed to produce the final outcome. Other frequently employed architectures were long-short term memory (LSTM) networks or recurrent neural networks (RNNs) and LSTMs with an attention mechanism. Some teams experimented with multichannel neural networks in a multi-task learning setting.

5 Results

5.1 Metrics

The official metric used in this shared task is the macro-averaged F1 score. This metric was also used in previous CLPsych shared tasks that classified online posts into one of four labels (Milne et al., 2016; Milne, 2017); as a way of defining a single figure of merit, macro-averaging treats each class as contributing equally to performance, which helps avoid performance on a single class dominating the result when there is class imbalance (cf. Table 1).

In addition, we adopt two metrics introduced in those previous shared tasks, derived from systems' four-way classifications: *urgent* is the accuracy in making the binary distinction between a, b vs. c, d, and *flagged* is the accuracy in distinguishing b, c, d from a. The reasoning behind these metrics lies in real-world use cases one might encounter. A system that is good at identifying *urgent* posts can be viewed as a first step in potentially time-sensitive triage (erring on the side of caution by including moderate as well as severe risk), while a system that is good at *flagged* distinctions helps avoid wasting valuable human effort on no-risk cases.[7]

For each of the three tasks we report official rankings based on the primary system identified by the team. Additionally, in Section 6 we report on the best run in terms of macro-F1 score, whether primary or not.

Tables 4, 5, and 6 provide the results of the primary runs of the participating teams for each of the three tasks, ranked by highest macro-F1 score. For tasks A and C, the CLaC team (Mohammadi et al., 2019) ranks first with a combination of conventional and neural models: an SVM is employed at the end of the pipeline, where it acts as a meta-classifier on top of a set of CNN, Bi-LSTM, Bi-RNN and Bi-GRU neural networks. However, for both of those tasks, the primary runs do not generate the best *unofficial* macro-F1 score on the test set: a different variation on the CLaC approach, in which SVM uses as input both the neural features and the predicted class probabilities from the SVM, yields the best macro-F1 score, 0.533 for task A as compared with 0.481 for the primary system. On the other hand, the CAMH system, which uses a stacked parallel CNN with LIWC and a universal sentence encoder (Cer et al., 2018), produced the best unofficial F1 score for task C: 0.278 as compared to 0.268 for the CLaC primary system.

For task B the best official score is 0.457, obtained by the HLAB team, where the system used logistic regression with features from Suicide-Watch and non-SuicideWatch language that were processed separately. The best unofficial F1 score (0.504) is also obtained by HLAB system, using BERT features generated separately from Suicide-Watch and non-SuicideWatch posts.

6 Discussion

In comparing the results of tasks A and B, we note that systems, especially the top systems, perform comparably in terms of predicting the severe risk label (d). This suggests that, in general, information about all the other Reddit posts by a user does not necessarily add noise that hurts the performance, but rather, in some instances, it might have positive impact. Surprisingly identifying severe risk posts in task C yields good results given

that the set of available posts excludes Suicide-Watch and other mental health subreddits. However, the overall F1 score is low, which suggests that future work should focus on correctly classifying the non-severe risk labels (c and b). Across tasks, classifying label b has a low performance, which is mainly due to its smaller training size in comparison to the other labels. Additionally, and as expected, all systems are better at predicting the two extreme labels (d and a) as opposed to the medium-risk labels(c and b).

As a way to augment the training data and to benefit from other available datasets, Hevia et al. (2019) experimented with including ReachOut data from the CLPsych 2016 and 2017 shared tasks (Milne et al., 2016; Milne, 2017). Unfortunately, adding this dataset resulted in slightly worse performance. Although both datasets adopt a four-way scale, the annotation guidelines are different and there is no guaranteed one-to-one mapping between the two.

One of the interesting findings from the different systems is that severe-risk users appear to use a distinct vocabulary in comparison to the rest of the labels. This would support the intuition of building separate language models for Suicide-Watch and non-SuicideWatch, or special features that can benefit from emotion and mental-health related lexicons.

Interestingly, we note that most submitted systems over-predict label d when the correct label is c. This confirms the value of reporting the *urgent* F1 score, noting that, in some instances, the distinction between the two labels is hard even for the crowdsourcers (Shing et al., 2018). Additionally, a number of the false positives observed concern users seeking advice for a relative or a friend as opposed to themselves. This suggests that building models specifically to separate such cases would be of value.

7 Ethical considerations

Mental health is a sensitive subject area, and work on technology for mental health using social media has broad implications. Benton et al. (2017) and Chancellor et al. (2019) provide thoughtful and comprehensive consideration of ethical issues. Informed by their discussions we focus here on several key ethical considerations for this shared task and how we handled them.

---

[7]Similarly to the previous shared tasks with four-way labeling, we exclude the no-risk label a in evaluation for screening task C. However, macro-F1 score is calculated over all four labels for tasks A and B.

team	accuracy	macro-f1	(flagged) f1	(urgent) f1	(d) f1	(c) f1	(b) f1	(a) f1
CLaC	0.504	**0.481**	**0.922**	0.776	0.543	0.4	0.244	0.737
ASU	0.544	0.477	0.882	0.826	0.655	0.281	**0.316**	0.656
SBU-HLAB	0.56	0.459	0.842	0.839	**0.692**	0.235	0.25	0.658
IDLab	0.544	0.445	0.852	0.789	0.673	0.292	0.167	0.649
CAMH	0.528	0.435	0.897	0.783	0.623	0.327	0.083	0.708
TTU	0.504	0.402	0.902	0.844	0.6	0.14	0.2	0.667
Affective Computing	**0.592**	0.378	0.92	**0.862**	0.685	0.065	0	**0.762**
CMU	0.472	0.373	0.876	0.773	0.545	0.302	0	0.646
JXUFE	0.464	0.364	0.882	0.779	0.571	0.217	0.087	0.582
UniOvi-WESO	0.512	0.312	0.897	0.821	0.633	0.062	0	0.553
USI-UPF	0.376	0.291	0.753	0.707	0.475	**0.408**	0	0.281
IBM data science	0.432	0.178	0.861	0.788	0.594	0	0	0.118

Table 4: Official results of task A primary systems ordered by macro-F1 score

team	accuracy	macro-f1	(flagged) f1	(urgent) f1	(d) f1	(c) f1	(b) f1	(a) f1
SBU-HLAB	**0.56**	**0.457**	0.821	**0.816**	**0.699**	0.245	0.25	0.634
CAMH	0.512	0.413	**0.91**	0.812	0.598	0.226	0.105	**0.721**
TsuiLab	0.408	0.37	0.789	0.603	0.506	0.264	0.205	0.507
Chen et al.	0.424	0.358	0.83	0.738	0.478	0.14	0.182	0.633
CLaC	0.416	0.339	0.843	0.718	0.549	0.185	0.069	0.554
USI-UPF	0.336	0.311	0.743	0.667	0.439	0.089	**0.417**	0.299
ASU	0.368	0.261	0.765	0.691	0.536	0.151	0	0.358
JXUFE	0.36	0.259	0.798	0.694	0.508	**0.298**	0	0.231
uOttawa	0.448	0.253	0.787	0.71	0.596	0	0	0.418
IBM data science	0.416	0.212	0.82	0.738	0.566	0	0	0.28
TTU	0.416	0.148	0.848	0.775	0.591	0	0	0

Table 5: Official results of task B primary systems ordered by macro-F1 score

7.1 Participants and research oversight

Social media posts are a window into people's thoughts and often into details of their lives. This has enormous value in understanding and predicting mental health, but it stands in tension with concerns about privacy, and formalized ethical standards only address these issues to a limited extent. The dataset used in this shared task was derived from previously existing, publicly available material on Reddit, and we obtained an Institutional Review Board (IRB) determination that work using the material constitutes "secondary research for which consent is not required", including the ability to share the dataset with other researchers, under the U.S. Federal Policy for the Protection of Human Subjects.[8] However, we also took several additional steps regarding participant protection and research oversight.

First, although a key characteristic of Reddit is its focus on anonymity (Gutman, 2018), users retain the ability to volunteer identifying information. As discussed in Section 2.3, therefore, we implemented additional, conservative measures for automatic de-identification to reduce the possibility of including any identifying information in either metadata or text data. In informal review of two sets of 100 randomly sampled postings from our training data, after de-identification — one from all postings and the other just from SuicideWatch — we found zero instances of personally identifying information in either text or metadata.

In addition, in order for teams to participate in the shared task, we required them (a) to provide evidence that their *own* organization's IRB (or equivalent ethical review board) had reviewed and approved their research activity using the dataset, (b) to provide a data management plan including provisions for appropriate protection of the data, and (c) to affirm that all team members had read Benton et al. (2017) and were committed to its broad ethical principles.[9] Mindful of Chancellor et al.'s call to include key stakeholders in the research process, the design of participant applications and their reviewing took place in consultation with clinical and domain experts at the American Association of Suicidology.

---

[8] `https://www.hhs.gov/ohrp/regulations-and-policy/regulations/common-rule/index.html`

[9] Teams' papers in this proceedings may or may not explicitly have mentioned IRB or ethical review, but it can be presumed in all cases to have been done.

team	accuracy	macro-f1	(flagged) f1	(urgent) f1	(d) f1	(c) f1	(b) f1
CLaC	0.673	**0.268**	0.671	**0.625**	**0.527**	**0.189**	0.087
CAMH	0.613	0.226	**0.673**	0.599	0.497	0.048	**0.133**
SBU-HLAB	**0.69**	0.176	0.587	0.554	0.465	0.065	0
IBM data science	0.435	0.165	0.554	0.455	0.329	0.097	0.069
ASU	0.597	0.159	0.63	0.575	0.396	0.082	0
USI-UPF	0.5	0.136	0.377	0.297	0.291	0.115	0
uOttawa	0.52	0.129	0.541	0.485	0.386	0	0
TTU	0.222	0.118	0.542	0.489	0.353	0	0

Table 6: Official results of task C primary systems ordered by macro-F1 score

7.2 The role of predictive models

Social media's window into the "clinical whitespace" (Coppersmith et al., 2018) offers the potential to identify and intervene with people who do not or cannot receive attention through conventional healthcare interactions. At the same time, algorithmic prediction of suicidality creates new challenges, such as creating potentially stigmatizing labels for false or even true positives, or generating an overwhelming number of new cases requiring intervention.

We cannot hope to address these issues in a single shared task, but we did have them in mind when designing it. Our view, informed by research in other domains, is that the most substantial, rapid progress on a problem takes place when a community is constructed around a common task with common data, even when the task and data are not perfect. (As is the case here, for example, in starting with crowdsourced judgments; see Section 2.2.) The way to understand tradeoffs and consequences involving false negatives and false positives is to build systems that make predictions, and then to involve clinicians and other practitioners in discussion of what those systems do, and how this relates to the real-world need — which makes CLPsych, as the venue for this shared task, just as important as the shared task itself.

8 Conclusions

The CLPsych 2019 shared task succeeded in its primary aims, which were to elicit community interest and effort in the problem of suicidality assessment using social media, and to lay solid foundations for work on this problem that will ultimately lead to deployable technology. The best results here show strong performance in culling out, among users who have posted to Reddit's SuicideWatch forum, those who are in urgent need of attention, and, conversely, in distinguishing people who might need attention from those who are at no risk. We also see a solid start on the even more challenging problem of identifying users in need of attention from more ordinary posts that do *not* come from SuicideWatch. On evaluation of finer grained, four-way classification we find that the medium risk categories (low and moderate, as opposed to no risk or severe risk) are more challenging for systems, just as they are more difficult for human judges (Shing et al., 2018).

We aim to address some of the limitations of the present shared task in the near future. Although crowdsourced judgments permit easily repeatable evaluations, we also hope to facilitate community-level evaluation against expert judgments. We are also working on the creation of secure community infrastructure for research on sensitive mental health data, in order to reduce practical obstacles and reduce data privacy concerns by bringing researchers to the data, rather than disseminating data out to researchers. Our ultimate goal is to create an environment where rapid progress can be achieved by combining the benefits of large scale, publicly available, annotated data, as explored here, with private social media and associated outcomes obtained using fully consented, donated data (e.g. via OurDataHelps.org, Coppersmith et al. (2018)).

Acknowledgments

The second author was supported in part by an Amazon Machine Learning Research Award and a UMB-UMCP Research and Innovation Seed Grant. The authors gratefully acknowledge important discussions with Kate Loveys, Kate Niederhoffer, Rebecca Resnik, April Foreman, Jonathan Singer, Beau Pinkham, and Tony Wood, as well as Amazon and the University of Maryland Center for Health-related Informatics and Bioimaging (CHIB) for their generous sponsorship of CLPsych this year.

References

Kristen Allen, Shrey Bagroy, Alex Davis, and Tamar Krishnamurti. 2019. ConvSent at CLPsych 2019 task a: Using post-level sentiment features for suicide risk prediction on Reddit. In *Proceedings of the Sixth Workshop on Computational Linguistics and Clinical Psychology*.

Ashwin Karthik Ambalavanan, Pranjali Dileep Jagtap, Soumya Adhya, and Murthy Devarakonda. 2019. Using contextual representations for suicide risk assessment from internet forums. In *Proceedings of the Sixth Workshop on Computational Linguistics and Clinical Psychology*.

Christos Baziotis, Nikos Pelekis, and Christos Doulkeridis. 2017. Datastories at semeval-2017 task 4: Deep lstm with attention for message-level and topic-based sentiment analysis. In *Proceedings of the 11th International Workshop on Semantic Evaluation (SemEval-2017)*, pages 747–754, Vancouver, Canada. Association for Computational Linguistics.

Adrian Benton, Glen Coppersmith, and Mark Dredze. 2017. Ethical research protocols for social media health research. In *Proceedings of the First ACL Workshop on Ethics in Natural Language Processing*, pages 94–102.

Semere Kiros Bitew, Giannis Bekoulis, Johannes Deleu, Lucas Sterckx, Klim Zaporojets, Thomas Demeester, and Chris Develder. 2019. Predicting suicide risk from online postings in Reddit the UGent-IDLab submission to the CLPysch 2019 shared task A. In *Proceedings of the Sixth Workshop on Computational Linguistics and Clinical Psychology*.

David M Blei, Andrew Y Ng, and Michael I Jordan. 2003. Latent dirichlet allocation. *Journal of machine Learning research*, 3(Jan):993–1022.

Olivier Bodenreider. 2004. The unified medical language system (umls): integrating biomedical terminology. *Nucleic acids research*, 32(suppl_1):D267–D270.

Daniel Cer, Yinfei Yang, Sheng-yi Kong, Nan Hua, Nicole Limtiaco, Rhomni St John, Noah Constant, Mario Guajardo-Cespedes, Steve Yuan, Chris Tar, et al. 2018. Universal sentence encoder. *arXiv preprint arXiv:1803.11175*.

Stevie Chancellor, Michael L Birnbaum, Eric D Caine, Vincent Silenzio, and Munmun De Choudhury. 2019. A taxonomy of ethical tensions in inferring mental health states from social media. In *Proceedings of the 2nd ACM Conference on Fairness, Accountability, and Transparency (FAT* '19)*.

Lushi Chen, Abeer Aldayel, Nikolay Bogoychev, and Tao Gong. 2019. Similar minds post alike: Assessment of suicide risk by hybrid language and behavioral model. In *Proceedings of the Sixth Workshop on Computational Linguistics and Clinical Psychology*.

Glen Coppersmith, Mark Dredze, and Craig Harman. 2014. Quantifying mental health signals in Twitter. In *Proceedings of the Workshop on Computational Linguistics and Clinical Psychology: From Linguistic Signal to Clinical Reality*, pages 51–60.

Glen Coppersmith, Ryan Leary, Patrick Crutchley, and Alex Fine. 2018. Natural language processing of social media as screening for suicide risk. *Biomedical Informatics Insights*, 10:1178222618792860.

Darcy J Corbitt-Hall, Jami M Gauthier, Margaret T Davis, and Tracy K Witte. 2016. College students' responses to suicidal content on social networking sites: an examination using a simulated Facebook newsfeed. *Suicide and life-threatening behavior*, 46(5):609–624.

Jacob Devlin, Ming-Wei Chang, Kenton Lee, and Kristina Toutanova. 2018. Bert: Pre-training of deep bidirectional transformers for language understanding. *arXiv preprint arXiv:1810.04805*.

Ethan Fast, Binbin Chen, and Michael S Bernstein. 2016. Empath: Understanding topic signals in large-scale text. In *Proceedings of the 2016 CHI Conference on Human Factors in Computing Systems*, pages 4647–4657. ACM.

Joseph C Franklin, Jessica Ribeiro, Kathryn Fox, Kate Bentley, Evan Kleiman, Xieyining Huang, Katherine Musacchio, Adam Jaroszewski, Bernard Chang, and Matthew Nock. 2016. Risk factors for suicidal thoughts and behaviors: A meta-analysis of 50 years of research. *Psychological bulletin*, 143. DOI 10.1037/bul0000084.

Devin Gaffney and J Nathan Matias. 2018. Caveat emptor, computational social science: Large-scale missing data in a widely-published Reddit corpus. *PloS one*, 13(7):e0200162.

Rachel Gutman. 2018. Reddit's case for anonymity on the Internet. *The Atlantic*.

Alejandro González Hevia, Rebeca Cerezo Menéndez, and Daniel Gayo-Avello. 2019. Analyzing the use of existing systems for the CLPsych 2019 shared task. In *Proceedings of the Sixth Workshop on Computational Linguistics and Clinical Psychology*.

Matthew Honnibal and Ines Montani. 2017. spaCy 2: Natural language understanding with bloom embeddings, convolutional neural networks and incremental parsing. *To appear*.

Micah Iserman, Taleen Nalabandian, and Molly E. Ireland. 2019. Dictionaries and decision trees for the 2019 CLPsych shared task. In *Proceedings of the Sixth Workshop on Computational Linguistics and Clinical Psychology*.

Thomas E Joiner, Jr, Rheeda L Walker, M David Rudd, and David A Jobes. 1999. Scientizing and routinizing the assessment of suicidality in outpatient practice. *Professional psychology: Research and practice*, 30(5):447.

Matthew Matero, Akash Idnani, Youngseo Son, Salvatore Giorgi, Huy Vu, and Mohammadzaman Zamani. 2019. Suicide risk assessment with multi-level dual-context language and BERT. In *Proceedings of the Sixth Workshop on Computational Linguistics and Clinical Psychology*.

Catherine M McHugh, Amy Corderoy, Christopher James Ryan, Ian B Hickie, and Matthew Michael Large. 2019. Association between suicidal ideation and suicide: meta-analyses of odds ratios, sensitivity, specificity and positive predictive value. *BJPsych open*, 5(2).

David N. Milne, Glen Pink, Ben Hachey, and Rafael A. Calvo. 2016. CLPsych 2016 shared task: Triaging content in online peer-support forums. In *Proceedings of the Third Workshop on Computational Linguistics and Clinical Psychology*, pages 118–127, San Diego, CA, USA. Association for Computational Linguistics.

D.N. Milne. 2017. Triaging content in online peer-support: an overview of the 2017 CLPsych shared task. Available online at http://clpsych.org/shared-task-2017.

Saif M Mohammad. 2017. Word affect intensities. *arXiv preprint arXiv:1704.08798*.

Elham Mohammadi, Hessam Amini, and Leila Kosseim. 2019. CLaC at clpsych 2019: Fusion of neural features and predicted class probabilities for suicide risk assessment based on online posts. In *Proceedings of the Sixth Workshop on Computational Linguistics and Clinical Psychology*.

Michelle Morales, Danny Belitz, Natalia Chernova, Prajjalita Dey, and Thomas Theisen. 2019. An investigation of deep learning systems for suicide risk assessment. In *Proceedings of the Sixth Workshop on Computational Linguistics and Clinical Psychology*.

Matthew K. Nock, Franchesca Ramirez, and Osiris Rankin. 2019. Advancing Our Understanding of the Who, When, and Why of Suicide Risk. *JAMA Psychiatry*, 76(1):11–12.

Matthew E Peters, Mark Neumann, Mohit Iyyer, Matt Gardner, Christopher Clark, Kenton Lee, and Luke Zettlemoyer. 2018. Deep contextualized word representations. *arXiv preprint arXiv:1802.05365*.

Dina Popovic, Eduard Vieta, Jean-Michel Azorin, Jules Angst, Charles L Bowden, Sergey Mosolov, Allan H Young, and Giulio Perugi. 2015. Suicide attempts in major depressive episode: evidence from the bridge-ii-mix study. *Bipolar disorders*, 17(7):795–803.

Esteban A. Ríssola, Diana Ramírez-Cifuentes, Ana Freire, and Fabio Crestani. 2019. Suicide risk assessment on social media: USI-UPF at the CLPsych 2019 shared task. In *Proceedings of the Sixth Workshop on Computational Linguistics and Clinical Psychology*.

Victor Ruiz, Lingyun Shi, Jorge Guerra, Wei Quan, Neal Ryan, Candice Biernesser, David Brent, and Fuchiang Tsui. 2019. CLPsych2019 shared task: Predicting users suicide risk levels from their Reddit posts on multiple forums. In *Proceedings of the Sixth Workshop on Computational Linguistics and Clinical Psychology*.

Maarten Sap, Gregory Park, Johannes Eichstaedt, Margaret Kern, David Stillwell, Michal Kosinski, Lyle Ungar, and Hansen Andrew Schwartz. 2014. Developing age and gender predictive lexica over social media. In *Proceedings of the 2014 Conference on Empirical Methods in Natural Language Processing (EMNLP)*, pages 1146–1151.

H Andrew Schwartz, Johannes C Eichstaedt, Margaret L Kern, Lukasz Dziurzynski, Stephanie M Ramones, Megha Agrawal, Achal Shah, Michal Kosinski, David Stillwell, Martin EP Seligman, et al. 2013. Personality, gender, and age in the language of social media: The open-vocabulary approach. *PloS one*, 8(9):e73791.

Han-Chin Shing, Suraj Nair, Ayah Zirikly, Meir Friedenberg, Hal Daumé III, and Philip Resnik. 2018. Expert, crowdsourced, and machine assessment of suicide risk via online postings. In *Proceedings of the Fifth Workshop on Computational Linguistics and Clinical Psychology: From Keyboard to Clinic*, pages 25–36. Association for Computational Linguistics.

Yla R Tausczik and James W Pennebaker. 2010. The psychological meaning of words: LIWC and computerized text analysis methods. *Journal of language and social psychology*, 29(1):24–54.

CLaC at CLPsych 2019:
Fusion of Neural Features and Predicted Class Probabilities for Suicide Risk Assessment Based on Online Posts

Elham Mohammadi, Hessam Amini and Leila Kosseim
Computational Linguistics at Concordia (CLaC) Lab
Department of Computer Science and Software Engineering
Concordia University, Montréal, Québec, Canada
`first.last@concordia.ca`

Abstract

This paper summarizes our participation to the CLPsych 2019 shared task, under the name *CLaC*. The goal of the shared task was to detect and assess suicide risk based on a collection of online posts. For our participation, we used an ensemble method which utilizes 8 neural sub-models to extract neural features and predict class probabilities, which are then used by an SVM classifier. Our team ranked first in 2 out of the 3 tasks (tasks A and C).

1 Introduction

The CLPsych 2019 shared task (Zirikly et al., 2019) focuses on the prediction of a person's degree of suicide risk based on a collection of their Reddit posts (Shing et al., 2018). It is a multi-class classification task where a subject can be assigned to one of the four categories of *no* (class *a*), *low* (class *b*), *moderate* (class *c*), or *severe* risk (class *d*), and consists of three different tasks:

Task A aims at suicide risk prediction based solely on the posts written on the Suicide Watch subreddit[1].

Task B focuses on making the same prediction by taking into account a person's posts on Suicide Watch, as well as their posts on other subreddits.

Task C has the goal of estimating suicide risk by looking at a subject's posts on different subreddits, but excluding Suicide Watch.

The first two tasks are dedicated to assessing risk; while Task C aims at screening. We participated in all 3 tasks[2] under the team name *CLaC* and ranked first in tasks A and C.

2 System Overview

Our system is composed of 8 neural network submodels, each with a specific type of input word embedding and hidden layer. The extracted neural features and softmax probabilities from all 8 neural networks are combined by a fusion component and the resulting features are used in the final SVM classifier. Figure 1 illustrates the overall architecture of the system. Each component is explained in the following sections.

2.1 Word Embeddings

As shown in Figure 1, GloVe (Pennington et al., 2014) and ELMo (Peters et al., 2018) have been used as pretrained word embeddings. The 300d GloVe word embedder has been pretrained on 840B tokens of web data from Common Crawl. For ELMo, the original 1024d version, pretrained on the 1 Billion Word Language Model Benchmark (Chelba et al., 2014) has been used.

2.2 Hidden Layers

Four different types of hidden layers have been used: a Convolutional Neural Network (CNN) (LeCun et al., 1999), a Bidirectional vanilla Recurrent Neural Network (Bi-RNN), a Bidirectional Long Short-term Memory network (Bi-LSTM) (Hochreiter and Schmidhuber, 1997), and a Bidirectional Gated Recurrent Unit network (Bi-GRU) (Cho et al., 2014).

2.3 Pooling

In order to create a vector representation for each post, three different types of pooling were applied to the output of the hidden layer. In the rest of the paper, these will be referred to as *AVG*, *MAX*, and *ATTN*.

---

[1] `https://www.reddit.com/r/SuicideWatch`

[2] This research was recognized as an IRB exempt by Concordia University's research ethics board.

Proceedings of the Sixth Workshop on Computational Linguistics and Clinical Psychology, pages 34–38
Minneapolis, Minnesota, June 6, 2019. ©2019 Association for Computational Linguistics

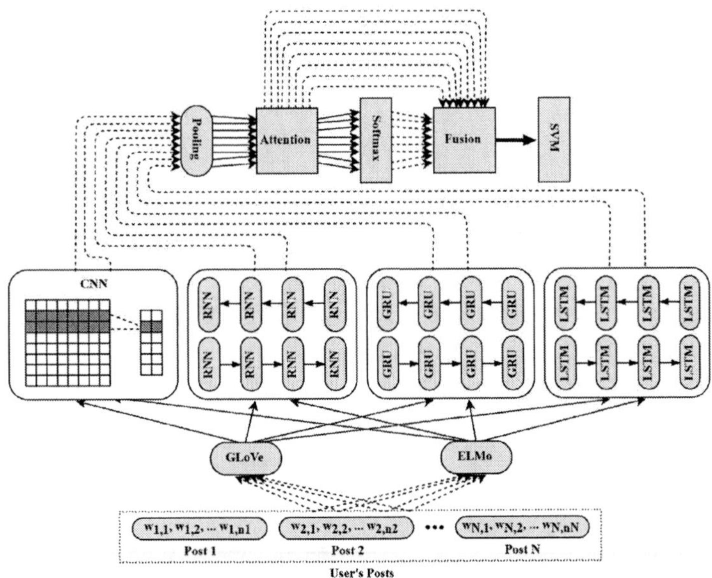

Figure 1: Architecture of the Model. The number of arrows between components correspond to the number of sub-models that move in that flow. The solid lines represent neural connections; while the dotted lines show the flow of data without the existence of a neural connection. The bold arrow between the Fusion and SVM correspond to the flow of data that exists only in the final model.

	Task A				Task B				Task C			
	#HL / #K	#HN / KH	Pooling Type	Max Post Length	#HL / #K	#HN / KH	Pooling Type	Max Post Length	#HL #K	#HN / KH	Pooling Type	Max Post Length
CNN–GloVe	2	300	MAX	400	2	200	AVG	400	2	100	MAX	400
CNN–ELMo	1	400	MAX	400	2	100	MAX	400	2	100	MAX	400
Bi-RNN–GloVe	2	64	MAX	400	2	32	ATTN	200	2	32	ATTN	200
Bi-RNN–ELMo	2	32	MAX	400	1	64	ATTN	200	1	64	ATTN	400
Bi-LSTM–GloVe	2	32	AVG	400	1	64	ATTN	200	2	32	ATTN	200
Bi-LSTM–ELMo	2	32	AVG	400	1	64	ATTN	200	1	64	ATTN	200
Bi-GRU–GloVe	2	64	MAX	400	2	32	ATTN	200	2	32	ATTN	400
Bi-GRU–ELMo	2	64	MAX	400	1	64	ATTN	200	1	64	ATTN	400

Table 1: Hyperparameters used for each sub-model. *#HL*: number of hidden layers, *#HN*: number of hidden nodes in each layer, *#K*: number of kernels (for the CNNs), *KH*: kernel height (for the CNNs).

AVG pooling simply averages the output vectors of the hidden layers. *MAX* pooling is applied on the resulting vectors after applying Concatenated Rectified Linear Unit (CReLU) on the output vectors of the hidden layers (i.e. ReLU applied on the concatenation of each output vector and its negative). *ATTN* is an attention mechanism (Bahdanau et al., 2014) applied to the output vectors of the hidden layers. While *ATTN* may not be considered a pooling method, we do so in order to differentiate between *ATTN* and the attention mechanism presented in Section 2.4. Since *ATTN*'s functioning is similar to the attention mechanism used to calculate the weighted average of the representations for a user's posts, its mechanism will be explained in detail in Section 2.4.

2.4 The Attention Mechanism

It was hypothesized that all posts by a user do not contribute equally to signal her/his mental state. In order to take into account the posts of each user based on their importance in detecting suicide risk, an attention mechanism was used. This mechanism automatically assigns weights to each post from a user, then calculates the weighted average of the representations of all the posts, and uses this average as a representation of the user. Equation 1 shows how the output of the attention mechanism is computed.

$$U = \sum_{i'=1}^{N} p_{i'} \omega_{i'} \qquad (1)$$

where $p_{i'}$ stands for the representation of the i'-

	Task A					Task B					Task C				
# of Neural Features	174					80					925				
SVM's Hyperparameter	kernel	degree	γ	C	class weight	kernel	degree	γ	C	class weight	kernel	degree	γ	C	class weight
Run 1	poly	1	auto	3.0	yes	sigmoid	–	scale	0.8	no	poly	3	scale	0.1	yes
Run 2	poly	4	scale	0.1	no	poly	2	scale	0.5	yes	sigmoid	–	scale	0.4	yes
Run 3	poly	1	auto	0.3	yes	sigmoid	–	scale	0.2	no	poly	2	scale	0.2	yes

Table 2: Hyperparameters used in the submitted runs. The column *degree* refers to the degree of the polynomial kernels. The values of *auto* and *scale* for γ refer to when the parameter γ is set to 1/number-of-features and 1/(number-of-features×variance-of-features), respectively. The value of *class weight* indicates whether weights proportional to the inverse of the number of samples from classes are applied to the parameter C.

th post by a user, $\omega_{i'}$ refers to the weight assigned to the post, and U corresponds to the vector representation for that specific user.

In order to calculate the corresponding weights for the posts, a single n-to-1 fully connected layer is first applied to the representation of each post, where n corresponds to the size of the document representation. The final weights are calculated by applying a softmax to the concatenation of the results of applying the fully-connected layer on the representations of all posts from a user. Equations 2 and 3 show how the weights are calculated:

$$\nu_i = p_i \times w \tag{2}$$

$$\omega = Softmax([\nu_1, \nu_2, \nu_3, \dots, \nu_N]) \tag{3}$$

where w corresponds to the weights in the neural layer, and ν_i refers to the resulting scalar, after feeding p_i (the representation of the i-th post) to the fully-connected layer.

As stated in Section 2.3, the overall mechanism of *ATTN* is similar to the attention mechanism applied to a user's posts. The only difference resides in the level of their functioning: the attention mechanism is applied to the post representations, whereas *ATTN* is applied to the outputs of the hidden layer, at (multiple-)token-level.

2.5 The Sub-models' Optimization Technique

PyTorch (Paszke et al., 2017) was used to develop and train the neural sub-models. At the end of each sub-model, a fully-connected classification layer was used, followed by a softmax activation function. Each sub-model was trained separately on the training data and optimized using the validation data.

The Adam optimizer (Kingma and Ba, 2014) with a learning rate of 5×10^{-4} was used as the optimization technique. Cross-entropy was used as the loss function, and in order to handle the imbalanced class distribution, weights were assigned to each class proportional to the inverse of the number of samples in that class. Due to limitation in computational resources, mini-batches with a maximum size of 32 were applied at the post level for each user.

2.6 The Fusion Component

The fusion component is responsible for creating a final vector representation for each user from the neural features and the predicted probability distributions over classes.

The neural features of the user representations are the result of each sub-model's attention component. In the fusion components, these user representations are first concatenated, and later, the mutual information between each neural feature and the final classes is calculated (using the Scikit-learn library (Pedregosa et al., 2011)). A subset of these features that have the highest mutual information with the final classes are then selected as the final neural features.

The fusion component also uses the predicted probability distributions of the classes for each user from the softmax output of all sub-models. The final user representations are generated by concatenating the neural features and the predicted probability distributions from all sub-models, to be fed to the SVM (see Figure 1).

2.7 The Support Vector Classifier

As shown in Figure 1, the final classifier is an SVM (Cortes and Vapnik, 1995), which uses as input the final user representations generated by the fusion component. The SVM was trained on the samples from the training data, and the validation dataset was used to find the best set of hyperparameters. We used the Scikit-learn library (Pedregosa et al., 2011) for developing and training

Run #	Task A			Task B			Task C		
	macro	flagged	urgent	macro	flagged	urgent	macro	flagged	urgent
1	**0.481**	**0.922**	**0.776**	0.359	0.857	0.714	0.250	0.675	0.610
2	0.416	0.918	0.851	0.381	0.815	0.732	0.239	0.667	0.616
3	0.533	0.922	0.838	**0.339**	**0.843**	**0.718**	**0.268**	**0.671**	**0.625**

Table 3: F1 scores of each run on the shared task test dataset. The results from the primary runs (the ones considered in the ranking) are highlighted in bold.

the SVM model. The final hyperparameters of the SVM classifiers are presented in Section 2.8.

2.8 Final Submitted Models

Before training the model and its sub-models, posts from 33% of the users in the training dataset were randomly selected in a stratified fashion, in order to be used for validation.

When feeding the posts to the sub-models, only the first 200 or 400 tokens were used[3], depending on which limit yielded a better performance at validation time, and the rest were disregarded.

The training process of each sub-model was stopped when the performance on the validation data was at its maximum (for each task, we used the main evaluation metric for that specific task; see Section 3). The validation data was also used in order to find the best set of hyperparameters of the models for each task.

The full model utilizes 8 different sub-models, each one with a unique input word embedding (GloVe or ELMo) and hidden layer type (CNN, Bi-RNN, Bi-LSTM or Bi-GRU). Table 1 shows the hyperparameters of the sub-models for each task, where each sub-model is named by its type of hidden layer and input word embedding.

For each task, we submitted three different runs:

Run 1 where the SVM classifier only uses the neural features.

Run 2 where the SVM classifier only uses the predicted probability of classes.

Run 3 where both the neural features and predicted probabilities are used by the SVM classifier.

Table 2 summarizes the hyperparameters used in each run.

3 Results and Discussion

Table 3 presents a summary of the results of the three runs in each of the three tasks, based on three evaluation metrics:

macro: Macro-averaged F1 on classes a, b, c, d for tasks A and B, and macro-averaged F1 on classes b, c, d for task C. This was the official metric for this shared task, on which we optimized our systems.

flagged: F1 for *flagged* versus *non-flagged*, where *flagged* includes classes b, c, d, and *non-flagged* consists of class a.

urgent: F1 for *urgent* versus *non-urgent*, where *urgent* includes classes c and d, and *non-urgent* consists of classes a and b.

In tasks A and C, the highest macro-averaged F1 was achieved by run 3, and for Task B, the highest F1 was achieved by run 2. This shows the effectiveness of using both the neural features and the predicted probabilities for the final SVM classifier.

In all three tasks, the best flagged F1 was achieved by run 1, showing that using only the neural features leads to better performance when distinguishing between no-risk users (class a) and users that require attention (classes b, c, d).

4 Conclusion

In this paper, we proposed a model based on an ensemble technique that uses a fusion of neural features and predicted probability distribution over classes from 8 neural sub-models, with an SVM as a final classifier. Our first rank in tasks A and C of CLPsych 2019 shared task shows that this technique can be useful in the task of suicide risk assessment. Moreover, it was found that using both neural features and predicted probability of classes generally led to a better performance.

Acknowledgments

The authors would like to thank the anonymous reviewers for their comments on an earlier version of this paper.

This work was financially supported by the Natural Sciences and Engineering Research Council of Canada (NSERC).

---

[3]The average size of posts across all tasks is $\sim$78 tokens.

References

Dzmitry Bahdanau, Kyunghyun Cho, and Yoshua Bengio. 2014. Neural machine translation by jointly learning to align and translate. *Computing Research Repository*, arXiv:1409.0473.

Ciprian Chelba, Tomas Mikolov, Mike Schuster, Qi Ge, Thorsten Brants, Phillipp Koehn, and Tony Robinson. 2014. One billion word benchmark for measuring progress in statistical language modeling. In *15th Annual Conference of the International Speech Communication Association (INTERSPEECH 2014)*, Singapore.

Kyunghyun Cho, Bart van Merrienboer, Caglar Gulcehre, Dzmitry Bahdanau, Fethi Bougares, Holger Schwenk, and Yoshua Bengio. 2014. Learning phrase representations using RNN encoder–decoder for statistical machine translation. In *Proceedings of the 2014 Conference on Empirical Methods in Natural Language Processing (EMNLP 2014)*, pages 1724–1734, Doha, Qatar. Association for Computational Linguistics.

Corinna Cortes and Vladimir Vapnik. 1995. Support-vector networks. *Machine Learning*, 20(3):273–297.

Sepp Hochreiter and Jürgen Schmidhuber. 1997. Long short-term memory. *Neural Computation*, 9(8):1735–1780.

Diederik P. Kingma and Jimmy Ba. 2014. Adam: A method for stochastic optimization. *Computing Research Repository*, arXiv:1412.6980.

Yann LeCun, Patrick Haffner, Léon Bottou, and Yoshua Bengio. 1999. Object recognition with gradient-based learning. In *Shape, contour and grouping in computer vision*, pages 319–345. Springer.

Adam Paszke, Sam Gross, Soumith Chintala, Gregory Chanan, Edward Yang, Zachary DeVito, Zeming Lin, Alban Desmaison, Luca Antiga, and Adam Lerer. 2017. Automatic differentiation in PyTorch. In *NIPS 2017 Autodiff Workshop*, Long Beach, California, USA.

Fabian Pedregosa, Gaël Varoquaux, Alexandre Gramfort, Vincent Michel, Bertrand Thirion, Olivier Grisel, Mathieu Blondel, Peter Prettenhofer, Ron Weiss, Vincent Dubourg, Jake Vanderplas, Alexandre Passos, David Cournapeau, Matthieu Brucher, Matthieu Perrot, and Édouard Duchesnay. 2011. Scikit-learn: Machine learning in Python. *Journal of Machine Learning Research*, 12(Oct):2825–2830.

Jeffrey Pennington, Richard Socher, and Christopher Manning. 2014. GloVe: Global vectors for word representation. In *Proceedings of the 2014 Conference on Empirical Methods in Natural Language Processing (EMNLP 2014)*, pages 1532–1543, Doha, Qatar. Association for Computational Linguistics.

Matthew Peters, Mark Neumann, Mohit Iyyer, Matt Gardner, Christopher Clark, Kenton Lee, and Luke Zettlemoyer. 2018. Deep contextualized word representations. In *Proceedings of the 2018 Conference of the North American Chapter of the Association for Computational Linguistics: Human Language Technologies (NAACL-HLT 2018)*, pages 2227–2237, New Orleans, Louisiana, USA. Association for Computational Linguistics.

Han-Chin Shing, Suraj Nair, Ayah Zirikly, Meir Friedenberg, Hal Daumé III, and Philip Resnik. 2018. Expert, crowdsourced, and machine assessment of suicide risk via online postings. In *Proceedings of the Fifth Workshop on Computational Linguistics and Clinical Psychology: From Keyboard to Clinic*, pages 25–36, New Orleans, Louisiana, USA. Association for Computational Linguistics.

Ayah Zirikly, Philip Resnik, Özlem Uzuner, and Kristy Hollingshead. 2019. CLPsych 2019 shared task: Predicting the degree of suicide risk in Reddit posts. In *Proceedings of the Sixth Workshop on Computational Linguistics and Clinical Psychology: From Keyboard to Clinic*, Minneapolis, Minnesota, USA.

Suicide Risk Assessment with Multi-level Dual-Context Language and BERT

Matthew Matero[1], Akash Idnani[1], Youngseo Son [1]
Salvatore Giorgi[2], Huy Vu[1], Mohammadzaman Zamani[1]
Parth Limbachiya[1], Sharath Chandra Guntuku[2], H. Andrew Schwartz[1]

[1] Stony Brook University   [2] University of Pennsylvania

mmatero@cs.stonybrook.edu

Abstract

Mental health predictive systems typically model language as if from a single context (e.g. Twitter posts, status updates, or forum posts) and often limited to a single level of analysis (e.g. either the message-level or user-level). Here, we bring these pieces together to explore the use of open-vocabulary (BERT embeddings, topics) and theoretical features (emotional expression lexica, personality) for the task of suicide risk assessment on support forums (the CLPsych-2019 Shared Task). We used *dual context* based approaches (modeling content from suicide forums separate from other content), built over both traditional ML models as well as a novel dual RNN architecture with user-factor adaptation. We find that while affect from the suicide context distinguishes with no-risk from those with "any-risk", personality factors from the non-suicide contexts provide distinction of the levels of risk: low, medium, and high risk. Within the shared task, our dual-context approach (listed as SBU-HLAB in the official results) achieved state-of-the-art performance predicting suicide risk using a combination of suicide-context and non-suicide posts (Task B), achieving an F1 score of 0.50 over hidden test set labels.

1 Introduction

Suicidal behavior is conceptualized by the thoughts, plans, and acts an individual makes toward intentionally ending their own life (Nock et al., 2008). With deaths by suicide increasing substantially (Curtin et al., 2016), researchers are turning to automated analysis of user generated content to potentially provide methods for early detection of suicide risk severity (Coppersmith et al., 2018; De Choudhury et al., 2016; Shing et al., 2018). If an automated process could detect elevated risk in a person, personalized (potentially digital and early) interventions could be provided to the individual to alleviate the risk.

Importantly, suicide risk assessment follows a growing body of work which has provided language-based models for measuring theoretically related psychological constructs: valence and arousal (Preoţiuc-Pietro et al., 2016; Mohammad, 2018), depression (Schwartz et al., 2014; Eichstaedt et al., 2018), and stress (Guntuku et al., 2019). However, few have evaluated the role of such theoretical models alongside standard open-vocbaulary features (e.g. ngrams, embeddings, topics), or integrated both message-level assessment (e.g. emotional valence) along with user-level assessment (e.g., personality).

In this study, we investigate a series of dual context (treating suicide forum posts separate from other forum posts) and multi-level approaches (user-level assessments of demographics and personality as well as aggregates of message-level features) for suicide risk prediction. **Our contributions** include: (1) proposal and evaluation of a *dual-context* modeling approach where language in a suicide-specific context is treated separate from language from other forums, (2) a novel deep learning architecture (*DualDeepAtt*) that both (a) applies dual-context modeling to GRU cells and attention layers and (b) adds a user-factor adaptation layer, (3) comparison of individual theoretically related linguistic assessments, (4) evaluation of models based on theoretically-motivated features versus models based on open-vocabulary features with multiple approaches to aggregating message-level features.

2 Data

The dataset was collected from Reddit, released as the CLPsych 2019 Shared Task (Zirikly et al., 2019), where collections of users' posts were annotated into 4 suicide risk categories (no risk, low, moderate, severe) and then aggregated into sin-

Proceedings of the Sixth Workshop on Computational Linguistics and Clinical Psychology, pages 39–44
Minneapolis, Minnesota, June 6, 2019. ©2019 Association for Computational Linguistics

gle labels representing their highest suicide risk across all collections (Shing et al., 2018). All users had posted in r/SuicideWatch and had at least 10 posts total across the platform. The task of suicide risk prediction was sub-divided into 3 sub-tasks, each based on different levels of data. The first task (Task A) consisted of users' posts from r/SuicideWatch annotated for suicide risk level. The second (Task B) consisted of the same users as in Task A and included their entire Reddit post history (including their r/SuicideWatch posts). The third task (Task C) consisted of users' entire Reddit post history apart from posts in r/SuicideWatch. Additionally Task C includes a set of 'control users' who are labeled as no risk[1]. Task A and B shared the same number of users(Training = 496, Test = 128), while Task C had 993 training and 248 test.

Ethics Statement: This research was evaluated by an institutional review board and deemed exempt.

3   Open and Theoretical Features

We extracted three sets of linguistic features: 1) theoretical dimensions, 2) open-vocabulary, and 3) meta-features (post statistics, forum names). Language features have been shown to be predictive of several mental health outcomes (Guntuku et al., 2017). We extracted open-vocabulary and theoretical dimensions from both *message-level* (post body, title) and *user-level* (collections of posts) features. Depending on predictive modeling choice, message-level features can then be aggregated to user-level through various mechanisms: RNN with attention, or explicit aggregation – mean, minimum, and maximum.

Theoretical dimensions. Our theoretical dimensions ranged from capturing message-level user states (able to change) to user-level traits (slow changing). The *Messsage-level states*, calculated separately for both the title and content, included **affect** and **intensity** (Preoţiuc-Pietro et al., 2016) as well as **valence, arousal,** and **dominance** (Mohammad, 2018). These features were generated per-message and aggregated to the users. *User-level traits* included language-based inferences of demographics **age/gender** (Sap

et al., 2014), assessments of **big-5 personality** traits (Schwartz et al., 2013) as well as trait **anxiety, anger,** and **depression** (Schwartz et al., 2014).

Open-Vocabulary Features. We also included higher dimensional features meant to capture open ended content. This included dimensionally reduced **BERT embeddings** – originally a 768-dimensional representation is extracted from a pre-trained model (Devlin et al., 2019) for post contents and titles (separately). Given the training sizes, we decided to further reduce these dimensions down to 50 and 20 dimensions for body and title respectively, using non-negative matrix factorization (NMF) (Févotte and Idier, 2011). Following successful use of topics for mental health modeling in the past (Eichstaedt et al., 2018), we also inferred 25 **LDA Topics** (Blei et al., 2003) trained using Gibb's Sampling over suicide watch posts excluding words used more frequently outside of the forum.

Meta-features. We also included various user-level **post statistics**: average 1-gram length, average 1-grams per post, and total 1-grams, as well as **subreddit** features: a 39 dimensional feature vector was derived from popular subreddits. We began with the 1973 subreddits that were mentioned by at least 0.5% of training users, and use NMF to reduce to 20 dimensions. The remaining 19 dimensions are subreddits that were most distinctive, in training, of high risk users.

4   Correlation and Distribution Analysis

To uncover the associations between the theoretical dimensions and suicide risk level, we perform a correlation analysis for Task B data, shown in table 1. Those scoring higher in the female dimension were associated with higher suicide risk scores and age had no significant effect. Prior epidemiological studies (Mościcki, 1997) have showed that nearly 80% of suicide completers are men, whereas the majority of lifetime attempters are women.

Among personality factors, being agreeable, conscientious, and extroverted were associated with lower suicide risk while higher neuroticism was positively correlated with higher suicide risk. Prior studies have found similar associations in other samples through traditional surveys (Velting, 1999) establishing that language on social media

---

[1]Control users are those who have no r/SuicideWatch or other mental health subreddit posts

Dimension	r	Dimension	r
Age	–	Agreeableness	-.14
Gender	.14	Conscientious.	-.14
Anger	.32	Extroversion	-.17
Anxiety	.33	Neuroticism	.32
Depression	.32	Openness	–

Table 1: Pearson correlations (r) between theoretical linguistic dimensions and suicide risk level over the training data. Gender was continuously coded (larger indicating more likely female). Correlations are significant at $p < .01$ multi-test corrected.

Figure 1: Topics correlated with higher risk (blue, top 4 rows) and lower risk (red, bottom row), treating risk as a continuous value. All correlations significant at $p < .05$, Benjamini-Hochberg corrected.

forums could be a good proxy for measuring suicidal ideation. Corroborating these findings, users with high anger, anxiety and depression scores were associated with higher suicide risk.

We also analyze the correlations between r/SuicideWatch topic dimensions, as shown in figure 1. Here, we showcase certain topics that correlate well with risk level and the words expressed in that topic.

Additionally, for certain features we explore their distributions over users of differing risk levels. From our correlation analysis, we pick emotional stability, the reverse encoding of neuroticism, depression and affect scores. For affect, we examine only user's posts from Task A (r/SuicideWatch), while we look at all available posts for depression and emotional stability.

In Figure 2 we show emotional stability, de-

pression, and mean affect scores of users belonging to each risk level. For emotional stability, as the value gets lower the less stability a person expresses, which holds across the risk levels with no risk users having higher stability values and less variance compared to high risk users. A similar pattern is expressed for depression scores, where high risk users trend towards higher values. There is also a slower decline for high risk users causing a longer tail on the distribution compared to other risk levels. Lastly, we see that while affect scores distinguish no risk from others, they do not provide a separation among the degrees of risk. The affect model was message-level and distributions here were for mean over their suicide watch messages. Also, those who are deemed low risk have the highest variance, while moderate and high risk users show very similar distributions.

5 Dual-Context Predictive Modeling

Our predictive approaches attempted to model language from a suicide context (that from suicide watch) separately from other forum posts – *dual-context*. We used a range of regularized logistic regression and attention-based RNN architectures for Tasks A and B, and logistic regression alone for Task C. All non-neural models were implemented via the DLATK Python package (Schwartz et al., 2017).

Task A. The logistic regression model used open-vocabulary, theoretical, and meta-features as input (termed as '*OpenTheory*'). We also evaluated the performance of BERT embeddings alone (termed as '*Bert*'). The neural model used an LSTM with hierarchical post-level attention (Yang et al., 2016). We fed it the concatenation of open-vocabulary features, Affect, Intensity, and VAD NRC Lexicon scores of each SuicideWatch post. The model was run on all posts of each user in the time order of their posting to make a prediction on the risk level of each user. This model is referred to as *DeepAtt*.

Task B. For Task B, we were able to experiment with the dual-context model. Our logistic regression based approach, termed as '*DualOpenTheory*' takes in features from SuicideWatch and non-SuicideWatch language that were processed separately. Similar to the previous task, we evaluate a '*DualContextBert*' model that uses BERT features from both SuicideWatch and separately

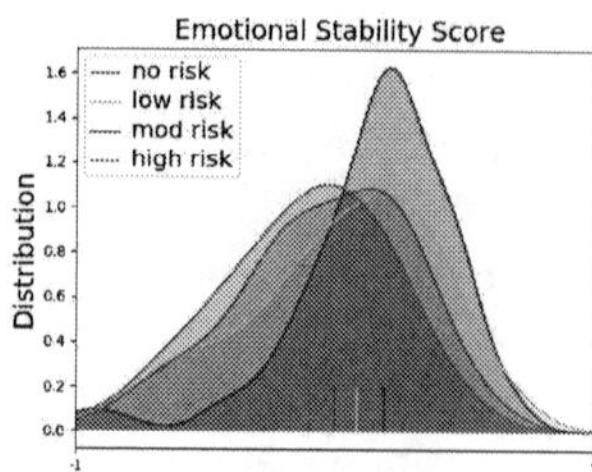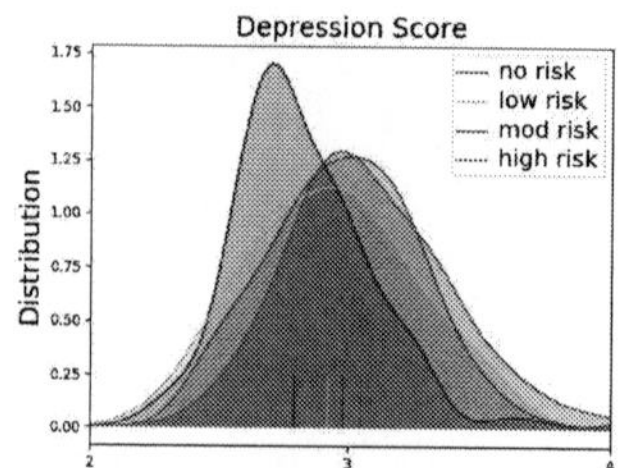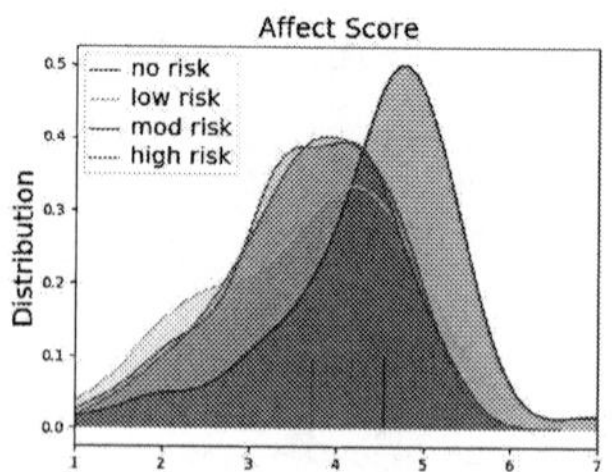

Figure 2: Density estimations, separated by risk level, of user emotional stability (left), depression score (middle), and mean message affect (right). Emotional stability and depression were calculated across non-suicide context while affect was from suicide context (from r/SuicideWatch posts). While affect provided some separation of no risk from any risk, emotional stability and depression distinguish all levels of risk. Across all three theoretical dimensions, there was less variance across no risk users.

non-SuicideWatch messages. Task B also enabled us to use subreddit features among the meta features (non r/SuicideWatch subreddits were assumed unavailable for Task A). For logistic regression models while the data is processed separately, only one model is trained on the joint feature sets.

For the neural dual-context model, visualized in figure 3, we used two separate GRU cells (termed as 'DualDeepAtt'); one takes the same input features of our Task A model from SuicideWatch posts, and the other runs by taking subreddit info feature vector in addition to the same input features, processed on non-SuicideWatch posts, of the SuicideWatch GRU cell (SuicideWatch subreddit info is already taken into account by having a separate GRU cell). We used the separate attention weights for SuicideWatch (SW) GRU hidden vectors and non-SuicideWatch (NSW) GRU hidden vectors as following:

$$[\overrightarrow{v_{SW}}; \overrightarrow{v_{NSW}}] = \left[\sum \alpha_{sw}\overrightarrow{h_{sw}}; \sum \alpha_{nsw}\overrightarrow{h_{nsw}}\right]$$

Then, we applied user-factor adaptation (Lynn et al., 2017) to the concatenation of the sum of hidden vectors with attentions of the SW GRU cell and the NSW GRU cell as following:

$$\overrightarrow{fv} = [F_0\times[\overrightarrow{v_{SW}}; \overrightarrow{v_{NSW}}]; \ldots; [F_N\times[\overrightarrow{v_{SW}}; \overrightarrow{v_{NSW}}]]$$

Here, we used age, gender, and latent factors of users with the following transformation: $F_i = \frac{X_i-min(X_i)}{max(X_i)-min(X_i)}$ For latent factors, we derived 3 user-level latent factors from the history of Reddit posts of the users, which are equivalent to the "user-embed" in (Lynn et al., 2017) as they found

these factors from language just as effective as personality factors.

Finally, we concatenate the user-level feature vector with the factorized output vector ($[\overrightarrow{fv}; \overrightarrow{UserFeatures}]$). Here, we used Anger, Anxiety, Depression scores, average word lengths, total word counts of each user for user features.

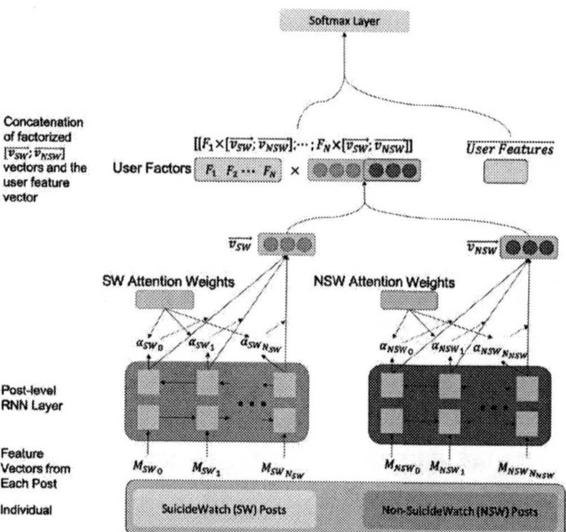

Figure 3: Dual-context, RNN-attention, use-factor adaptation architecture used in Task B. The left RNN handles features related to suicide watch posts and the right RNN handles non-SuicideWatch. User factors are multiplied into the concatenated vector for adaptation, as well as simply concatenated before softmax layer

Task C. We build logistic regression models using a) BERT embeddings alone: 'Bert'; b) open-vocabulary, theoretical dimensions, meta-features, and subreddit latent factors 'OpenTheoryUser'; and c) same as b but without user traits of personality, age/gender, and anxiety, anger, depression scores 'OpenTheorySubr'.

6 Results

We compare our models performance during training using 10-fold cross-validation as well as 3 models for each task using the designated test set. Across each task the models that take advantage of both open vocabulary and theoretical constructs outperform others.

6.1 Task A

A combination of open-vocabulary, theoretical dimensions, and meta-features performed best at predicting suicide risk based on annotated SuicideWatch posts. Table 2 shows the results on the cross-validation setting we employed in the training set and the performance released on the test set. While the logistic regression models had similar performance across train and test sets, the neural models outperformed others on the test set.

In models designed for Task A when performing message to user level aggregations we performed average, minimum, and maximum and concatenated the vectors. This outperformed aggregations using average or minimum/maximum together.

	Train		Test	
Model	Acc	F1	Acc	F1
Open	.55	.44	-	-
Theory	.47	.32	-	-
OpenTheory	.54	.40	-	-
OpenTheory w/ Min, Max	**.57**	**.46**	.56	.46
DeepAtt	.53	.44	**.59**	**.50**
Bert w/ Min,Max	.55	.42	.53	.40

Table 2: Task A: Suicide Risk Prediction Performance (measured by Accuracy and F1-scores). Best performing models are highlighted. Meta features for Task A only contains post statistics as all posts come from SuicideWatch.

6.2 Task B

We found a large improvement from using the dual-context type approach, shown in table 3. Overall, the OpenTheory approach performed best on the training set and also achieving similar performance on the test set. However, the *dual-context* BERT embeddings based logistic regression outperformed other approaches on the test set. DualDeepAtt was not far behind but likely was hindered by the limited amount of training, relative to parameters for the task.

	Train		Test	
Model	Acc	F1	Acc	F1
Open	.54	.44	-	-
Theory	.48	.33	-	-
Single Context OpenTheory	.50	.35	-	-
Dual Context OpenTheory	**.58**	**.47**	.56	.46
DualDeepAtt	.47	.41	.51	.44
DualContextBert	.53	.43	**.57**	**.50**

Table 3: Task B: Suicide Risk Prediction Performance (measured by Accuracy and F1-scores). Best performing models are highlighted.

6.3 Task C

Task C proved the most difficult for our models. The dual-context approach did not apply and our approach modeled such that a majority of users were no risk while the test F1 only evaluated over those deemed to have some risk. Still, A combination of open vocabulary and theoretical features outperform other approaches. Here, our best performing model was *OpenTheoryUser* (scoring accuracy of .69 and F1 of .18), which accounted for all user level traits and a mean aggregation of message-level open-vocabulary features.

7 Conclusion

We presented new approaches for identifying suicide risk among users on support based forums, focused largely on (a) utilizing dual-contexts of language, (b) message and user multi-level models, and (c) exploring both theoretical dimensions and open vocabulary features. We also compared aggregation techniques and proposed a novel RNN architecture for processing dual context data. We found dual-context models yielded significant gains and while theoretical dimensions of language related in the expected direction (more depressive and anxious language correlated with higher risk), a combination of BERT-based features and theoretical dimensions was best when building predictive models.

References

David M Blei, Andrew Y Ng, and Michael I Jordan. 2003. Latent dirichlet allocation. *Journal of machine Learning research*, 3(Jan):993–1022.

Glen Coppersmith, Ryan Leary, Patrick Crutchley, and Alex Fine. 2018. Natural language processing of social media as screening for suicide risk. *Biomedical informatics insights*, 10:1178222618792860.

Sally C Curtin, Margaret Warner, and Holly Hedegaard. 2016. Increase in suicide in the united states, 1999–2014.

Munmun De Choudhury, Emre Kiciman, Mark Dredze, Glen Coppersmith, and Mrinal Kumar. 2016. Discovering shifts to suicidal ideation from mental health content in social media. In *Proceedings of the 2016 CHI conference on human factors in computing systems*, pages 2098–2110. ACM.

Jacob Devlin, Ming-Wei Chang, Kenton Lee, and Kristina Toutanova. 2019. Bert: Pre-training of deep bidirectional transformers for language understanding. In *Proceedings of NAACL-HLT 2019: The Annual Meeting of the North American Association for Computational Linguistics*.

Johannes C Eichstaedt, Robert J Smith, Raina M Merchant, Lyle H Ungar, Patrick Crutchley, Daniel Preoţiuc-Pietro, David A Asch, and H Andrew Schwartz. 2018. Facebook language predicts depression in medical records. *Proceedings of the National Academy of Sciences*, 115(44):11203–11208.

Cédric Févotte and Jérôme Idier. 2011. Algorithms for nonnegative matrix factorization with the β-divergence. *Neural computation*, 23(9):2421–2456.

Sharath Chandra Guntuku, Anneke Buffone, Kokil Jaidka, Johannes Eichstaedt, and Lyle Ungar. 2019. Understanding and measuring psychological stress using social media.

Sharath Chandra Guntuku, David B Yaden, Margaret L Kern, Lyle H Ungar, and Johannes C Eichstaedt. 2017. Detecting depression and mental illness on social media: an integrative review. *Current Opinion in Behavioral Sciences*, 18:43–49.

Veronica Lynn, Youngseo Son, Vivek Kulkarni, Niranjan Balasubramanian, and H Andrew Schwartz. 2017. Human centered nlp with user-factor adaptation. In *Proceedings of the 2017 Conference on Empirical Methods in Natural Language Processing*, pages 1146–1155.

Saif Mohammad. 2018. Obtaining reliable human ratings of valence, arousal, and dominance for 20,000 english words. In *Proceedings of the 56th Annual Meeting of the Association for Computational Linguistics (Volume 1: Long Papers)*, volume 1, pages 174–184.

Eve K Mościcki. 1997. Identification of suicide risk factors using epidemiologic studies. *Psychiatric Clinics of North America*, 20(3):499–517.

Matthew K Nock, Guilherme Borges, Evelyn J Bromet, Christine B Cha, Ronald C Kessler, and Sing Lee. 2008. Suicide and suicidal behavior. *Epidemiologic reviews*, 30(1):133–154.

Daniel Preoţiuc-Pietro, H Andrew Schwartz, Gregory Park, Johannes Eichstaedt, Margaret Kern, Lyle Ungar, and Elisabeth Shulman. 2016. Modelling valence and arousal in facebook posts. In *Proceedings of the 7th Workshop on Computational Approaches to Subjectivity, Sentiment and Social Media Analysis*, pages 9–15.

Maarten Sap, Gregory Park, Johannes Eichstaedt, Margaret Kern, David Stillwell, Michal Kosinski, Lyle Ungar, and Hansen Andrew Schwartz. 2014. Developing age and gender predictive lexica over social media. In *Proceedings of the 2014 Conference on Empirical Methods in Natural Language Processing (EMNLP)*, pages 1146–1151.

H Andrew Schwartz, Johannes Eichstaedt, Margaret L Kern, Gregory Park, Maarten Sap, David Stillwell, Michal Kosinski, and Lyle Ungar. 2014. Towards assessing changes in degree of depression through facebook. In *Proceedings of the Workshop on Computational Linguistics and Clinical Psychology: From Linguistic Signal to Clinical Reality*, pages 118–125.

H Andrew Schwartz, Johannes C Eichstaedt, Margaret L Kern, Lukasz Dziurzynski, Stephanie M Ramones, Megha Agrawal, Achal Shah, Michal Kosinski, David Stillwell, Martin EP Seligman, et al. 2013. Personality, gender, and age in the language of social media: The open-vocabulary approach. *PloS one*, 8(9):e73791.

H Andrew Schwartz, Salvatore Giorgi, Maarten Sap, Patrick Crutchley, Lyle Ungar, and Johannes Eichstaedt. 2017. Dlatk: Differential language analysis toolkit. In *Proceedings of the 2017 Conference on Empirical Methods in Natural Language Processing: System Demonstrations*, pages 55–60.

Han-Chin Shing, Suraj Nair, Ayah Zirikly, Meir Friedenberg, Hal Daumé III, and Philip Resnik. 2018. Expert, crowdsourced, and machine assessment of suicide risk via online postings. In *Proceedings of the Fifth Workshop on Computational Linguistics and Clinical Psychology: From Keyboard to Clinic*, pages 25–36.

Drew M Velting. 1999. Suicidal ideation and the five-factor model of personality. *Personality and Individual Differences*, 27(5):943–952.

Zichao Yang, Diyi Yang, Chris Dyer, Xiaodong He, Alex Smola, and Eduard Hovy. 2016. Hierarchical attention networks for document classification. In *Proceedings of the 2016 Conference of the North American Chapter of the Association for Computational Linguistics: Human Language Technologies*, pages 1480–1489.

Ayah Zirikly, Philip Resnik, Özlem Uzuner, and Kristy Hollingshead. 2019. CLPsych 2019 shared task: Predicting the degree of suicide risk in Reddit posts. In *Proceedings of the Sixth Workshop on Computational Linguistics and Clinical Psychology: From Keyboard to Clinic*.

Using natural conversations to classify autism with limited data: Age matters

Michael Hauser[1]   Evangelos Sariyanidi[1]   Birkan Tunc[1,2]   Casey J. Zampella[1]

Edward S. Brodkin[2]   Robert T. Schultz[1,2,3]   Julia Parish-Morris[1,2]

[1] Center for Autism Research, Children's Hospital of Philadelphia
[2] Department of Psychiatry, University of Pennsylvania
[3] Department of Pediatrics, University of Pennsylvania

Abstract

Spoken language ability is highly heterogeneous in Autism Spectrum Disorder (ASD), which complicates efforts to identify linguistic markers for use in diagnostic classification, clinical characterization, and for research and clinical outcome measurement. Machine learning techniques that harness the power of multivariate statistics and non-linear data analysis hold promise for modeling this heterogeneity, but many models require enormous datasets, which are unavailable for most psychiatric conditions (including ASD). In lieu of such datasets, good models can still be built by leveraging domain knowledge.

In this study, we compare two machine learning approaches: the first approach incorporates prior knowledge about language variation across middle childhood, adolescence, and adulthood to classify 6-minute naturalistic conversation samples from 140 age- and IQ-matched participants (81 with ASD), while the other approach treats all ages the same. We found that individual age-informed models were significantly more accurate than a single model tasked with building a common algorithm across age groups. Furthermore, predictive linguistic features differed significantly by age group, confirming the importance of considering age-related changes in language use when classifying ASD. Our results suggest that limitations imposed by heterogeneity inherent to ASD and from developmental change with age can be (at least partially) overcome using domain knowledge, such as understanding spoken language development from childhood through adulthood.

1   Introduction

Autism Spectrum Disorder (ASD) is a neurobiologically-based condition characterized by social communication impairments and restricted, repetitive patterns of behaviors and interests [1]. Although ASD is a neurodevelopmental disorder, it is currently diagnosed using behavior alone, including spoken language. For the roughly 70 percent of individuals with ASD that have average to above-average verbal abilities [2], language is an important pathway to social connections. For clinicians and care providers, spoken language can provide a window into internal cognitive and social processing. Given that primary diagnostic tools for ASD often rely on language-mediated semi-structured interviews and play activities to elicit behaviors found in the condition [3], measuring and quantifying subtle differences in spoken language between individuals with ASD and matched typically developing (TD) controls is important for improving diagnostic speed and reliability. Furthermore, since the emergence of spoken language before age 5 is a critical predictor of later functional outcomes in ASD [4, 5, 6], characterizing spoken language development is crucial for understanding long-term developmental outcomes.

Behavioral heterogeneity in ASD is a persistent challenge for researchers and clinicians. In fact, generalizability from one individual to the next is so low that it is often said, "If you have met one person with autism, you have met one person with autism". Wide phenotypic variability has made it difficult to draw reliable statistical conclusions about ASD, and indeed, has made it challenging to study the disorder at all [7]. Significant variability is similarly present in the verbal domain, with the spoken language skills of individuals with ASD ranging from severely impaired to verbally gifted [8]. As an illustration, a recent narrative study found that intra-group variability (ASD alone) was greater than inter-group variability (between ASD and TD) [9].

Recent attempts to leverage machine learning for understanding and classifying individuals with

Proceedings of the Sixth Workshop on Computational Linguistics and Clinical Psychology, pages 45–54
Minneapolis, Minnesota, June 6, 2019. ©2019 Association for Computational Linguistics

ASD have grappled with this phenotypic variability [10, 11]. Unfortunately, many of the most exciting machine learning models (e.g., models that are able to capture nonlinear dependencies across many dimensions), require large, well-characterized training datasets to function correctly, which are rare in ASD (and are particularly scarce for children). These two constraints in ASD research (wide variability in high dimensional spaces, and lack of large datasets), suggest that it may be useful to proactively incorporate information that psychiatrists and linguists deem important, thus guiding machine learning models to learn relevant dependencies while ignoring irrelevant ones.

2 Language in ASD

Prior research suggests that language is a valuable metric that can be used to distinguish individuals with ASD from TD controls. For example, the NEPSY narrative retelling test, in which a child listens to and retells a story while being evaluated on how many key story elements were remembered, has been explored for its utility in supporting ASD identification [12]. In an analysis of 97 children aged 4-8 years, Prud'hommeaux and colleagues found that children with ASD were more likely than TD controls to veer off topic and incorporate their own specialized interests into the narrative. Similarly, another study showed that TD children are more likely to use similar words and semantic concepts to those given in the narrative, while children with ASD will retell the narrative with different words and concepts related to their own specialized interests [9]. Although promising, these and other studies that focus on one-sided language samples, rather than more ecologically valid conversations, miss a potential source of informative variance in language in ASD: the conversational partner.

Typically, natural conversations involve dynamic adjustments on a variety of levels that facilitate rapport and communication; this is called "linguistic accommodation" or "alignment" [13]. Increased accommodation is associated with perceptions of better conversation [14], but most prior research on language in ASD has used samples from structured or semi-structured elicitation tasks - or conversations conducted with an autism specialist - rather than natural conversations [15]. Thus, it is unknown whether and how typical (non-expert) speakers adjust their conversational behaviors to accommodate social communication differences in ASD, and whether the extent of accommodation changes over the course of development. To explore this new area, the machine learning models employed in this study include include dyadic features derived from a natural conversation (such as turn-taking rates) and interlocutor (conversation partner) features, as well as features from individuals with ASD.

3 Developmental Changes in Conversation

Individuals with and without ASD continue to develop socially and cognitively throughout childhood, adolescence, and into early adulthood. For example, although Theory of Mind (or the ability to take another person's perspective) emerges in early childhood [16], it becomes increasingly sophisticated throughout typical adolescence and early adulthood [17]. Thus, age-related differences in conversation (which is inherently social) are likely to be found.

Physical and emotional changes between childhood and adolescence (e.g., puberty [18]) increase the likelihood that people's preferred topic of conversation might change over time as well. Whereas young children may be more likely to talk about family and school, older children may be more focused on peer relationships [19], and adults might naturally gravitate toward talking about occupations or romantic partners. Unfortunately, few studies have explored natural conversation across development, and normative expectations for brief conversations are poorly understood across developmental phases and ages.

4 Current Study

The purpose of the current study is to test whether separating a large sample of individuals with and without ASD into different age groups, namely middle childhood (8 to 11), adolescence (12 to 17) and adulthood (18 and up), increases the accuracy and reliability of a simple machine learning classification model for classifying ASD vs. TD, despite inevitable trade-offs in sample size.

Given the likelihood that natural conversation differs between children and adolescents in a variety of measurable ways (e.g., preferred topics), and that adolescents also converse differently than adults, we hypothesized that diagnostic classification accuracy would improve significantly when conducted within each age group separately, as compared to the combined sample. This is in contrast to generally accepted doctrine in machine learning (i.e., that more data is better), since in our study we divide our larger dataset into three smaller datasets.

We further tested whether the specific features that best distinguished diagnostic groups differed significantly by age. Based on prior research and clinical observation, we hypothesized that the relative predictive value of specific features would differ across development.

5 Methods

5.1 Participants

One hundred forty individuals participated in the present study (ASD: N=81, TD: N=59). Participants were categorized by age into three subgroups (see Table 1): middle childhood (8-11 years), adolescence (12-17 years) and adulthood (18-50 years). Diagnoses were confirmed (ASD group) or ruled out (TD group) using the Clinical Best Estimate process [20] informed by the Autism Diagnostic Observation Schedule - Second Edition (ADOS-2) [3] and adhering to DSM-V criteria for ASD [21]. To control for non-age related phenotypic heterogeneity, age subgroups were matched on Full Scale IQ estimates (WASI-II) [22], verbal and nonverbal IQ estimates, and sex ratio (Table 1). Participants with ASD were also matched across age subgroups on autism symptom severity, based on ADOS-2 Calibrated Severity Scores [23] and scores on the Social Communication Questionnaire (SCQ) [24]. All participants were native English speakers.

5.2 Procedure

All aspects of this study were approved by the Institutional Review Boards of the Children's Hospital of Philadelphia and the University of Pennsylvania. All adult participants and parents of minor children provided written informed consent for participation. The primary experimental task for this study was a slightly modified version of the Contextual Assessment of Social Skills (CASS) [25]. The CASS is a semi-structured assessment of conversational ability designed to mimic real-life first-time encounters. Participants engaged in two three-minute face-to-face conversations with two different confederates (research staff, blind to participant diagnostic status and unaware of the dependent variables of interest). In the first conversation (Interested condition), the confederate demonstrated social interest by engaging both verbally and non-verbally in the conversation. In the second conversation (Bored condition), the confederate demonstrated boredom and disengagement both verbally (e.g., one-word answers, limited follow-up questions) and non-verbally (e.g., neutral affect, limited eye-contact and gestures). Prior to each conversation, study staff provided the following prompt to the participants and confederates before leaving the room: "Thank you both so much for coming in today. Right now, you will have three minutes to talk and get to know each other, and then I will come back into the room."

CASS confederates included 42 undergraduate students or BA-level research staff (12 males, 30 females, all native English speakers). Fourteen confederates interacted with the ASD group, 7 with the TD group, and 21 with both groups. Confederates were semi-randomly selected, based on availability and clinical judgment. Confederate sex ratios did not differ by diagnostic group (p=n.s.). In order to provide opportunities for participants to initiate and develop the conversation, and in accordance with CASS confederate instructions [25], confederates in both conditions were trained to wait 10 seconds before initiating the conversation and to speak for no more than 50% of the time. If conversational lapses occurred, confederates were trained to wait 5 seconds before re-initiating the conversation. No specific conversational topic prompts were provided to either speaker.

Audio/video recordings of CASS conversations were obtained using a specialized "TreeCam", built in-house (Figure 1), placed between the participant and confederate (seated facing one another) on a floor stand. The TreeCam has two HD video cameras pointing in opposite directions to allow simultaneous recording of participant and confederate, as well as directional microphones to record audio. For these analyses, the language sample began when the first word of the CASS was uttered, after study staff left the room, and ended when study staff re-entered.

Table 1: Sex ratio, mean age (in years) and mean IQ scores for ASD and TD children (8-11 years), adolescents (12-17 years), and adults (18-50 years), and measures of autism symptoms for ASD participants.

Dx	N	Age group	N	Sex (f/m)	Age	Full-scale IQ	Verbal IQ	Non-verbal IQ	ADOS CSS	SCQ
		Children	22	8, 14	9.98	105	103	105	7.32	19.81
ASD	81	Adolescents	24	7, 17	14.62	102	103	101	6.58	17.38
		Adults	35	5, 30	26.73	104	108	99	7.06	17.23
		Children	19	8, 11	9.58	103	104	102	.	.
TD	59	Adolescents	12	6, 6	14.17	103	101	103	.	.
		Adults	28	5, 23	28.42	109	110	106	.	.

Note: Diagnostic groups did not significantly differ on sex ratio, age, or IQ within age bins, and age bins did not differ from one another on these variables (all p=ns). In the ASD group, age bins did not differ significantly from one another on ADOS-2 calibrated severity scores (CSS) or on SCQ scores (all p=ns). Five participants with ASD had missing scores on the SCQ (1 child, 4 adults).

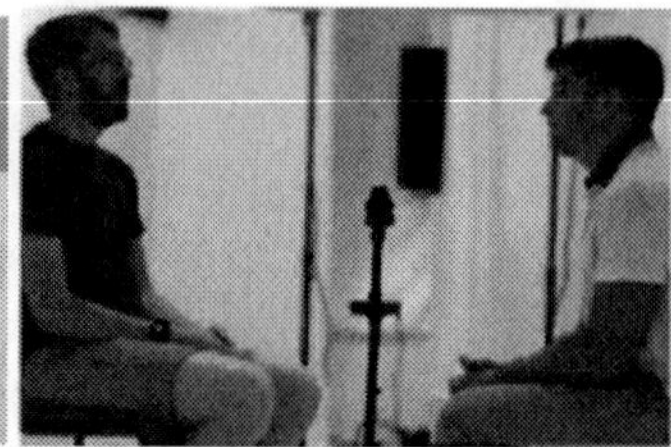

(a) The TreeCam audio/video capture device. (b) Illustration of the task environment. Participants and confederates sat face-to-face while engaging in a "get to know each other" dialogue, with the TreeCam placed in between.

Figure 1: Experimental setup of the TreeCam device, as well as participants and confederates.

5.3 Audio Data Processing

Audio streams were extracted from audio/video recordings, and saved in lossless .flac format. A team of reliable annotators produced time-aligned, verbatim, orthographic transcripts of audio recordings in the transcription software XTrans [26]. Each recording was processed by two junior annotators and one senior annotator, all of whom were undergraduate students and native English speakers. Before becoming junior annotators for this cohort, each team member received at least 10 hours of training in Quick Transcription [27] modified for use with clinical interviews of participants with ASD [10, 11, 28]. In addition, annotators achieved reliability (defined as >90% in common with a Gold Standard transcript) on segmenting (marking speech start and stop times) and transcribing (writing down words and sounds produced, using the modified Quick Transcription specification) before beginning independent annotation. Training files included audio recordings of conversations between individuals with and with-

out autism that were not used in this study.

For CASS recordings, one reliable junior annotator segmented utterances into pause groups, while the second transcribed words produced by each speaker. A senior annotator then thoroughly reviewed and corrected each file. All senior annotators had at least 6 months of prior transcription experience. Final language data were exported from XTrans as tab-delimited files that were batch imported into R. Annotations marking non-speech sounds like laughter, indicators of language errors like stutters, and punctuation were removed, while other disfluencies (including filled pauses and whole-word repetitions) remained.

5.4 Speech/Language Features

One hundred twenty-three features were calculated for each speaker (participant, confederate) in the Bored condition and the Interested condition separately, using base R [29], qdap [30], and Linguistic Inquiry and Word Count (LIWC) software [31]. There were six main feature groups: pause/overlap metrics (12), segment/turn metrics (6), speaking rate/word complexity metrics (9), LIWC categories (80), lexical entropy/diversity measures (5), and parts of speech (9). Formality and polarity (2) were also computed at the conversation level for each speaker, using all words produced by a given speaker in each condition, leading to a total of 123 linguistic features. Differences between speakers were calculated within each condition (Participant Interested - Confederate Interested, Participant Bored - Confederate Bored) and within each speaker across conditions (Participant Interested - Participant Bored, Confederate Interested - Confederate Bored), yielding $8 \times 123 = 984$ features.

LIWC [31] is a commonly used software for analyzing text-based natural language data. LIWC relies on a dictionary of words that are grouped by semantic similarity into lexical categories. These word-language lexica are designated by a majority vote by human judges, as are which words that fall into each, or multiple, of these lexica. This type of text analysis has been used successfully to analyze various mental disorders [32], as well as to characterize personality traits from transcribed language or written text [33].

Lexical features are included in the current study as they have proven informative in prior ASD research. For example, the words produced by interviewing psychologists correlate significantly with ASD symptom severity [34]. Bone and colleagues conducted their analysis across a wide age range (3.58 to 13.17 years), and interlocutors were autism experts, but their research nonetheless suggests that word choice by conversational partners could be a potentially sensitive marker of ASD phenotype. In the current study, confederate word choice is captured.

Difference metrics were included in our feature set for two primary reasons. First, the original intent of the CASS task was to probe how individuals with ASD handle variations in conversational context, as compared to TD peers. Thus, within-speaker differences across two contexts (Bored interlocutor, Interested interlocutor) are pertinent relative to the original design. Second, interlocutor differences within a given condition were included as a general measure of linguistic accommodation; to study how closely the speaking rates, pause rates, and preferred conversational topics of the two speakers align. Research shows that greater linguistic accommodation is associated with social success [35] and also suggests that reduced accomodation in ASD in childhood [36] may improve by adulthood [37].

We recognize that for linear models, introducing new features as linear combinations of old features (such as the difference between the Interested and Bored conditions) is algebraically equivalent to not introducing these features at all. However, by introducing these additional features, we are guiding the model to learn dependencies that clinicians deem important and have functional value in real-world social contexts. This is especially true when using an automated feature selection technique, such as the f-value employed here, as these techniques limit the number of dimensions that can be used by a model. In the current study, rather than requiring our model to learn to take the difference across two dimensions, we are giving the model this knowledge *a priori*, and thus allowing the model to learn to use this difference with only one dimension. This type of reasoning forms the motivation for sparse coding (see below).

6 Results and Discussion

6.1 Model Design

Linear logistic regression, also known as the Maximum Entropy classifier or the softmax classifier, was used to classify ASD vs. TD. Features were down-selected before being input into the model by identifying dimensions with the highest f-value (largest mean separation between groups). The model was trained and tested according to leave one out, with an internal 5-fold cross validation to determine what percentage of the total features are kept from the f-value, selected from 0.5%, 1%, 2%, 5%, 10% or 20%. The top scoring f-test values can be seen in Figure 3 for the different age ranges. We used an ℓ_2-regularization penalty in the cost function in order to smooth out model coefficients. Our models were implemented in the Python library SciKit-Learn [38].

We use logistic regression so as to have an interpretable linear model. With more complex nonparametric and/or non-linear models, it is more difficult to understand the contribution of different variables on the model performance. We did not use a sparsity constraint in the model, such as an ℓ_1 penalty, since we are already imposing sparsity on the feature space by downsampling the feature dimension to those features with large f-values.

When designing the model, one may consider using age or gender as a covariate that automatically adjusts the model parameters, within for example a hierarchical Bayes network [39]. There are at least two difficulties with doing this in a purely data driven way. First, it introduces many additional parameters into the model one would need to learn, which on limited data is suboptimal in a statistical sense. Second, such hierarchical models are nonlinear, and thus difficult to interpret, which was an important design criteria for our model. Instead, we chose to use domain knowledge from developmental psychology to strictly define different models for different developmental age groups.

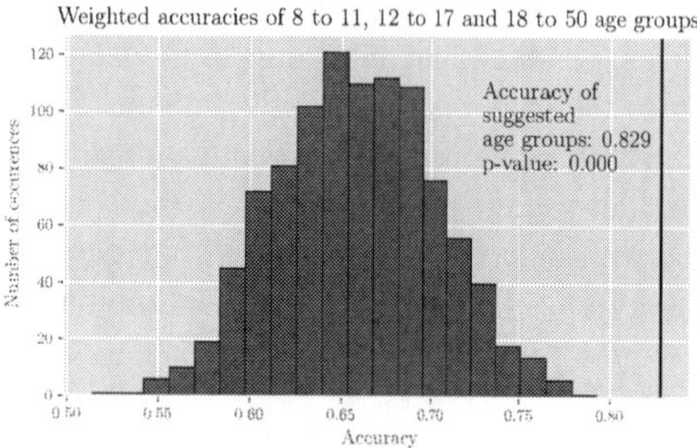

Figure 2: Comparison of the classification accuracy (weighted average of the three age groups) of the actual age-based split against 1000 randomized splits (not based on age) where sample sizes and proportions of classes in each sample were kept same as the actual split. The black vertical line shows the actual accuracy, and the red histogram shows the distribution of accuracy for random splits. The proportion of the distribution to the right of the vertical line defines the p-value.

Table 2: Classification accuracy for the three individual age groups and the entire sample. The weighted average (based on sample size) accuracy of the three age-specific models is 0.829 (p < 0.001, see Figure 2).

Age Range of Model	Accuracy
8 to 11	0.756
12 to 17	0.806
18 to 50	0.889
Weighted average	0.829
8 to 50	0.686

6.2 Classification Accuracy

Classification accuracy for three age-specific models, as well as the accuracy of a model for all ages together (8 and older), are shown in Table 2. Age-specific models outperformed the single model. The weighted average of the three age-specific models, weighted according to number of samples in each age group, was 0.829. In contrast, the single model for all ages achieved an accuracy of 0.686. Thus, our age-informed approach resulted in a 20.8% relative increase in accuracy, $p < 0.001$ (Figure 2). Again, this is notable as it contrasts with the standard doctrine in machine learning that training a model on more data is better; in our case we trained three models on roughly a third of the data each, yielding improved results.

6.3 Distinguishing Features by Age Group

Different linguistic features emerged as important for distinguishing between TD and ASD partici-

pants in each age group, as seen in Figure 3.

In the 8 to 11 age group, overall pronouns and personal pronouns predicted diagnostic status, such that children with ASD produced smaller proportions of pronouns than matched TD peers. In particular, the first person plural pronoun "we" was used relatively less frequently by the ASD group, suggesting that children with ASD were less likely to describe themselves as associating with others during conversation. Children in the ASD group also tended to use more out-of-dictionary words than TD children (i.e., they produced a smaller percentage of words that were in the LIWC dictionary, relative to their total word production), which could be due to children with ASD talking about specialized, idiosyncratic interests or simply using low-frequency words or phrases. Finally, children with ASD spoke more slowly, measured in words per minute with breath pauses removed, than matched TD children, and used comparatively fewer verbs (Figure 3a).

Top linguistic features that predicted diagnosis in the 12 to 17 age group are shown in Figure 3b. The Bored condition emerged as particularly important for distinguishing between TD and ASD adolescents, as did confederate word choice. Pronouns were predictive in this age group as well. Specifically, the second person personal pronoun "you" was produced relatively more often by TD teens in relation to confederates in the Bored condition. This could indicate more attempts by the TD group to engage with an obviously bored conversational partner, and relatively diminished effort put forth by teens with ASD. Confederates speaking with autistic teens used words associated with less authenticity, but greater clout, than when speaking with TD peers, and responded more often to TD participants with negations (perhaps in response to increased questions/comments about themselves, as indicated by greater use of "you" by TD teens).

Finally, linguistic features that differentiated between conversation samples from adults with and without ASD are shown in Figure 3c. Interestingly, these features were primarily temporal; for example, top features included the number of overlapping pauses (interruptions) in the conversation, as well as the rate of pauses per minute. This suggests that whereas topics of conversation might be comparable in ASD and TD adults (i.e., similar tendencies to discuss occupations or romantic

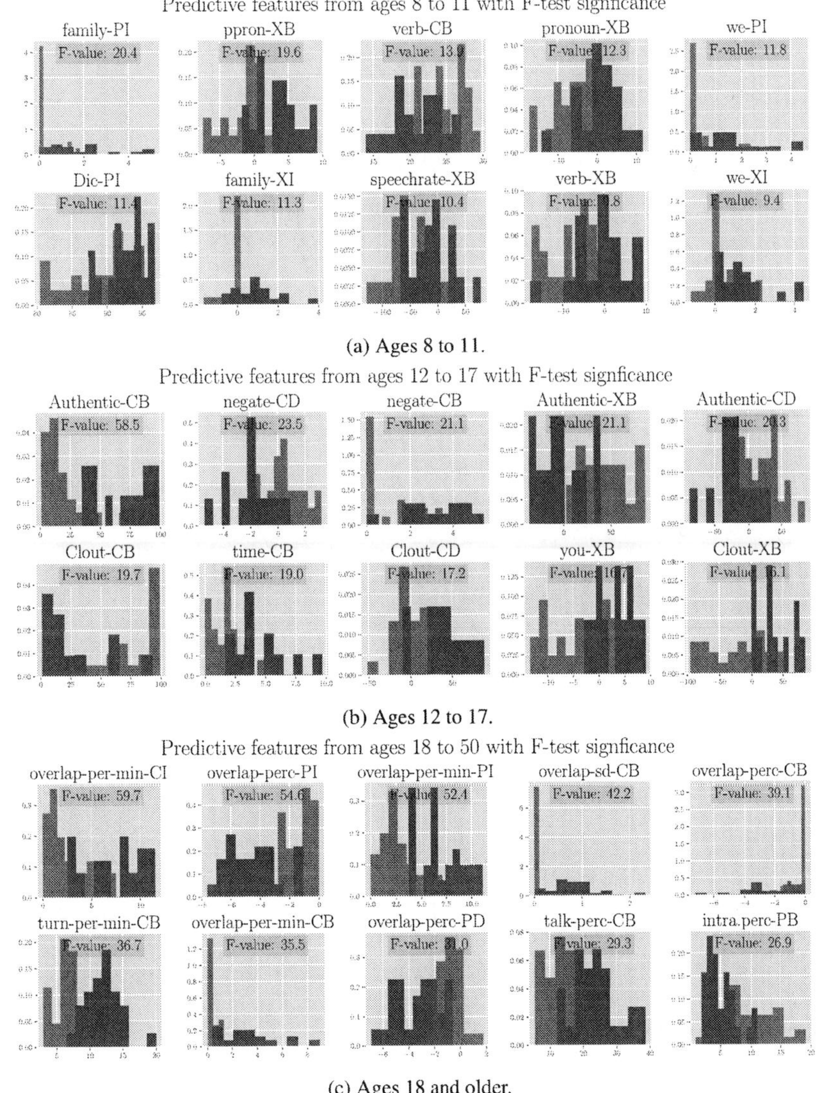

Figure 3: Histograms of the top 10 most discriminant features (ranked by f-test value) for the different age ranges considered, namely middle childhood, adolescence, and adulthood. In all figures, red is the ASD sample, and blue is the TD sample. Acronyms: PI = participant:interested, PB = participant:bored, PD = participant:difference (interested-bored), CI = confederate:interested, CB = confederate:bored, CD = confederate:difference (interested-bored), XI = cross:interested (participant-confederate), XB = cross:bored (participant-confederate).

partners), the way in which conversations occur may include awkward pauses, interruptions, and other temporal atypicalities that could negatively impact conversation quality.

The linguistic features identified in our machine learning analysis are consistent with prior research, as well as with observations about ASD made by clinicians and linguists. Importantly, our analysis goes a step further by quantifying the *extent* to which each of these features is important for distinguishing diagnostic groups at each age.

6.4 Feature Consistency Across Age Groups

The purpose of this subsection is to quantify which predictive speech/language features change by age group (i.e., how many predictive features remain predictive regardless of age). To do this, we measured change in the f-value.

Suppose we have age groups $(8, 11)$ and $(12, 17)$, and would like to compare changes in

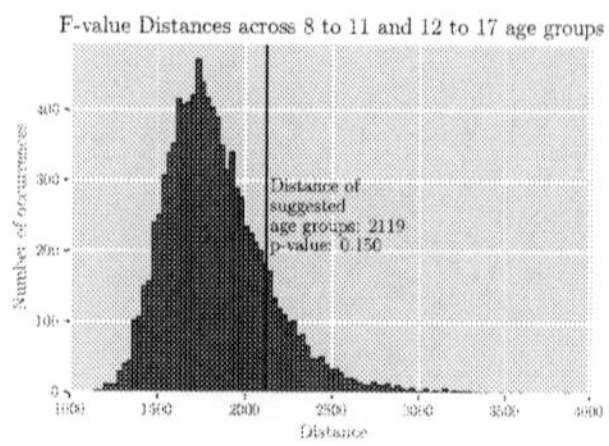
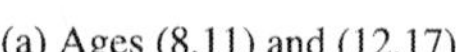
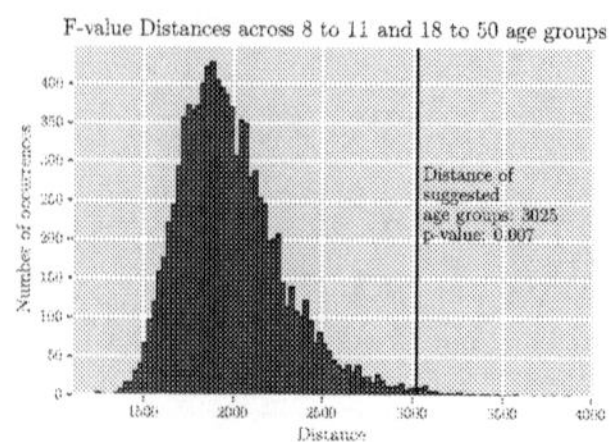

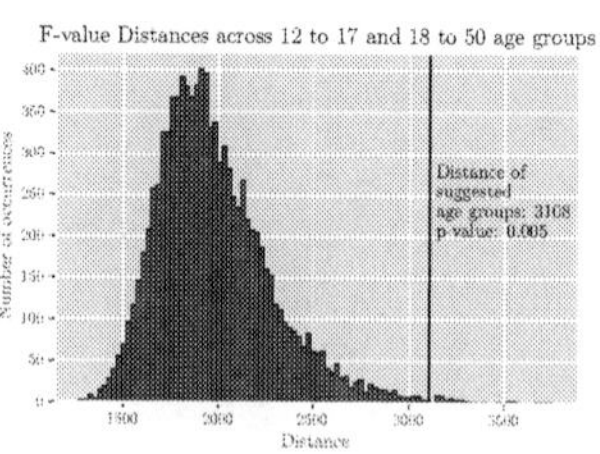

(a) Ages (8,11) and (12,17)   (b) Ages (8,11) and (18,50)   (c) Ages (12,17) and (18,50)

Figure 4: f-value distances ($\|f_i - f_j\|_1$) of the actual age-based split against 1000 randomized splits (not based on age) where sample sizes and proportions of classes were kept same as the actual split. The black vertical lines show the actual distance, and the red histograms show the distributions of distances for random splits.

f-values between $f_{(8,11)}$ and $f_{(12,17)}$. Since each participant is associated with $8 \times 123 = 984$ features, as mentioned in Section 6.1, then $f_{(8,11)}$ and $f_{(12,17)}$ are both 984-dimensional vectors, with each dimension containing the f-value of its corresponding feature. Measuring distances across dimensions does not make sense in this case, as each of the individual f-values are calculated in one dimension independently of each other. Thus, we use the ℓ_1-norm, sometimes referred to as the Manhattan distance, when measuring these distances, i.e. $\|f_{(8,11)} - f_{(12,17)}\|_1$.

Given that the magnitudes $\|f_{(8,11)}\|_1 = 1505$, $\|f_{(12,17)}\|_1 = 1848$ and $\|f_{(18,50)}\|_1 = 3035$, we see that the changes in magnitude of the feature importance from one age group to another are proportionally very large, and in fact often exceed, the magnitude of the features themselves. This tells us that the specific linguistic features that are important for distinguishing between ASD and TD, as defined by the f-test, vary enormously across age groups, especially when considered against the scale of the linguistic features themselves (Figures 4a- 4c, and Table 3).

Table 3: Measuring the extent to which the feature importance changes with the ℓ_1-norm, according to each feature's f-value, depending on which age group is under consideration. The p-value corresponds to distances developed from the null hypothesis where no age groups are considered, while ensuring correct proportions of ages and classes are kept.

Measurement	Value	p-value
$\|f_{(8,11)} - f_{(12,17)}\|_1$	2119	0.150
$\|f_{(8,11)} - f_{(18,50)}\|_1$	3025	0.007
$\|f_{(12,17)} - f_{(18,50)}\|_1$	3108	0.005

7 Discussion

In this study, we demonstrated that machine learning models for classifying and characterizing ASD improve significantly after incorporating domain knowledge. Specifically, we showed that models accounting for developmental changes in spoken language and conversation are more accurate for distinguishing ASD from typical development, relative to models resting on the assumption that language patterns during natural conversation remain consistent across ages. We further showed that linguistic features most strongly predicting ASD vary significantly across age groups, suggesting that specific atypicalities in the ways that individuals with ASD use language (versus TD controls) are not static across development.

These findings highlight the value of machine learning models that are clinically informed, particularly for understanding highly heterogeneous conditions like ASD. Developing separate models for different age groups (i.e., middle childhood, adolescence, and adulthood), we were able to significantly improve the models' classification performance and reliability, despite reductions in sample size. This bodes well for future applications of machine learning for studying psychiatric conditions. Future research will incorporate pitch-related features, extend classification to non-ASD psychiatric conditions, and explore the use of more complex nonlinear models for classification and prediction in larger sample sizes.

8 Conclusions and Future Work

This study has implications for our clinical understanding of ASD across the lifespan. We have identified sets of precise, objective linguistic features that are highly predictive of ASD at three different developmental stages. These features provide specific, developmentally-informed intervention targets that could be used to improve language and conversation skills in individuals with ASD. We anticipate that additional features identified through machine learning in other domains could similarly inform future efforts to develop targeted clinical interventions.

For future work, we would like to use these techniques in a longitudinal study for measuring treatment progress. This can be done by tracking feature values of an individual as they change through time. Additionally, we would like to use these techniques to see if they can be used to differentiate between other mental health disorders, such as anxiety, depression and obsessive compulsive disorder.

Acknowledgements

This work was supported by NIMH grant R34MH104407, Services to Enhance Social Functioning in Adults with Autism Spectrum Disorder (E.S. Brodkin, PI); by the National Center for Research Resources, Grant UL1RR024134, now the National Center for Advancing Translational Sciences, Grant UL1TR000003 (MPIs: E.S. Brodkin and R.T. Schultz); by the Intellectual and Developmental and Disabilities Research Center at the Childrens Hospital of Philadelphia and the University of Pennsylvania, NICHD U54HD86984 (MPIs: M. Robinson and R.T. Schultz); and by the Institute for Translational Medicine and Therapeutics (ITMAT) Transdisciplinary Program in Translational Medicine and Therapeutics (MPIs: E.S. Brodkin and R.T. Schultz), and by generous gifts from the Eagles Charitable Foundation and the Allerton Foundation to R.T. Schultz.

References

[1] Fifth Edition, American Psychiatric Association, et al. Diagnostic and statistical manual of mental disorders. *Arlington: American Psychiatric Publishing*, 2013.

[2] Helen Tager-Flusberg and Connie Kasari. Minimally verbal school-aged children with autism spectrum disorder: The neglected end of the spectrum. *Autism Research*, 6(6):468–478, 2013.

[3] C Lord, M Rutter, P DiLavore, S Risi, K Gotham, and S Bishop. Autism diagnostic observation schedule–2nd edition (ados-2). *Los Angeles, CA: Western Psychological Corporation*, 2012.

[4] Christopher Gillberg and Suzanne Steffenburg. Outcome and prognostic factors in infantile autism and similar conditions: A population-based study of 46 cases followed through puberty. *Journal of autism and developmental disorders*, 17(2):273–287, 1987.

[5] Patricia Howlin, Susan Goode, Jane Hutton, and Michael Rutter. Adult outcome for children with autism. *Journal of child psychology and psychiatry*, 45(2):212–229, 2004.

[6] André Venter, Catherine Lord, and Eric Schopler. A follow-up study of high-functioning autistic children. *Journal of child psychology and psychiatry*, 33(3):489–597, 1992.

[7] Meng-Chuan Lai, Michael V Lombardo, Bhismadev Chakrabarti, and Simon Baron-Cohen. Subgrouping the autism spectrum": Reflections on dsm-5. *PLoS biology*, 11(4):e1001544, 2013.

[8] Deborah K Anderson, Catherine Lord, Susan Risi, Pamela S DiLavore, Cory Shulman, Audrey Thurm, Kathleen Welch, and Andrew Pickles. Patterns of growth in verbal abilities among children with autism spectrum disorder. *Journal of consulting and clinical psychology*, 75(4):594, 2007.

[9] Masoud Rouhizadeh, Emily PrudHommeaux, Jan Van Santen, and Richard Sproat. Measuring idiosyncratic interests in children with autism. In *Proceedings of the conference. Association for Computational Linguistics. Meeting*, volume 2015, page 212. NIH Public Access, 2015.

[10] Julia Parish-Morris, Christopher Cieri, Mark Liberman, Leila Bateman, Emily Ferguson, and Robert T Schultz. Building language resources for exploring autism spectrum disorders. In *LREC... International Conference on Language Resources & Evaluation:[proceedings]. International Conference on Language Resources and Evaluation*, volume 2016, page 2100. NIH Public Access, 2016.

[11] Julia Parish-Morris, Mark Liberman, Neville Ryant, Christopher Cieri, Leila Bateman, Emily Ferguson, and Robert Schultz. Exploring autism spectrum disorders using hlt. In *Proceedings of the third workshop on computational lingusitics and clinical psychology*, pages 74–84, 2016.

[12] Emily Prudhommeaux and Masoud Rouhizadeh. Automatic detection of pragmatic deficits in children with autism. In *The... Workshop on Child, Computer and Interaction*, volume 2012, page 1. NIH Public Access, 2012.

[13] Cindy Gallois and Howard Giles. Communication accommodation theory. *The international encyclopedia of language and social interaction*, pages 1–18, 2015.

[14] Stanford W Gregory, Kelly Dagan, and Stephen Webster. Evaluating the relation of vocal accommodation in conversation partners' fundamental frequencies to perceptions of communication quality. *Journal of Nonverbal Behavior*, 21(1):23–43, 1997.

[15] Daniel Bone, Chi-Chun Lee, Matthew P Black, Marian E Williams, Sungbok Lee, Pat Levitt, and Shrikanth Narayanan. The psychologist as an interlocutor in autism spectrum disorder assessment: Insights from a study of spontaneous prosody. *Journal of Speech, Language, and Hearing Research*, 57(4):1162–1177, 2014.

[16] Henry M Wellman. *The child's theory of mind.* The MIT Press, 1992.

[17] Iroise Dumontheil, Ian A Apperly, and Sarah-Jayne Blakemore. Online usage of theory of mind continues to develop in late adolescence. *Developmental science*, 13(2):331–338, 2010.

[18] Deborah Yurgelun-Todd. Emotional and cognitive changes during adolescence. *Current opinion in neurobiology*, 17(2):251–257, 2007.

[19] B Bradford Brown. Adolescents' relationships with peers. *Handbook of adolescent psychology*, pages 363–394, 2004.

[20] Catherine Lord, Eva Petkova, Vanessa Hus, Weijin Gan, Feihan Lu, Donna M Martin, Opal Ousley, Lisa Guy, Raphael Bernier, Jennifer Gerdts, et al. A multisite study of the clinical diagnosis of different autism spectrum disorders. *Archives of general psychiatry*, 69(3):306–313, 2012.

[21] American Psychiatric Association et al. *Diagnostic and statistical manual of mental disorders (DSM-5®)*. American Psychiatric Pub, 2013.

[22] David Wechsler. *WASI-II: Wechsler abbreviated scale of intelligence.* PsychCorp, 2011.

[23] Vanessa Hus, Katherine Gotham, and Catherine Lord. Standardizing ados domain scores: Separating severity of social affect and restricted and repetitive behaviors. *Journal of autism and developmental disorders*, 44(10):2400–2412, 2014.

[24] Michael Rutter, Anthony Bailey, and Cathrine Lord. *The social communication questionnaire: Manual.* Western Psychological Services, 2003.

[25] Allison B Ratto, Lauren Turner-Brown, Betty M Rupp, Gary B Mesibov, and David L Penn. Development of the contextual assessment of social skills (cass): A role play measure of social skill for individuals with high-functioning autism. *Journal of Autism and Developmental Disorders*, 41(9):1277–1286, 2011.

[26] Meghan Lammie Glenn, Stephanie M Strassel, and Haejoong Lee. Xtrans: A speech annotation and transcription tool. In *Tenth Annual Conference of the International Speech Communication Association*, 2009.

[27] Owen Kimball, Chai-Lin Kao, Teodoro Arvizo, John Makhoul, and Rukmini Iyer. Quick transcription and automatic segmentation of the fisher conversational telephone speech corpus. In *RT04 Workshop*, 2004.

[28] Julia Parish-Morris, Mark Y Liberman, Christopher Cieri, John D Herrington, Benjamin E Yerys, Leila Bateman, Joseph Donaher, Emily Ferguson, Juhi Pandey, and Robert T Schultz. Linguistic camouflage in girls with autism spectrum disorder. *Molecular autism*, 8(1):48, 2017.

[29] RDC Team et al. R: A language and environment for statistical computing. *R foundation for statistical computing, Vienna, Austria*, 2008.

[30] Tyler W Rinker. qdap: Quantitative discourse analysis package. *University at Buffalo/SUNY*, 2013.

[31] James W Pennebaker, Ryan L Boyd, Kayla Jordan, and Kate Blackburn. The development and psychometric properties of liwc2015. Technical report, 2015.

[32] Stanley D Rosenberg and Gary J Tucker. Verbal behavior and schizophrenia: The semantic dimension. *Archives of General Psychiatry*, 36(12):1331–1337, 1979.

[33] James W Pennebaker and Laura A King. Linguistic styles: Language use as an individual difference. *Journal of personality and social psychology*, 77(6):1296, 1999.

[34] Manoj Kumar, Rahul Gupta, Daniel Bone, Nikolaos Malandrakis, Somer Bishop, and Shrikanth S Narayanan. Objective language feature analysis in children with neurodevelopmental disorders during autism assessment. In *INTERSPEECH*, pages 2721–2725, 2016.

[35] Kate Muir, Adam Joinson, Rachel Cotterill, and Nigel Dewdney. Characterizing the linguistic chameleon: Personal and social correlates of linguistic style accommodation. *Human Communication Research*, 42(3):462–484, 2016.

[36] Zoë Louise Hopkins. *Language alignment in children with an autism spectrum disorder.* PhD thesis, University of Sussex, 2016.

[37] Katie E Slocombe, Ivan Alvarez, Holly P Branigan, Tjeerd Jellema, Hollie G Burnett, Anja Fischer, Yan Hei Li, Simon Garrod, and Liat Levita. Linguistic alignment in adults with and without aspergers syndrome. *Journal of Autism and Developmental Disorders*, 43(6):1423–1436, 2013.

[38] F. Pedregosa, G. Varoquaux, A. Gramfort, V. Michel, B. Thirion, O. Grisel, M. Blondel, P. Prettenhofer, R. Weiss, V. Dubourg, J. Vanderplas, A. Passos, D. Cournapeau, M. Brucher, M. Perrot, and E. Duchesnay. Scikit-learn: Machine learning in Python. *Journal of Machine Learning Research*, 12:2825–2830, 2011.

[39] Christopher M Bishop. *Pattern recognition and machine learning.* springer, 2006.

The importance of sharing patient-generated clinical speech and language data

Kathleen C. Fraser
National Research Council Canada
Ottawa, Canada
`kathleen.fraser@nrc-cnrc.gc.ca`

Nicklas Linz
German Research Center
for Artificial Intelligence (DFKI)
Saarbrücken, Germany
`nicklas.linz@dfki.de`

Hali Lindsay
German Research Center
for Artificial Intelligence (DFKI)
Saarbrücken, Germany
`hali.lindsay@dfki.de`

Alexandra König
Memory Clinic at Nice University Hospital,
University of Côte d'Azur, and INRIA
Nice, France
`alexandra.konig@inria.fr`

Abstract

Increased access to large datasets has driven progress in NLP. However, most computational studies of clinically-validated, patient-generated speech and language involve very few datapoints, as such data are difficult (and expensive) to collect. In this position paper, we argue that we must find ways to promote data sharing across research groups, in order to build datasets of a more appropriate size for NLP and machine learning analysis. We review the benefits and challenges of sharing clinical language data, and suggest several concrete actions by both clinical and NLP researchers to encourage multi-site and multi-disciplinary data sharing. We also propose the creation of a collaborative data sharing platform, to allow NLP researchers to take a more active responsibility for data transcription, annotation, and curation.

1 Introduction

The Workshop on Computational Linguistics and Clinical Psychology (CLPsych) has brought together a strong community of NLP researchers and clinical experts, working on areas as diverse as the early detection of dementia through speech analysis, characterization of the properties of autistic children's language, identifying signs of depression and anxiety from written text, and many more. One theme that has emerged over time is the importance of clinically validated data, and at the same time, the difficulty in obtaining such data.

For example, and drawing only from the past proceedings of this workshop, numerous researchers have explicitly mentioned the small size of their dataset as a limitation of the work (Jarrold et al., 2014; Glasgow and Schouten, 2014; Fraser et al., 2014; Lamers et al., 2014; Bullard et al., 2016; Parish-Morris et al., 2016; Guo et al., 2017; Iter et al., 2018). These researchers point out that the consequences of such small datasets can include a lack of diversity in and representativeness of the training data, models which do not converge to a stable solution, unknown generalizability to other datasets, difficulty in interpreting the results, and limited clinical utility.

Other work has sought to overcome these limitations by using data scraped from social media or web forums (Coppersmith et al., 2014, 2015; Mitchell et al., 2015). While solving some problems, this approach introduces others, including uncertainty around the accuracy of the diagnosis and, crucially, the lack of a clinically-confirmed healthy control group (Coppersmith et al., 2014). Furthermore, such methods of data collection likely exclude many populations, including children and the elderly.

Here, we argue that large, clinically-validated datasets of patient-generated speech and langauge are imperative if we want to move the field forward, and that one way to create such datasets is to join together as a community and commit to finding better ways to share data.

2 Background

The issue of data sharing arises in many fields, including NLP more generally (where sharing corpora is strongly encouraged) and medical research (where data openness varies by domain). Clinical

Proceedings of the Sixth Workshop on Computational Linguistics and Clinical Psychology, pages 55–61
Minneapolis, Minnesota, June 6, 2019. ©2019 Association for Computational Linguistics

NLP sits at the intersection of these two fields, and thus faces its own unique challenges to data sharing (Chapman et al., 2011).

In NLP, data openness has long been recognized as the key to reproducible research and fair comparison between competing systems. One example of this is the popularity of the "shared task", in which systems from different research groups are trained, validated, and tested on the same data, allowing precise comparison across systems and leading to steady improvements in areas such as machine translation, speaker identification, parsing, information retrieval, etc. (Liberman and Cieri, 1998). In many areas of NLP, recent improvements in performance and generalizability have been reported due to the availability of larger and larger corpora (Jozefowicz et al., 2016; Koehn and Knowles, 2017).

The value of data sharing has been recognized in other scientific fields, where it has permitted the accumulation of massive data sets in areas such as astronomy and climatology. For example, while it is not possible for any one telescope to see all parts of the sky simultaneously, by sharing data with each other, astronomers can collectively build an accurate picture of the night sky (Borgman, 2012). The medical community has also identified important benefits to sharing data, as well as several critical practical and ethical challenges (Souhami, 2006; Hansson et al., 2016; Figueiredo, 2017).

In the following sections, we outline the benefits and challenges of data sharing as it applies specifically to patient-generated speech and text, within the context of NLP research.

3 Arguments for sharing data

Rationales for sharing data may vary for different stakeholders in the academic process (i.e., researchers, funding agencies, study participants).

When it comes to the computational study of clinical speech data, two broad groups of researchers are involved in the data sharing process: clinical researchers, who actively collect speech and language data, and computational linguistics researchers, who analyse and build models from the data. Both groups of researchers may be motivated by the fact that sharing data advances the state of research and innovation (Borgman, 2012; Figueiredo, 2017; Campbell et al., 2002; Fischer and Zigmond, 2010). Through the aggregation of multiple local studies, researchers are able to create a combined data set bigger than any single lab could reasonably collect (Borgman, 2012; Fischer and Zigmond, 2010), thus creating a more complete representation of reality. Proposals of innovative speech and language measures are more likely to attract the interest of the medical community when the conclusions are backed by a large study population. These large datasets can also support the application of complex computational modelling techniques, such as deep learning, that are not typically effective for small data.

Data sharing can also be used as a tool to reproduce and verify previous research (Borgman, 2012; Liberman and Cieri, 1998), which helps to validate findings for use in a clinical setting. Furthermore, data sharing can also have a professional benefit to researchers, as it fulfills the requirements of some granting agencies (e.g., NIH and NSF) (Borgman, 2012; Fischer and Zigmond, 2010), and can increase the citation rates and impact of researchers' studies (Piwowar et al., 2007; Figueiredo, 2017).

Societal interest in data sharing, and thereby that of funding agencies, is motivated differently. Since funding bodies often support research using tax revenue, there is interest in making results, including data, of publicly-funded research available to the public (Borgman, 2012; Figueiredo, 2017; Pennebaker, 2004). Additionally, data sharing has been found to increase the overall quality of the produced research. It maximizes the use of collected data, as it enables others to ask new questions of existing data (Borgman, 2012; Figueiredo, 2017; Fischer and Zigmond, 2010) and diversifies the perspective on these data (Fischer and Zigmond, 2010). Financially, sharing data leads to a greater return on public investment in research, since the production costs of data sets can be shared between different actors (Liberman and Cieri, 1998; Fischer and Zigmond, 2010) and it avoids the generation of duplicate data sets (Figueiredo, 2017; Liberman and Cieri, 1998; Fischer and Zigmond, 2010).

Participants in studies, including patient and healthy controls, might be motivated by the multiple benefits to society listed above. Participants are also often motivated by making a contribution to new, improved or safer medical treatments and want their participation to have the widest possible impact (Hansson et al., 2016). They are often willing to share de-identified personal data and do not necessarily see it as an invasion of their privacy (Hansson et al., 2016). The willingness to share data may be even greater in patient populations, since results from research may directly benefit themselves or other with the diseases (Souhami, 2006; Hansson et al., 2016).

4 Challenges to sharing data

Despite the many benefits, there are also challenges within scientific communities that can prevent the sharing of data, including ethical and legal considerations, practical barriers, and the desire for researchers to protect and manage access to the data that support their research programs.

A primary concern regarding the sharing of patient data is personal privacy and security (Souhami, 2006; Childs et al., 2011), which is magnified in the case of clinical speech and language data that will be linked by necessity to personal health data (e.g., medical diagnosis, cognitive test results). Audio and visual data may not be possible to fully anonymize, and are also considered personal information. Study participants in general are wary of being identified by insurance providers, employers or other third parties as the risk of exposure of personal information may result in social or psychological harm (Hansson et al., 2016). This can lead to inaccurate self-reporting or even the avoidance of medical care if a person believes that the disclosure of certain information (e.g. drug use) will be revealed to others, resulting in harm or persecution. Additionally, even if participants gave consent for the initial data collection, obtaining consent for the secondary use of data may be impossible, as patients may be deceased or have relocated (Souhami, 2006).

For these reasons, in some cases it may not be ethically or legally permissible to share clinical data, and legal measures are in place to protect the privacy of patients and research participants. For example, in the United States medical information is protected under the Health Insurance Portability and Accountability Act (HIPAA) and the Health Information Technology for Economic and Clinical Health Act (HITECH Act) (Annas, 2003; Blumenthal, 2010); similar regulations exist in countries around the world. These policies mean that data collected by clinicians acting in their clinical capacities may be subject to stricter regulation than data in traditional academic research. Non-compliance with federal regulations can result in fines or loss of license. Additionally, many clinicians (including psychologists[1] and psychiatrists[2]) are bound by a professional code of ethics which may preclude the sharing of patient data.

Data sharing can be difficult on a practical level. Often, data collected at separate sites are not formatted for consistent and comparable sharing (Borgman, 2012). In some cases, audio or video data may not even exist as a digital file (MacWhinney, 2007). Limited financial and personnel resources may prevent the labour-intensive preparation and documentation of clinical speech and language data into convenient, transmittable formats (Campbell et al., 2002; Borgman, 2012). Different research projects may involve different speech/language tasks, different recording conditions, different diagnostic criteria, and different clinical populations, which may limit the extent to which datasets can be combined across projects.

In addition to these challenges are personal considerations within the research community itself. Allowing others to work on private datasets could expose errors within the data or in previous publications (Childs et al., 2011). A real example of this can be found in the social psychology literature, where the re-analysis of data from the implicit association test challenged the conclusions of the original study (Blanton et al., 2009; McConnell and Leibold, 2009). Data sharing efforts typically do not factor into tenure or promotional considerations (Borgman, 2012), and there is a perceived lack of reward or credit for the considerable time and effort required (Fischer and Zigmond, 2010; Borgman, 2012). This is compounded by the reality that one's research may be considered less novel or innovative, since allowing access to data resources would allow other researchers to publish similar work on the same data (Figueiredo, 2017; Childs et al., 2011; Campbell et al., 2002).

Other concerns relate to the inability to control the applications of the data and the possibility of misuse or misinterpretation (Campbell et al., 2002; Figueiredo, 2017). Research protocols describe the purpose of the data collection, e.g. improving care and providing timely intervention, and clinicians may be wary of outside parties using these data for more profit-oriented objectives.

---

[1] https://www.apa.org/ethics/code/principles.pdf

[2] https://www.psychiatry.org/psychiatrists/practice/ethics

5 Examples of successful data sharing

We now briefly discuss two case studies in successful data sharing, while acknowledging that many other models exist and may also be appropriate to our community (for example, shared tasks).

One successful example of a data repository in NLP is the Linguistic Data Consortium, or LDC (Liberman and Cieri, 1998). The LDC manages dozens of widely-used speech and language corpora, including TIMIT, Gigaword, the Penn Treebank, and many other foundational datasets in NLP. As of 2018, it has distributed more than 140,000 copies of datasets to over 4,000 organizations (Cieri et al., 2018). Originally supported by grants, the LDC has been sustained by membership fees and data sales since 2015. It also has a scholarship program to provide free data access to researchers who do not have the resources to pay for a membership (DiPersio and Cieri, 2016). Particularly relevant to our discussion here, the LDC has recently started to move in the direction of creating clinical databases, including for autism and neurodegenerative disorders (Cieri et al., 2018).

In the clinical speech research realm, one successful initiative has been the TalkBank Project, including AphasiaBank and DementiaBank (MacWhinney, 2007; Forbes et al., 2012). The project is supported by grants, and members of the TalkBank consortium are expected, wherever possible, to contribute data of their own. AphasiaBank has a standard protocol of tasks that facilitates comparison and aggregation of data across individual research projects. Furthermore, demographic and neuropsychological test data are also given for the participants, and all audio, video, and transcription files use a common format. Individual datasets in the database are protected according to the sensitivity of the data and the terms of the consent. The project has its own code of ethics, and provides guidelines for research ethics board applications and consent form templates. While AphasiaBank was started by and for researchers, it has become an important resource for clinicians and educators as well (Forbes et al., 2012).

Both platforms can be used as good examples for how sharing patient-generated clinical speech and language data can be realized. In particular, they create a separation between the work of creating the data from the work of maintaining and distributing the data (Cieri et al., 2018). They have also managed the issues of security and data privacy, and have created standards for data formatting and data collection.

However, contributions to TalkBank (and the limited clinical datasets on the LDC) appear to be made mostly by clinical researchers, which still places most of the burden of preparing, documenting, transcribing, and annotating the data on their shoulders. A more collaborative model of data sharing, which involves various contributions from both clinical and computational researchers, may encourage greater participation.

6 Recommendations

Based on the literature and examples above, we offer a preliminary (and surely incomplete) set of recommended best practices to promote collaboration and data sharing. Some actions that can be taken by *researchers who are collecting data* that will aid data sharing include:

- Having a long-term data management plan in place from the initial stages of a project, and including it in the funding proposal.

- Obtaining open and transparent consent from participants, that allows sharing and re-use of the data and realistically describes the benefits and harms of data sharing.

- Reviewing archival consent forms to determine if the original terms allow sharing to any degree.[3]

- Collecting data that can be anonymized to the greatest extent possible (e.g., eliciting speech on more general topics rather than personal histories, where appropriate).

- Where it is necessary to collect data of a more personal nature (as will be the case in many situations arising in couples and family therapy, or in relation to mental health conditions), considering automated or manual approaches to anonymizing the data, including offering participants the chance to anonymize their own data.

- Using file formats and transcription protocols that are common in the field, as well as a standardized protocol of tasks and meta-data (e.g. demographic information).

---

[3]For example, see `https://talkbank.org/share/irb/` for some guidelines on this topic.

Some actions that can be taken by *researchers who intend to make use of shared data* that will encourage and support data sharing include:

- Making other kinds of contributions to shared repositories, including: digitized versions of archival data, transcriptions, scripts for data processing and feature extraction, spreadsheets of extracted information, etc.

- Incentivizing data sharing through citations, acknowledgements, collaborations, and respectful use of the data and adherence to the relevant codes of ethics.

- Creating resources/platforms to lower the technical barriers to data sharing, and improve security and privacy of data.

- Communicating openly with the data owners, both to promote trust and to increase awareness of the kinds of emerging technologies that can benefit research in the field.

7   Conclusion and next steps

Access to larger datasets would undoubtedly improve the accuracy, generalizability, and clinical utility of computer models of patient-generated speech and language. However, clinical data is expensive and time-consuming to collect. Therefore, we argue that increased data sharing across research groups may be the only way to collect datasets of the size needed for robust machine learning, and to establish the population norms and empirical validation that will be required to allow NLP technologies to be recognized and used in clinical practice.

Existing platforms like the LDC and TalkBank are one option, particularly for sharing existing data sets. However, other models of data sharing may also be appropriate. Specifically, we propose a collaborative platform to support the continuous aggregation of data in a multi-disciplinary setting, where different parties can contribute according to their expertise (e.g., clinicians collect data, NLP researches transcribe or curate data). This shifts some of the responsibility from the clinical researchers to the computational researchers, while increasing the total value of the resulting data resource for everyone.

As a first step towards this goal, we advocate for the creation of a multi-disciplinary working group, consisting of clinicians and clinical researchers, patient organizations, and NLP researchers. This group should carefully review the feasibility of the recommendations made in the previous section, gauge interest in such a project from the various stakeholders, define the concrete requirements of a platform that would enable multi-disciplinary data collection and sharing, and determine how it could be prototyped and sustained through funding, over a longer period of time. It is essential that clinicians take a leading role in defining the concrete objectives and orientation of this group, ensuring that clinical research goals and improved patient outcomes are the main focus.

References

George J Annas. 2003. HIPAA regulations—a new era of medical-record privacy? *New England Journal of Medicine*, 348(15):1486–1490.

Hart Blanton, James Jaccard, Jonathan Klick, Barbara Mellers, Gregory Mitchell, and Philip E Tetlock. 2009. Strong claims and weak evidence: Reassessing the predictive validity of the IAT. *Journal of Applied Psychology*, 94(3):567–582.

David Blumenthal. 2010. Launching HITECH. *New England Journal of Medicine*, 362(5):382–385.

Christine L Borgman. 2012. The conundrum of sharing research data. *Journal of the American Society for Information Science and Technology*, 63(6):1059–1078.

Joseph Bullard, Cecilia Ovesdotter Alm, Xumin Liu, Qi Yu, and Rubén Proano. 2016. Towards early dementia detection: Fusing linguistic and non-linguistic clinical data. In *Proceedings of the Third Workshop on Computational Linguistics and Clinical Psychology*, pages 12–22.

Eric G Campbell, Brian R Clarridge, Manjusha Gokhale, Lauren Birenbaum, Stephen Hilgartner, Neil A Holtzman, and David Blumenthal. 2002. Data withholding in academic genetics: Evidence from a national survey. *Journal of the American Medical Association*, 287(4):473–480.

Wendy W Chapman, Prakash M Nadkarni, Lynette Hirschman, Leonard W D'avolio, Guergana K Savova, and Ozlem Uzuner. 2011. Overcoming barriers to NLP for clinical text: The role of shared tasks and the need for additional creative solutions. *Journal of the American Medical Informatics Association*, 18(5):540–543.

Becky Childs, Gerard Van Herk, and Jennifer Thorburn. 2011. Safe Harbour: Ethics and accessibility in sociolinguistic corpus building. *Corpus Linguistics and Linguistic Theory*, 7(1):163–180.

Christopher Cieri, Mark Liberman, Stephanie Strassel, Denise DiPersio, Jonathan Wright, and Andrea

Mazzucchi. 2018. From 'solved problems' to new challenges: A report on LDC activities. In *Proceedings of the Eleventh International Conference on Language Resources and Evaluation (LREC-2018)*, pages 3265–3269.

Glen Coppersmith, Mark Dredze, and Craig Harman. 2014. Quantifying mental health signals in Twitter. In *Proceedings of the Workshop on Computational Linguistics and Clinical Psychology: From Linguistic Signal to Clinical Reality*, pages 51–60.

Glen Coppersmith, Mark Dredze, Craig Harman, and Kristy Hollingshead. 2015. From ADHD to SAD: Analyzing the language of mental health on Twitter through self-reported diagnoses. In *Proceedings of the 2nd Workshop on Computational Linguistics and Clinical Psychology: From Linguistic Signal to Clinical Reality*, pages 1–10.

Denise DiPersio and Christopher Cieri. 2016. Trends in HLT research: A survey of LDC's data scholarship program. In *Proceedings of the Language Resources and Evaluation Conference (LREC)*, pages 1614–1618.

Ana Sofia Figueiredo. 2017. Data sharing: Convert challenges into opportunities. *Frontiers in Public Health*, 5:327.

Beth A Fischer and Michael J Zigmond. 2010. The essential nature of sharing in science. *Science and Engineering Ethics*, 16(4):783–799.

Margaret M Forbes, Davida Fromm, and Brian MacWhinney. 2012. AphasiaBank: A resource for clinicians. In *Seminars in Speech and Language*, volume 33, pages 217–222. Thieme Medical Publishers.

Kathleen C Fraser, Graeme Hirst, Naida L Graham, Jed A Meltzer, Sandra E Black, and Elizabeth Rochon. 2014. Comparison of different feature sets for identification of variants in progressive aphasia. In *Proceedings of the Workshop on Computational Linguistics and Clinical Psychology: From Linguistic Signal to Clinical Reality*, pages 17–26.

Kimberly Glasgow and Ronald Schouten. 2014. Assessing violence risk in threatening communications. In *Proceedings of the Workshop on Computational Linguistics and Clinical Psychology: From Linguistic Signal to Clinical Reality*, pages 38–45.

Jia-Wen Guo, Danielle L Mowery, Djin Lai, Katherine Sward, and Mike Conway. 2017. A corpus analysis of social connections and social isolation in adolescents suffering from depressive disorders. In *Proceedings of the Fourth Workshop on Computational Linguistics and Clinical Psychology — From Linguistic Signal to Clinical Reality*, pages 26–31.

Mats G Hansson, Hanns Lochmüller, Olaf Riess, Franz Schaefer, Michael Orth, Yaffa Rubinstein,

Caron Molster, Hugh Dawkins, Domenica Taruscio, Manuel Posada, et al. 2016. The risk of re-identification versus the need to identify individuals in rare disease research. *European Journal of Human Genetics*, 24(11):1553.

Dan Iter, Jong Yoon, and Dan Jurafsky. 2018. Automatic detection of incoherent speech for diagnosing schizophrenia. In *Proceedings of the Fifth Workshop on Computational Linguistics and Clinical Psychology: From Keyboard to Clinic*, pages 136–146.

William Jarrold, Bart Peintner, David Wilkins, Dimitra Vergryi, Colleen Richey, Maria Luisa Gorno-Tempini, and Jennifer Ogar. 2014. Aided diagnosis of dementia type through computer-based analysis of spontaneous speech. In *Proceedings of the Workshop on Computational Linguistics and Clinical Psychology: From Linguistic Signal to Clinical Reality*, pages 27–37.

Rafal Jozefowicz, Oriol Vinyals, Mike Schuster, Noam Shazeer, and Yonghui Wu. 2016. Exploring the limits of language modeling. *arXiv preprint arXiv:1602.02410*.

Philipp Koehn and Rebecca Knowles. 2017. Six challenges for neural machine translation. *arXiv preprint arXiv:1706.03872*.

Sanne MA Lamers, Khiet P Truong, Bas Steunenberg, Franciska de Jong, and Gerben J Westerhof. 2014. Applying prosodic speech features in mental health care: An exploratory study in a life-review intervention for depression. In *Proceedings of the Workshop on Computational Linguistics and Clinical Psychology: From Linguistic Signal to Clinical Reality*, pages 61–68.

Mark Liberman and Christopher Cieri. 1998. The creation, distribution and use of linguistic data: The case of the Linguistic Data Consortium. In *Proceedings of the 1st International Conference on Language Resources and Evaluation (LREC)*, pages 159–164.

Brian MacWhinney. 2007. The TalkBank Project. In *Creating and Digitizing Language Corpora*, pages 163–180. Springer.

Allen R McConnell and Jill M Leibold. 2009. Weak criticisms and selective evidence: Reply to Blanton et al.(2009). *Journal of Applied Psychology*, 94(3):583–589.

Margaret Mitchell, Kristy Hollingshead, and Glen Coppersmith. 2015. Quantifying the language of schizophrenia in social media. In *Proceedings of the 2nd Workshop on Computational Linguistics and Clinical Psychology: From Linguistic Signal to Clinical Reality*, pages 11–20.

Julia Parish-Morris, Mark Liberman, Neville Ryant, Christopher Cieri, Leila Bateman, Emily Ferguson, and Robert Schultz. 2016. Exploring autism spectrum disorders using HLT. In *Proceedings of the*

Third Workshop on Computational Linguistics and Clinical Psychology, pages 74–84.

James W Pennebaker. 2004. Theories, therapies, and taxpayers: On the complexities of the expressive writing paradigm. *Clinical Psychology: Science and Practice*, 11(2):138–142.

Heather A Piwowar, Roger S Day, and Douglas B Fridsma. 2007. Sharing detailed research data is associated with increased citation rate. *PloS One*, 2(3):e308.

Robert Souhami. 2006. Governance of research that uses identifiable personal data. *The BMJ*, 333(7563):315–316.

Depressed Individuals Use Negative Self-Focused Language When Recalling Recent Interactions with Close Romantic Partners but Not Family or Friends

Taleen Nalabandian and **Molly E. Ireland**

Department of Psychological Sciences, Texas Tech University, Lubbock, Texas
{taleen.nalabandian,molly.ireland}@ttu.edu

Abstract

Depression is characterized by a self-focused negative attentional bias, which is often reflected in everyday language use. In a prospective writing study, we explored whether the association between depressive symptoms and negative, self-focused language varies across social contexts. College students ($N = 243$) wrote about a recent interaction with a person they care deeply about. Depression symptoms positively correlated with negative emotion words and first-person singular pronouns (or negative self-focus) when writing about a recent interaction with romantic partners or, to a lesser extent, friends, but not family members. The pattern of results was more pronounced when participants perceived greater self-other overlap (i.e., interpersonal closeness) with their romantic partner. Findings regarding how the linguistic profile of depression differs by type of relationship may inform more effective methods of clinical diagnosis and treatment.

1 Introduction

Depression is often characterized by a negative attentional bias, wherein depressed individuals view themselves and their surrounding environment negatively (Beck, 1967). For example, when listening to a string of words, depressed individuals are more likely to identify negative (rather than neutral) homophones (e.g., *weak* rather than *week*; Wenzlaff & Eisenberg, 2001). Depressed individuals also selectively recall negative more than positive experiences (Dalgleish & Werner-Seidler, 2014). Further, people who are currently depressed associate more negative and fewer positive traits with not only themselves, but also their parents and romantic partners (Gara et al., 1993).

With depression affecting millions worldwide (WHO, 2018) and depression rates increasing for adolescents and young adults in particular (Twenge, Joiner, Rogers, & Martin, 2017), researchers across multiple fields are focused on finding more effective methods of early diagnosis and treatment. Research at the intersection of clinical psychology and computational linguistics has extensively examined depressed individuals' language use as an alternative to more traditional self-report methods of measuring depressive symptomology. Self-reports can be particularly limited when assessing mental health conditions, such as depression, which tend to be stigmatized (Crocker & Major, 1989) and may involve biased self-perceptions (Beck, 1967; Beevers, 2005; c.f. Moore & Fresco, 2012). Given the limitations of self-reports, it is necessary to supplement depression scales (e.g., Beck Depression Inventory-II, Center for Epidemiologic Studies Depression Scale Revised) with less explicit measures.

Language use may serve as an implicit, behavioral measure of depression. Many studies have found that high rates of first-person singular pronouns and negative emotion words correlate with higher levels of depression in a variety of contexts, such as public social media posts (De Choudhury, Counts, Horvitz, & Hoff, 2014; Eichstaedt et al., 2018; Schwartz et al., 2014), private expressive writing tasks (Rude, Gortner, & Pennebaker, 2004), and diagnostic clinical interviews (Zimmerman et al., 2016; see Holtzman, 2017 for a meta-analysis). Depressed individuals' use of negative emotion words coincides with their negative attentional bias (Beevers, 2005) and emotion regulation deficits (Joorman & Stanton, 2016), while their use of first-person singular pronouns corresponds with

62

Proceedings of the Sixth Workshop on Computational Linguistics and Clinical Psychology, pages 62–73
Minneapolis, Minnesota, June 6, 2019. ©2019 Association for Computational Linguistics

their tendency to ruminate (i.e., engage in repetitive negative thinking about the self; Watkins & Teasdale, 2001).

Despite the clinical importance of behavioral indicators of mental health, effect sizes for the associations between language and depressive symptoms tend to be modest, which limits the use of language as a primary clinical outcome or ground truth (Baddeley, Pennebaker, & Beevers, 2012; Holtzman, 2017). For example, recent research suggests that self-focused language in particular may be better understood as an indicator of vulnerability to stress (or neuroticism) rather than depression per se (Tackman et al., 2018). We propose that some questions about the stability of the links between language and mental health symptoms stem from differences in how individuals experience and express depressive symptoms across contexts. In the current study, we consider how linguistic indicators of depression—presumably reflecting depressive symptoms and self-regulatory processes—vary across written descriptions of recent interactions with family, friends, and romantic partners.

Not all language categories are created equal. People tend to be less conscious of their use of function words (i.e., words that define syntax and express *how* people communicate, such as articles and pronouns) than content words (i.e., words that reflect conversation topic or what people are saying, such as nouns and verbs; Tausczik & Pennebaker, 2010). Function words make up a miniscule portion (<.1%) of the total words in an individual's repertoire, yet they comprise over half of the words used in everyday conversation and writing (Chung & Pennebaker, 2007). In some instances, first-person singular pronouns (*I, me, my*) predict levels of depression to a greater degree than do negative emotion words (De Choudhury et al., 2014), perhaps because function words may be less easily regulated than content words (Bell, Brenier, Gregory, Girand, & Jurafsky, 2009; Garrod & Pickering, 2016).

For example, mothers with postpartum depression (a major depressive episode following childbirth) were more likely to use first-person singular pronouns in their Facebook posts than were non-postpartum depression mothers, but their use of negative emotion words did not differ (De Choudhury et al., 2014). Follow-up interviews with those mothers revealed that many of their concerns with respect to posting about their depression stemmed from possible judgment from friends. Thus, content words (e.g., negative emotion words), which people are more conscious of, may be more easily censored in everyday language use, whereas function words (e.g., first-person singular pronouns) and syntax are less easily censored as they are processed more rapidly (Segalowitz, & Lane, 2000), with less conscious attention and control (Pulvermüller, Shtyrov, Hasting, & Carlyon, 2008).

Other social factors may play a role in the rate at which depressed people use certain content and function word categories. For instance, depressed individuals may disclose more or less while talking with certain people in their daily lives (Altman & Taylor, 1973). Specifically, students with higher levels of depression were more likely to use negative language while having a conversation with a friend rather than a stranger (Segrin & Flora, 1998). Naturalistic recordings of everyday life also show that depressed individuals are more likely to use negative emotion words in conjunction with self-focused speech (e.g., "I feel guilty") as well as when speaking with romantic partners than others (e.g., coworkers; Baddeley et al., 2012). Perhaps depressed individuals feel less obligated to maintain a socially desirable front with and thus are more comfortable communicating negative affect to romantic partners. Alternately, close relationships may be a source of distress or depressive symptoms rather than a buffer against stress for some individuals in distressed relationships (Kiecolt-Glaser & Newton, 2001; Joyner, & Udry, 2000). Romantic breakups, which often follow a pattern of negative interactions with romantic partners (Gottman & Levenson, 2000), are a common trigger for adolescents' first depressive episodes (Monroe, Rhode, Seeley, & Lewinsohn, 1999).

Intimate relationships powerfully impact mental health, having the potential to both protect against and cause significant psychological distress. Close interpersonal relationships are typically viewed a hallmark of mental health, as they foster feelings of belongingness or satisfy the fundamental need to belong (i.e., people have a basic desire to develop long-term close relationships with others; Baumeister & Leary, 1995). Decreased feelings of belongingness are strongly associated with depressive symptoms (Choenarom, Williams, & Hagerty, 2005;

Hagerty & Williams, 1999). Furthermore, chronic self-focus is bidirectionally associated with loneliness (Cacioppo, Chen, & Cacioppo, 2017), and loneliness is a major risk factor for depression, independent of related constructs such as perceived social support and stress (Cacioppo, Hughes, Waite, Hawkley & Thisted, 2006). Thus, decreased belongingness may serve as a possible mechanism that links negative emotion word and first-person singular pronoun use with depression. The social construct of belongingness may help explain why depressed individuals tend to use negative self-focused language in the presence of those close to them.

1.1 Hypotheses

Earlier research has focused on examining depressed individuals' language use in the context of in-person conversations with intimate versus non-intimate others (Baddeley et al., 2012; Segrin & Flora, 1998). We determined to test whether these results would replicate when depressed individuals reflect on and write about—rather than speak with—their significant others. In particular, we predict that when asked to think about and describe the most recent interaction with a romantic partner, close friend, or family member, those with higher levels of depression will be more likely to use negative self-focused language in their written responses.

Furthermore, we hypothesize that depressed individuals' language use in their written recollections of their significant other should be dependent on their level of belongingness or interpersonal closeness. In other words, those with higher levels of depression will use more negative self-focused language to a greater degree if they indicate higher levels of belongingness or interpersonal closeness with their indicated significant other.

Finally, because anxiety is often co-morbid with depression and the two mental health conditions have significant symptomological overlap (i.e., both are characterized by negative affect and self-focus), it is important to determine whether any statistical effects are solely attributable to depression or may stem from anxiety as well (Tennen, Hall, & Affleck, 1995).

2 Method

Texas Tech University undergraduates enrolled in a general psychology course (N = 243; M_{age} = 19.7, SD_{age} = 2.94; 62.6% female) participated in an online survey for course credit. Three participants did not complete the depression scale and thus could not be included in the depression analyses. Upon providing their electronic consent, students were asked to take the time to reflect on one person in their life they deeply care about, such as a family member, a close friend, or a romantic partner. Once they successfully visualized this person in their mind, they were instructed to describe the last interaction they experienced with them in a detailed written response. Interactions were broadly defined, encompassing in-person as well as distant (e.g., over the phone or internet) encounters. Participants were asked to indicate the exact date of their interaction to ensure compliance with the request to write about the *most recent* interaction with a significant other. Less than 8% (n = 19) of the 243 participants identified dates that were significantly discrepant from the time of their participation in the study (>4 months, or roughly one semester). For each model reported below, our conclusions were identical when excluding those 19 participants from the sample. Following the writing task, participants completed various questionnaires in order to assess their mental state and demographic information. All questionnaires—including those on depression, anxiety, belongingness, and demographics—were administered after the writing task to avoid any potential carryover effects on individuals' recollections or language use.

2.1 Measures

Depression. The Center for Epidemiologic Studies Depression Scale Revised (CESD-R; Eaton, Smith, Ybarra, Muntaner, & Tien, 2004) was used to measure participants' depressive symptoms and categorize participants as having subclinical depression or not. The CESD-R includes 20 items, each of which belong to various symptom categories of depression: Dysphoria, anhedonia, appetite, sleep, thinking/concentration, worthlessness, fatigue, agitation, and suicidal ideation (Eaton et al., 2004). Participants were asked to indicate how often they felt depressive symptoms (e.g.,

"Nothing made me happy") over the past two weeks on a scale of 0 (*not at all or less than one day last week*) to 4 (*nearly every day for two weeks*; Eaton et al., 2004). Utilizing the CESD-style scoring system, where the two highest responses are given the same score of 3 (Eaton et al., 2004), 53.8% of the present sample had a score of less than 16 and 46.3% had a score of equal to or greater than 16, meeting the criteria for subclinical depression (*M* = 16.6, *SD* = 13.3).

Anxiety. In addition to the CESD-R, participants were given the Generalized Anxiety Disorder 7-Item (GAD-7; Spitzer, Kroenke, Williams, & Löwe, 2006) scale to assess their level of anxiety. Items comprised of GAD-7 are based on diagnostic criteria for generalized anxiety disorder, such as excessive anxiety (e.g., "Worrying too much about different things"), difficulty controlling anxiety (e.g., "Not being able to stop or control worrying"), and key symptoms associated with experiencing anxiety (e.g., "Becoming easily annoyed or irritable"; Spitzer et al., 2006). Participants were asked to rate how often they were experiencing each symptom on a scale of 0 (*not at all*) to 3 (*nearly every day*) within the last two weeks (Spitzer et al., 2006). The current sample had relatively low anxiety (*M* = 6.8, *SD* = 5.8). Nearly half (46.5%) of the sample reported little to no anxiety (scoring 0-4 on the GAD-7), 22.2% had mild anxiety (scoring 5-9), 19.8% had moderate anxiety (scoring 10-14), and 11.5% were severe (≥15).

Belongingness. Three separate scales were used to measure the exploratory mechanism of belongingness: The Need to Belong (NTB; Leary, Kelly, Cottrell, & Schreindorfer, 2013) scale, the Interpersonal Needs Questionnaire (INQ; Van Orden, Cukrowicz, Witte, & Joiner, 2012), and the Inclusion of Other in the Self (IOS; Aron, Aron, & Smollan, 1992) scale.

The NTB scale is a trait measure of belongingness consisting of ten items, wherein participants identify how strongly they agree or disagree (on a scale of 1 = *strongly disagree* to 5 = *strongly agree*) with statements concerning their desire for interpersonal interaction and acceptance from others (e.g., "I do not like being alone" and "I want other people to accept me"; Leary et al., 2013).

The INQ is a state measure of perceived burdensomeness and thwarted belongingness for which participants indicate how they feel each of 15 statements (e.g., "These days, the people in my life would be better off if I were gone" and "These days, I feel disconnected from other people") accurately represent their beliefs about themselves and others on a scale of 1 (*not at all true*) to 7 (*very true*; Van Orden et al., 2012).

The IOS scale is a single-item measure of interpersonal closeness (Aron et al., 1992). Participants are presented with seven pairs of circles with varying degrees of overlap (Aron et al., 1992). For each pair, one circle represents the self and one circle represents the other (Aron et al., 1992). Participants identify which circle pair correctly embodies their relationship with a specified other (Aron et al., 1992; Figure 1).

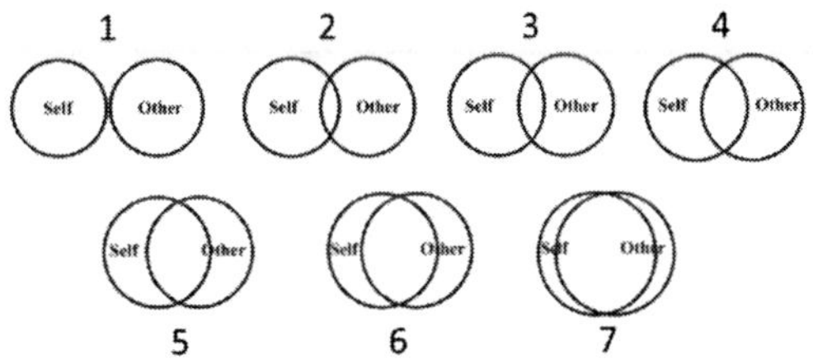

Figure 1: Levels of self and other overlap in the IOS scale (Aron et al., 1992).

In the present study, we asked participants to select the circles that best represented their relationship with the person they had previously described in the writing task.

2.2 Computerized Text Analysis

LIWC. Participant responses were analyzed with the Linguistic Inquiry and Word Count (LIWC; Pennebaker, Booth, Boyd, & Francis, 2015) software. LIWC is an objective measure that facilitates quantitative research in language. Users may import any given text(s) into the software, wherein LIWC outputs the frequency—specifically, the percentage—of word categories in each text. LIWC compares each text to its 125 psychological (affect, cognitive processes), topical (death, family), and grammatical (auxiliary verbs, personal pronouns) language categories. In the current study, we focused on rates of first-person singular pronouns (*I*, *me*, *my*) and negative emotion words (*stress*, *resent*, *lonely*). The negative emotion language category is made up of anxiety (*upset*, *worry*), anger (*hate*, *annoy*), and sadness (*cry*, *hurt*) words as well as some generic affective terms (*bad*, :(, *apath**) that do not easily fit into specific subcategories. With negative emotion words and first-person singular pronouns positively correlated ($r = .20$, $t(241) = 3.15$, $p = .002$, (95% CI [.07, .32]), we created a composite negative self-focus variable by averaging the standardized (i.e., z-scored) rates of negative emotion words and first-person singular pronouns.

2.3 Statistical Analyses

Regression analyses computed on R (version 3.5.2; R Core Team, 2018) assessed whether CESD-R levels of depression predicted negatively self-focused language use moderated by significant other (i.e., romantic partner, close friend, or family member). We also regressed language use on the interaction among depression, significant other, and belongingness (or interpersonal closeness) with separate models for each measure of belongingness (i.e., NTB scale, INQ, and IOS scale). Lastly, all models described were re-analyzed with GAD-7 levels of anxiety in place of CESD-R levels of depression.

Depression, anxiety, as well as perceived burdensomeness and thwarted belongingness (measured by the INQ) were all positively skewed and subsequently log transformed. The remaining variables were either categorical (e.g., interpersonal closeness measured by the IOS scale) or normally distributed (e.g., negative self-focused language, belongingness measured by the NTB scale) and did not require transformation. All variables analyzed were standardized.

3 Results

3.1 Depression

Consistent with our predictions, when writing about a loved one, significant other significantly moderated the association between depression and negative self-focused language, $b = .37$, $SE = .14$, $t(234) = 2.63$, $p = .009$, 95% CI [.09, .65]. Follow-up simple slope analyses revealed that those with higher levels of depression were significantly more likely to use negative self-focused language when writing about the last interaction they had with romantic partners ($b = .33$, $SE = .10$, $t(79) = 3.13$, $p = .002$, 95% CI [.12, .53]) or, to a lesser extent, friends ($b = .20$, $SE = .10$, $t(55) = 2.04$, $p = .046$, 95% CI [.004, .40]), but not family members, $b = -.05$, $SE = .10$, $t(100) = -0.45$, $p = .655$, 95% CI [-.25, .16] (Figure 2).

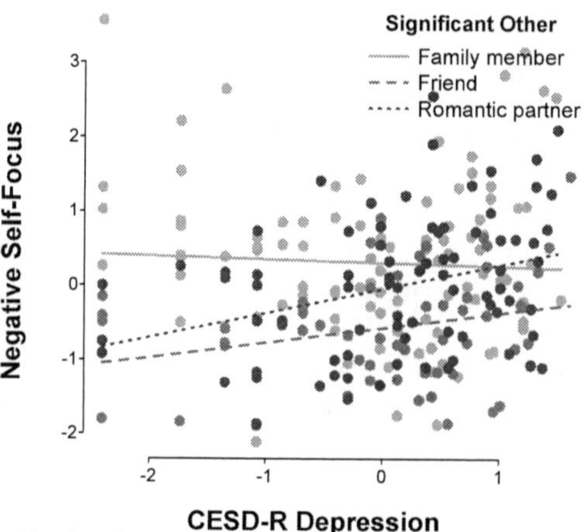

Figure 2: Depression predicting rates of negative self-focused language moderated by significant other.

Scales	b	SE	df	t	p	95% CI
IOS	.17	.09	228	1.87[†]	.063	-.01, .35
NTB	-.12	.15	228	-0.85	.397	-.41, .16
INQ	-.10	.14	228	-0.74	.459	-.38, .17

Table 1: Results for the three-way interaction effects of depression, significant other, and each measure of belongingness. $p < .1$[†]

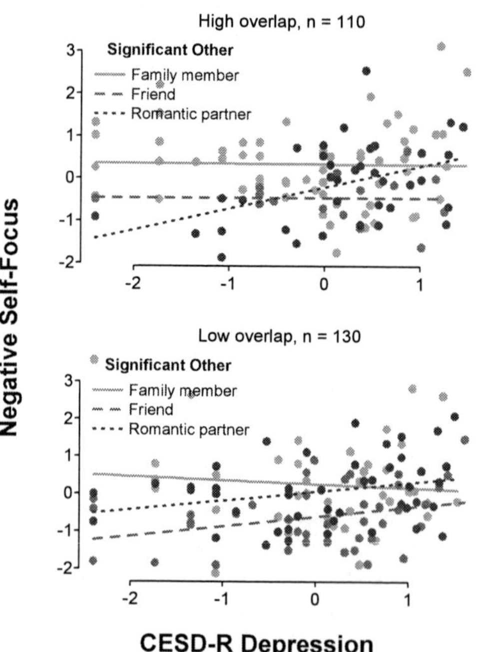

Figure 3: Depression predicting rates of negative self-focused language in recollections of significant others moderated by perceived self-other overlap.

SO	IOS	b	SE	df	t	p	95% CI
Partner	Low	.24	.13	43	1.86[†]	.070	-.02, .50
	High	.50	.18	34	2.86**	.007	.15, .86
Friend	Low	.26	.12	40	2.15*	.038	.02, .50
	High	-.001	.16	13	-0.01	.995	-.34, .34
Family	Low	-.10	.18	41	-0.54	.592	-.47, .27
	High	-.01	.12	57	-0.07	.947	-.25, .23

Table 2: Simple slope results for the three-way interaction effect of depression, significant other, and IOS. $p < .01$**, $p < .05$*, $p < .1$[†]

Partly consistent with our predictions, analyses revealed a marginal three-way interaction effect of depression, significant other, and inclusion of other in the self predicting negative self-focused language ($b = .17$, $SE = .09$, $t(228) = 1.87$, $p = .063$, 95% CI [-.01, .35]; Table 1). To assess the simple slopes of the interaction, we used a median split to convert IOS (median = 5) from a 7-level categorical variable to a 2-level categorical variable (i.e., Low IOS = scores of 5 and lower, High IOS = scores higher than 5). Simple slope analyses demonstrated that those with higher levels of depression were significantly more likely to use negative self-focused language when writing about an interaction with their romantic partner if they indicated high self-other overlap ($b = .50$, $SE = .18$, $t(34) = 2.86$, $p = .007$, 95% CI [.15, .86]; Figure 3).

Simple slope analyses also indicated that those with higher levels of depression were significantly more likely to use negative self-focused language when writing about a friend if they identified low self-other overlap ($b = .26$, $SE = .12$, $t(40) = 2.15$, $p = .038$, 95% CI [.02, .50]; Figure 3). All other simple slopes regarding the interaction effect for depression, significant other, and inclusion of other in the self were nonsignificant (all $ps > .05$; see Table 2). Similarly, the two remaining three-way interaction effects with belongingness (as measured by the NTB scale) as well as with perceived burdensomeness and thwarted belongingness (as measured by the INQ) as separate moderators were nonsignificant ($ps > .1$; Table 1).

3.2 Anxiety

To determine whether the findings might also extend to anxiety, we ran all the aforementioned models replacing CESD-R depression with GAD-7 anxiety. When writing about a loved one, significant other did not significantly moderate the association between anxiety and negative self-focused language, $b = -.02$, $SE = .14$, $t(237) = -.13$, $p = .898$, 95% CI [-.29, .26]. Three-way interaction effects with interpersonal closeness (as measured by the IOS scale; $b = .11$, $SE = .09$, $t(231) = 1.23$, $p = .218$, 95% CI [-.07, .29]) as well as with perceived burdensomeness and thwarted belongingness (as measured by the INQ; $b = -.12$, $SE = .14$, $t(230) = -.86$, $p = .389$, 95% CI [-.39, .15]) as separate moderators were not significant. Results showed a significant three-way interaction effect of anxiety, significant other, and belongingness (as measured by the NTB scale) predicting negative self-focused language ($b = -.32$, $SE = .16$, $t(231) = -2.02$, $p = .045$, 95% CI [-.62, -.01]). However, follow-up simple slope tests did not reach significance (all $ps > .05$), suggesting that the social mechanisms of negative self-focused language implicated in depression may not extend to anxiety. Alternatively, our sample may simply have had insufficient levels of anxiety. With roughly 32% of the sample identifying as moderately to severely anxious (compared with about half of the sample scoring as subclinically depressed), a lack of power could explain the null effects regarding anxiety.

4 Discussion

Due to stigma against mental illness and individuals' desire to be viewed positively, people may be reluctant to openly disclose depressive symptoms on self-report surveys or in daily interactions. Individuals with depression perceive themselves and the world around them in a negative light (Beck, 1967). Although this negative attentional bias is reflected in everyday language use in conversations with romantic partners (Baddeley et al., 2012) and friends (Segrin & Flora, 1998), depressed individuals tend to not use more negative language than others on average (e.g., in naturalistic recordings of students' conversations over the course of 2 week days; Mehl, 2006).

Extending findings from naturalistic recordings of spoken conversations, we found that depressed individuals are more likely to use negative self-focused language when writing about romantic partners and friends but not family. Such results are consistent with past research on depression and recall, which suggest that depressed individuals have a tendency to attend to (Beevers, 2005) and remember

(Dalgleish & Werner-Seidler, 2014) negative stimuli more than positive or neutral stimuli.

In addition, our analyses revealed that interpersonal closeness might serve as a potential mechanism to help understand depressed individuals' recall of and disclosure to romantic partners and friends. Specifically, depressed individuals perceiving a high overlap between themselves and their romantic partner as well as depressed individuals perceiving a low overlap between themselves and their friend were more likely to use negative self-focused language in their written recollections.

Perhaps depressed individuals view their romantic partners as an extension of themselves and, thus, feel more comfortable ruminating while thinking about them. For example, one participant scoring high on CESD-R depression (score = 54), interpersonal closeness (IOS = 7), and negative (4.01%) self-focused (7.66%) language describes their relationship with their partner as such:

> "We are two stubborn asses that have everything at our damn finger tips and too stupid, stubborn, and prideful to move forward … I'm so frustrated I think I'll have to buy a new keyboard when I'm done here."

In the first sentence, this participant confirms their interpersonal closeness, relaying how they perceive their partner as quite similar to themselves. In the second sentence, the participant demonstrates their negative self-focus, expressing their own frustration of the encounter. In cases like this, perhaps interpersonal closeness with a romantic partner exacerbates depressive symptomology, particularly if the partner shares their negative affective tendencies. Being exposed to negative self-relevant stimuli—such as seeing negative aspects of the self reflected in a romantic partner—triggers episodes of rumination, which in turn aggravates symptoms of depression (Beevers, 2005).

On the other hand, another participant scoring a bit lower on CESD-R depression (score = 29)—but still meeting criteria for subclinical depression—interpersonal closeness (IOS = 6), and negative (1.33%) self-focused (11.95%) language discusses how they feel comfortable disclosing to their close friend:

> "In the past, when I have felt like I could not talk to anyone else about my problems and the things that are causing me stress, I have always been able to vent my issues to him."

The participant's recollection of their close friend appears to embody a more adaptive style of coping than the previous participant's almost violent frustration with their romantic partner. Examining the discrepancy between these two participant responses reveals how interpersonal closeness with a significant other may be helpful for depressed individuals to a certain extent. Specifically, if the depressed individual perceives themselves as indistinguishable from their significant other because of shared negative experiences or traits, such interpersonal overlap may heighten depression by triggering rumination. In contrast, if the depressed individual perceives a strong self-other overlap because they feel that they may rely on that person for support, such interpersonal closeness may alleviate depressive symptomology.

Closeness, rather than the relationship type per se, may be responsible for differences in negativity across recalled interactions. Perceived interpersonal closeness tends to be stronger with romantic partners than with friends (Quintard, Jouffre, Croizet, & Bouquet, 2018), which may account for the significant interaction effect involving depressed individuals' high rates of negative self-focused language when recalling an experience with a friend they were less interpersonally close with. In other words, if perceived self-other overlap is inherently less between friends than romantic partners, then it stands to reason that the positive correlation between depression and negative self-focused language is robust for low rather than high IOS.

In any case, social support is heavily implicated as a proponent of relieving stress and promoting positive (mental and physical) health outcomes (Cohen & Wills, 1985). However, depressed individuals tend to withdraw from their social networks (Segrin, 2000; Segrin & Abramson, 1994). During depressive episodes—when social support is arguably needed most—individuals with depression may feel as though they do not belong and struggle to seek or obtain help (Schaefer, Kornienko, & Fox, 2011). Being able to rely on a significant other may lessen the degree of social repercussions of depression. Thus, differences in how depressed individuals use language with the people in their lives could potentially inform more effective methods of diagnosis and treatment of the disorder. Future research will explore social-cognitive mechanisms that may explain discrepancies in how depressive symptoms manifest in language use across social contexts.

4.1 Future Directions and Limitations

The present results converge with previous findings regarding everyday interactions with romantic partners (Baddeley et al., 2012). That is, people's recollections of recent interactions align with naturalistic data on how those conversations actually unfold. In particular, the rate at which negative self-focused language is used similarly across recollections and recordings of conversations with romantic partners provides further evidence of depressed individuals' negative attentional bias. However, our results are limited by the fact that—unlike Baddeley et al. (2012)—we cannot compare across interactions within person. It may be useful, in future studies, to use within-person designs to examine how the same person discusses family, friends, romantic partners, and acquaintances or colleagues.

The present study took a simplified approach to analyzing individuals' language use, focusing exclusively on two robust dictionary-based markers of depression: negative emotion words and first-person singular pronouns. We adopted that approach partly because the texts we analyzed were from a modest sample of individuals writing relatively short texts. In larger samples, it would be possible to apply more complex models of depressed and depression-prone language built, in part, on the results of larger social media studies or corpus analyses (Coppersmith, Dredze, Harman, & Hollingshead, 2015; Eichstaedt et al., 2018; Mowery et al., 2017; Resnik, Armstrong, Claudino, Nguyen, Nguyen, & Boyd-Graber, 2015; for a review, see Guntuku, Yaden, Kern, Ungar, & Eichstaedt, 2017). Such models could provide a more complete picture of the degree to which a depressed or at-risk individual "sounds" depressed—or uses linguistic features correlated with depression—across social contexts. Word or phrase-level analyses can be psychologically revealing in large samples ($N > $ ~5,000) but do not generalize well to smaller samples, where particular word-level indicators of depression

symptoms may only appear in a small percentage of total texts (Schwartz et al., 2013).

The aim of studying a nonclinical population was partly to advance research on preventing depression in individuals with subclinical depression or risk factors for depression. However, because our results are cross-sectional and correlational, it remains unclear whether participants' increased negative self-focus in recollections of interactions with romantic partners represents a risk factor for future depression, a cause of depressive symptoms, or an adaptive way of dealing with early depressive symptoms.

Selectively recalling or disclosing negative affect (or "venting") with romantic partners and masking depression symptoms from close friends and family may be an effective coping strategy, given that depression tends to cause friends to withdraw (Schaefer et al., 2011). To the degree that people are aware of the stigma against mental health conditions or depression, they may strategically disclose negative emotions to the people with whom they are most securely attached, which for a majority of adults is likely to be romantic partners more often than friends or family (Feeney, 2004). Indeed, although self-disclosure is overall healthy for individuals and relationships (Hendrick, 1981), the most personal disclosures—such as discussing depressive symptoms—are commonly reserved for one or two close friends or partners (Altman & Taylor, 1973; Saramäki et al., 2014).

To further understand how people interact with and think about various others in their lives, future research may focus on separately analyzing recollections of recent versus salient interactions (i.e., asking participants in the same study to recount the most recent and the most impactful or memorable interactions with family, friends, and romantic partners). For romantic partners in particular, it may be the case that currently-depressed individuals' most recent interactions are largely negative (reflecting their present mental state), but their most salient memories of that person will be positive to the degree that they feel close or securely attached with them.

Also of interest for future research is uncovering why recent recollections of family members do not seem to impact depressed individuals' language use. Depressed individuals may mask their negative self-focused symptoms during interactions with family so as to prevent them from worrying about them. Although depression is stigmatized across multiple social contexts (Halter, 2004), concealing depressive symptoms in order to protect family members may ironically be more prevalent in cultures that are more collectivist or place more importance on family, such as Latinx communities (Uebelacker et al., 2012).

An alternate explanation of our results is that family members may elicit less negative affect than do romantic partners. However, our preliminary (not yet published) results from a comparison of how depression forum users talk about their relationships across diverse forums on Reddit (based on posts containing variations of the phrase "my [social role]," e.g., "my dad") suggest that family members are described more negatively on social media than are friends or romantic partners. Based on those findings and the present results, we speculate that although depressed or depression-vulnerable individuals' everyday interactions with family members are low in negative affect, family members elicit at least as much negative affect as romantic partners or friends in general.

Our research may have relevance for therapeutic treatment of depression, especially in the context of family systems therapy or couple therapy. Observing how partners or family members interact, asking about recent interactions, and identifying potentially dysfunctional behaviors in these interactions are typically key parts of family systems and couple therapies, across therapeutic approaches (Barbato & D'Avanzo, 2008; Minuchin, 2013). Quantitative and qualitative text analyses have the potential to further inform how clients' symptoms vary across interactions with family and romantic partners, which in turn may help clinicians provide tailored advice on how to navigate important relationships in their lives.

Finally, the impact of our conclusions must be tempered by the fact that our results are from one relatively small, correlational study of writing by college students. Our trust in the present findings is buttressed by the fact that they align with previous work (e.g., Baddeley et al., 2012); however, future replications based on larger and more diverse samples are necessary before substantially building on these results. Other limitations include latent (unmeasured) variables, such as relationship length and the flexibility with which participants' most recent interactions were defined (remote vs. in-person). For instance, perceived belongingness or interpersonal closeness may be a function of how long the individuals have been romantic partners or friends—that is, longer relationships may predict stronger feelings of belongingness. Also, whether participants' interactions were over the phone, in person, or computer-mediated may play a role in what they are able to recall (e.g., in-person conversations may be more salient and thus allow for more vivid or accurate recollections). Future research should incorporate such variables into the current models.

4.2 Conclusion

A prospective, exploratory writing study assessed the association between interpersonal closeness, depression, and the language used to describe intimate relationships. We found that self-focused negativity positively correlates with self-reported depressive symptoms in recollections of recent interactions with close romantic partners, but not close family or friends.

Our results underline the importance of considering how symptoms of mental health conditions manifest differently across social contexts. Past mixed results regarding the linguistic signature of depression (Holtzman, 2017; Tackman et al., 2018) or, more broadly, positive and negative affect (Sun, Schwartz, Son, Kern, & Vazire, 2019), may be partly due to the self-regulatory exigencies of different relationships and social interactions. People do not experience mental health symptoms in a vacuum, but rather interact dynamically with their physical and social environments. Individuals take on different roles—and to some degree become different people, who may have different constellations of mental health symptoms and reveal those symptoms in different ways—across various social contexts.

The end goal of most computational linguistics research on mental health is arguably to not only identify linguistic features that correlate with some clinical outcome, but also to improve clinical diagnosis and treatment. We argue, and our results suggest, that we can only advance from the lab to reality, or predictive models to practice, by increasingly taking the nuances of person-situation interactions into consideration. We propose that research in this area should consider not only practical aspects of the environment, such as topics or social media platforms, but also social psychological variables, including individuals' relationships with and closeness to the people they are discussing.

References

Altman, I. & Taylor, D. A. (1973). *Social penetration: The development of interpersonal relationships.* Oxford, England: Holt, Rinehart & Winston.

Aron, A., Aron, E. N., & Smollan, D. (1992). Inclusion of other in the self scale and the structure of interpersonal closeness. *Journal of Personality and Social Psychology, 63,* 596-612. doi: 10.1037/0022-3514.63.4.596

Baddeley, J. L., Pennebaker, J. W., & Beevers, C. G. (2012). Everyday social behavior during a major depressive episode. *Social Psychological and Personality Science, 4,* 445-452. doi: 10.1177/1948550612461654

Barbato, A., & D'Avanzo, B. (2008). Efficacy of couple therapy as a treatment for depression: a meta-analysis. *Psychiatric Quarterly, 79,* 121-132. doi: 10.1007/s11126-008-9068-0

Baumeister, R. F. & Leary, M. R. (1995). The need to belong: Desire for interpersonal attachments as a fundamental human motivation. *Psychological Bulletin, 117,* 497-529. doi: 10.1037/0033-2909.117.3.497

Beck, A. T. (1967). *Depression: Clinical, experimental, and theoretical aspects.* New York, NY: Harper & Row.

Beevers, C. G. (2005). Cognitive vulnerability to depression: A dual process model. *Clinical Psychology Review, 25,* 975-1002. doi: 10.1016/j.cpr.2005.03.003

Bell, A., Brenier, J. M., Gregory, M., Girand, C., & Jurafsky, D. (2009). Predictability effects on durations of content and function words in conversational English. *Journal of Memory and Language, 60,* 92-111. doi: 10.1016/j.jml.2008.06.003

Cacioppo, J. T., Chen, H. Y., & Cacioppo, S. (2017). Reciprocal influences between loneliness and self-centeredness: A cross-lagged panel analysis in a population-based sample of African American, Hispanic, and Caucasian adults. *Personality and Social Psychology Bulletin, 43,* 1125-1135. doi: 10.1177/0146167217705120

Cacioppo, J. T., Hughes, M. E., Waite, L. J., Hawkley, L. C., & Thisted, R. A. (2006). Loneliness as a specific risk factor for depressive symptoms: Cross-sectional and longitudinal analyses. *Psychology and Aging, 21,* 140-151. doi: 10.1037/0882-7974.21.1.140

Choenarom, C., Williams, R. A., & Hagerty, B. M. (2005). The role of sense of belonging and social support on stress and depression in individuals with depression. *Archives of Psychiatric Nursing, 19,* 18-29. doi: 10.1016/j.apnu.2004.11.003

Chung, C. & Pennebaker, J. W. (2007). The psychological functions of function words. In K. Fiedler (Ed.), *Social Communication* (pp. 343-359). New York, NY: Psychology Press.

Cohen, S., & Wills, T. A. (1985). Stress, social support, and the buffering hypothesis. *Psychological Bulletin, 98,* 310-357. doi: 10.1037/0033-2909.98.2.310

Coppersmith, G., Dredze, M., Harman, C., & Hollingshead, K. (2015). From ADHD to SAD: Analyzing the language of mental health on Twitter through self-reported diagnoses. In *Proceedings of the 2nd Workshop on Computational Linguistics and Clinical Psychology* (pp. 1-10).

Crocker, J., & Major, B. (1989). Social stigma and self-esteem: The self-protective properties of stigma. *Psychological Review, 96,* 608-630. doi:10.1037/0033-295X.96.4.608

Dalgleish, T., & Werner-Seidler, A. (2014). Disruption in autobiographical memory processing in depression and the emergence of memory therapeutics. *Trends in Cognitive Sciences, 18,* 596-604. doi: 10.1016/j.tics.2014.06.010

De Choudhury, M., Counts, S., Horvitz, E. J., & Hoff, A. (2014). Characterizing and predicting postpartum depression from shared Facebook data. In *Proceedings of the 17th ACM conference on Computer supported cooperative work & social computing* (pp. 626-638).

Eaton, W. W., Smith, C., Ybarra, M., Muntaner, C., Tien, A. (2004). Center for Epidemiologic Studies Depression Scale: Review and Revision (CESD and CESD-R). In M. E. Maruish (Ed.), *The use of psychological testing for treatment planning and outcomes assessment: Instruments for adults* (pp. 363-377). Mahwah, NJ: Lawrence Erlbaum Associates Publishers.

Eichstaedt, J. C., Smith, R. J., Merchant, R. M., Ungar, L. H., Crutchley, P., Preotiuc-Pietro, D., Asch, D. A., & Schwartz, H. A. (2018). Facebook language predicts depression in medical records. *Proceedings of the National Academy of Sciences, 115,* 11203-11208. doi: 10.1073/pnas.1802331115

Feeney, J. A. (2004). Transfer of attachment from parents to romantic partners: Effects of individual and relationship variables. *Journal of Family Studies, 10,* 220-238. doi: 10.5172/jfs.327.10.2.220

Gara, M. A., Woolfolk, R. L., Cohen, B. D., Goldston, R. B., Allen, L. A., & Novalany, J. (1993). Perception of self and other in major depression. *Journal of Abnormal Psychology, 102,* 93-100.

Garrod, S., & Pickering, M. (2016). *Language processing* (1st ed.). London, England: Psychology press.

Gottman, J. M. & Levenson, R. W. (2000). The timing of divorce: Predicting when a couple will divorce over a 14-year period. *Journal of Marriage and Family, 62,* 737-745. doi: 10.1111/j.1741-3737.2000.00737.x

Guntuku, S. C., Yaden, D. B., Kern, M. L., Ungar, L. H., & Eichstaedt, J. C. (2017). Detecting depression and mental illness on social media: An integrative review. *Current Opinion in Behavioral Sciences, 18,* 43-49. doi: 10.1016/j.cobeha.2017.07.005

Hagerty, B. M. & Williams, R. A. (1999). The effects of sense of belonging, social support, conflict, and loneliness on depression. *Nursing Research, 48,* 215-219.

Halter, M. J. (2004). The stigma of seeking care and depression. *Archives of Psychiatric Nursing, 18,* 178-184. doi: 10.1016/j.apnu.2004.07.005

Hendrick, S. S. (1981). Self-disclosure and marital satisfaction. *Journal of Personality and Social Psychology, 40,* 1150-1159. doi: 10.1037/0022-3514.40.6.1150

Holtzman, N. S. (2017). A meta-analysis of correlations between depression and first person singular pronoun use. *Journal of Research in Personality, 68,* 63-68. doi: 10.1016/j.jrp.2017.02.005

Joormann, J., & Stanton, C. H. (2016). Examining emotion regulation in depression: a review and future directions. *Behaviour Research and Therapy, 86,* 35-49. doi: 10.1016/j.brat.2016.07.007

Joyner, K., & Udry, J. R. (2000). You don't bring me anything but down: Adolescent romance and depression. *Journal of Health and Social Behavior, 41,* 369-391.

Kiecolt-Glaser, J. K. & Newton, T. L. (2001). Marriage and health: His and hers. *Psychological Bulletin, 127,* 472-503. doi: 10.1037//0033-2909.127.4.472

Leary, M. R., Kelly, K. M., Cottrell, C. A., & Schreindorfer, L. S. (2013). Construct validity of the need to belong scale: Mapping the nomological network. *Journal of Personality Assessment, 95,* 610-624. doi: 10.1080/00223891.2013.819511

Mehl, M. R. (2006). The lay assessment of subclinical depression in daily life. *Psychological assessment, 18,* 340. doi: 10.1037/1040-3590.18.3.340

Minuchin, S. (2013). The family in therapy. In R. L. Smith & R. E. Montilla (Eds.). *Counseling and family therapy with Latino populations* (pp. 74-84). New York, NY: Routledge.

Monroe, S. M., Rohde, P., Seeley, J. R., & Lewinsohn, P. M. (1999). Life events and depression in adolescence: Relationship loss as a prospective risk factor for first onset of major depressive disorder. *Journal of Abnormal Psychology, 108,* 606–614.

Moore, M. T., & Fresco, D. M. (2012). Depressive realism: A meta-analytic review. *Clinical Psychology Review, 32,* 496-509. doi: 10.1016/j.cpr.2012.05.004

Mowery, D., Smith, H., Cheney, T., Stoddard, G., Coppersmith, G., Bryan, C., & Conway, M. (2017). Understanding depressive symptoms and psychosocial stressors on Twitter: A corpus-based study. *Journal of medical Internet research, 19,* e48. doi: 10.2196/jmir.6895

Pennebaker, J. W., Booth, R. J., Boyd, R. L., & Francis, M. E. (2015). Linguistic Inquiry and Word Count: LIWC2015. Austin, TX: Pennebaker Conglomerates (www.LIWC.net).

Pulvermüller, F., Shtyrov, Y., Hasting, A. S., & Carlyon, R. P. (2008). Syntax as a reflex: Neurophysiological evidence for early automaticity of grammatical processing. *Brain and language, 104,* 244-253. doi: 10.1016/j.bandl.2007.05.002

Quintard, V., Jouffre, S., Croizet, J. C., & Bouquet, C. A. (2018). The influence of passionate love on self-other discrimination during joint action. *Psychological Research,* 1-11. doi: 10.1007/s00426-018-0981-z

R Core Team. (2018). *R: A language and environment for statistical computing.* R Foundation for Statistical Computing, Vienna, Austria. url: r-project.org

Resnik, P., Armstrong, W., Claudino, L., Nguyen, T., Nguyen, V. A., & Boyd-Graber, J. (2015). Beyond LDA: Exploring supervised topic modeling for depression-related language in Twitter. In *Proceedings of the 2nd Workshop on Computational Linguistics and Clinical Psychology: From Linguistic Signal to Clinical Reality* (pp. 99-107).

Rude, S. S., Gortner, E. M., Pennebaker, J. W. (2004). Language use of depressed and depression-vulnerable college students. *Cognition and Emotion, 18,* 1121-1133. doi: 10.1080/02699930441000030

Saramäki, J., Leicht, E. A., López, E., Roberts, S. G., Reed-Tsochas, F., & Dunbar, R. I. (2014). Persistence of social signatures in human communication. *Proceedings of the National*

Academy of Sciences, 111, 942-947. doi: 10.1073/pnas.1308540110

Schaefer, D. R., Kornienko, O., & Fox, A. M. (2011). Misery does not love company: Network selection mechanisms and depression homophily. *American Sociological Review, 76,* 764-785. doi: 10.1177/0003122411420813

Schwartz, H. A., Eichstaedt, J. C., Kern, M. L., Dziurzynski, L., Ramones, S. M., Agrawal, M., Shah, A., Kosinski, M., Stillwell, D., Seligman, M. E. P., & Ungar, L. H. (2013). Personality, gender, and age in the language of social media: The open-vocabulary approach. *PloS one, 8,* e73791. doi: 10.1371/journal.pone.0073791

Schwartz, H. A., Eichstaedt, J., Kern, M. L., Park, G., Sap, M., Stillwell, D., Kosinski, M., & Ungar, L. (2014). Towards assessing changes in degree of depression through Facebook. In *Proceedings of the Workshop on Computational Linguistics and Clinical Psychology: From Linguistic Signal to Clinical Reality* (pp. 118-125).

Segalowitz, S. J., & Lane, K. C. (2000). Lexical access of function versus content words. *Brain and Language, 75,* 376-389. doi: 10.1006/brln.2000.2361

Segrin, C. (2000). Social skills deficits associated with depression. *Clinical Psychology Review, 20,* 379-403. doi: 10.1016/S0272-7358(98)00104-4

Segrin, C. & Abramson, L. Y. (1994). Negative reactions to depressive behaviors: A communication theories analysis. *Journal of Abnormal Psychology, 103,* 655-668. doi: 10.1037/0021-843X.103.4.655

Segrin, C., & Flora, J. (1998). Depression and verbal behavior in conversations with friends and strangers. *Journal of Language and Social Psychology, 17,* 492-503. doi: 10.1177/0261927X980174005

Spitzer, R. L., Kroenke, K., Williams, J. B., & Löwe, B. (2006). A brief measure for assessing generalized anxiety disorder: The GAD-7. *Archives of internal medicine, 166,* 1092-1097. doi: 10.1001/archinte.166.10.1092

Sun, J., Schwartz, H. A., Son, Y., Kern, M. L., & Vazire, S. (2019). The language of well-being: Tracking fluctuations in emotion experience through everyday speech. *Journal of Personality and Social Psychology.* doi: 10.1037/pspp0000244

Tackman, A. M., Sbarra, D. A., Carey, A. L., Donnellan, M. B., Horn, A. B., Holtzman, N. S., Edwards, T. S., Pennebaker, J. W., & Mehl, M. R. (2018). Depression, negative emotionality, and self-referential language: A multi-lab, multi-measure, and multi-language-task research synthesis. *Journal of Personality and Social Psychology.* doi: 10.1037/pspp0000187

Tausczik, Y. R. & Pennebaker, J. W. (2010). The psychological meaning of words: LIWC and computerized text analysis methods. *Journal of Language and Social Psychology, 29,* 24-54. doi: 10.1177/0261927X09351676

Tennen, H., Hall, J. A., & Affleck, G. (1995). Depression research methodologies in the *Journal of Personality and Social Psychology*: A review and critique. *Journal of Personality and Social Psychology, 68,* 870-884. doi: 10.1037/0022-3514.68.5.870

Twenge, J. M., Joiner, T. E., Rogers, M. L., & Martin, G. N. (2017). Increases in depressive symptoms, suicide-related outcomes, and suicide rates among US adolescents after 2010 and links to increased new media screen time. *Clinical Psychological Science, 6,* 3-17. doi: 10.1177/2167702617723376

Uebelacker, L. A., Marootian, B. A., Pirraglia, P. A., Primack, J., Tigue, P. M., Haggarty, R., Velazquez, L., Bowdoin, J. J., Kalibatseva, Z., & Miller, I. W. (2012). Barriers and facilitators of treatment for depression in a Latino community: A focus group study. *Community mental health journal, 48,* 114-126. doi: 10.1007/s10597-011-9388-7

Van Orden, K. A., Cukrowicz, K. C., Witte, T. K., & Joiner, T. E. (2012). Thwarted belongingness and perceived burdensomeness: Construct validity and psychometric properties of the Interpersonal Needs Questionnaire. *Psychological Assessment, 24,* 197-215. doi: 10.1037/a0025358

Watkins, E. & Teasdale, J. D. (2001). Rumination and overgeneral memory in depression: Effects of self-focus and analytic thinking. *Journal of Abnormal Psychology, 110,* 353-357. doi: 10.1037/0021-843X.110.2.333

Wenzlaff, R. M. & Eisenberg, A. R. (2001). Mental control after dysphoria: Evidence of a suppressed, depressive bias. *Behavior Therapy, 32,* 27-45. doi: 10.1016/S0005-7894(01)80042-3

Word Health Organization (WHO). (2018, March 22). Depression. Retrieved from https://www.who.int/news-room/fact-sheets/detail/depression

Zimmermann, J., Brockmeyer, T., Hunn, M., Schauenburg, H., & Wolf, M. (2016). First-person pronoun use in spoken language as a predictor of future depressive symptoms: Preliminary evidence from a clinical sample of depressed patients. *Clinical psychology & psychotherapy, 24,* 384-391. doi: 10.1002/cpp.2006

Linguistic Analysis of Schizophrenia in Reddit Posts

Jonathan Zomick
Psychology Department
Hofstra University
Hempstead, NY 11549
jzomick1@pride.hofstra.edu

Sarah Ita Levitan
Computer Science Department
Columbia University
New York, NY 10027
sarahita@cs.columbia.edu

Mark Serper
Psychology Department
Hofstra University
Mount Sinai School of Medicine
mark.r.serper@hofstra.edu

Abstract

We explore linguistic indicators of schizophrenia in Reddit discussion forums. Schizophrenia (SZ) is a chronic mental disorder that affects a person's thoughts and behaviors. Identifying and detecting signs of SZ is difficult given that SZ is relatively uncommon, affecting approximately 1% of the US population, and people suffering with SZ often believe that they do not have the disorder. Linguistic abnormalities are a hallmark of SZ and many of the illness's symptoms are manifested through language. In this paper we leverage the vast amount of data available from social media and use statistical and machine learning approaches to study linguistic characteristics of SZ. We collected and analyzed a large corpus of Reddit posts from users claiming to have received a formal diagnosis of SZ and identified several linguistic features that differentiated these users from a control (CTL) group. We compared these results to other findings on social media linguistic analysis and SZ. We also developed a machine learning classifier to automatically identify self-identified users with SZ on Reddit.

1 Introduction

Schizophrenia is a serious mental illness that affects roughly 1% of the US population (NIMH, 2019) and is reportedly one of the 25 top causes of disability around the world (Vos et al., 2015). Symptoms of the disorder are categorized as positive symptoms (e.g., delusions, hallucinations, disorganized thinking) or negative symptoms (e.g., diminished emotional expression, anhedonia, asociality) (APA, 2013). Individuals with SZ are at an elevated risk for suicide; an estimated 4-5% of people diagnosed with SZ die from suicide (Hor and Taylor, 2010; Carlborg et al., 2010). Early detection and diagnosis of the disorder has been speculated to improve long-term outcomes

for people suffering with SZ (Birchwood et al., 1997). However, early detection and diagnosis of SZ is challenging given that it is a relatively uncommon disease and diagnostic measures are reliant on self-report measures. Additionally, many people suffering from the disorder genuinely do not believe they have SZ (Rickelman, 2004).

Linguistic abnormalities are prominent symptoms of SZ (APA, 2013). Some of the linguistic markers associated with people with the illness include diminished emotional expression, incoherence, derailment, tangentiality, co-reference failure and lexical and syntactical errors (Rochester and Martin, 1979; Harvey and Serper, 1990; Hoekert et al., 2007; Covington et al., 2005; Kuperberg, 2010). Much of the research on language and SZ has focused on analyzing transcriptions of spoken language and handwritten samples, which tend to be small, manually collected datasets.

Some recent research has focused on analyzing language from social media posts (Birnbaum et al., 2017; Lyons et al., 2018; Coppersmith et al., 2015; Mitchell et al., 2015). With the advent of social media, many people who suffer from various forms of mental illness have found a sense of community and support, and these platforms offer a mode of expression for discussing their experiences openly online. Additionally, many online platforms allow users to post anonymously, giving users a sense of security and anonymity to discuss their experiences and struggles without the fear of being stigmatized or discriminated against (Balani and De Choudhury, 2015; Berry et al., 2017; Highton-Williamson et al., 2015).

There are many advantages to leveraging social media data for analyzing the linguistic characteristics of SZ. This open discussion enables the collection and annotation of social media posts of relatively uncommon disorders such as SZ. These corpora can be collected using automated or

Proceedings of the Sixth Workshop on Computational Linguistics and Clinical Psychology, pages 74–83
Minneapolis, Minnesota, June 6, 2019. ©2019 Association for Computational Linguistics

semi-automated methods, and enable analysis on a much larger scale. Regular social media use has risen above two billion users worldwide (Kemp, 2014), and youth comprise the largest and fastest growing demographic of social media users – over 90% of youth in the US reportedly engage in social media on a daily basis (Lenhart et al., 2015). Studying SZ among social media users can be useful for identifying early stages of the disorder, which is critical for early intervention.

Most of the research on social media posts and SZ has focused on Twitter data. In this paper we explore another popular social media platform: Reddit. Reddit is one of the fastest growing and widely used social media platforms, averaging over 330 million active monthly users, and as of 2018 was the fourth most visited website in the US (Hutchinson, 2018). Unlike Twitter, Reddit imposes no limits on the length of posts, enabling an analysis of longer language samples. In addition, Reddit is composed of subreddits, which are forums dedicated to specific topics. We leverage subreddits that are communities for individuals with SZ for identifying potential Reddit users with SZ, in order to collect a corpus of posts from these users (as described in Section 3).

These online posts provide a rich source of language data which we use to identify linguistic markers of SZ. We also use this data to train a machine learning classifier to automatically identify individuals with SZ using linguistic cues. Hopefully, an improved understanding of linguistic patterns unique to this population can assist in diagnostic procedures and be employed as an early detection mechanism.

The rest of this paper is organized as follows: Section 2 reviews relevant previous research, and 3 describes the dataset that we collected and the features that we use for analysis. In Section 4, we present the analysis of linguistic markers of SZ, and provide a detailed comparison of our findings with prior work. Section 5 presents the results of our machine learning classification of users with SZ. We discuss ethical considerations in Section 6 and conclude in Section 7.

2 Related Work

Some recent research has analyzed Twitter data of self-identified individuals with SZ with promising results. Mitchell et al. (2015) analyzed a variety of linguistic markers of SZ using tweets of self-identified individuals with SZ. Their features included lexicon-based and open-vocabulary approaches, and they discovered several significant signals for SZ. Further, they trained classifiers using these features and obtained an accuracy of 82%.

Coppersmith et al. (2015) used a similar approach to study 10 mental disorders, including SZ, and identified linguistic markers of each. They also leveraged the collected data to explore relationships between linguistic markers of multiple conditions, which is very difficult to analyze without a large-scale corpus. Birnbaum et al. (2017) also analyzed linguistic markers of SZ in Twitter data, and built a classifier to distinguish users with SZ from healthy controls. Importantly, they obtained clinician annotations of the data to validate the approach of annotating social media data based on self-disclosure of mental health conditions.

A limitation of analyzing Twitter data is that posts are constrained in character length so only very short samples of text are available for analysis. Furthermore, the character restrictions imposed by Twitter may affect users' linguistic expression and force users to communicate in ways that differ from their natural way of communicating. An alternative source of social media data are discussion board forums. Discussion board forums are not character-limited, and allow for focused conversations on topics within sub-forums. Lyons et al. (2018) analyzed several discussion board forums dedicated to mental disorders, including Reddit, and used posts from a financial discussion forum as a control. They studied linguistic features related to affective processes and personal pronoun usages, and found that these were effective at distinguishing between individuals with SZ and the control. In our work, we expand on this study by analyzing a larger set of linguistic features. We also collected a control group within the same platform to eliminate confounding factors such as stylistic and topical differences between discussion board forums.

Because all of these studies used overlapping feature sets, and in particular Linguistic Inquiry and Word Count (LIWC) features (Pennebaker et al., 2015b) (described in section 3), we had the opportunity to analyze markers of SZ across domains. We compare the results from our study of Reddit data with previously identified markers of SZ in the four studies described in this section.

This analysis allows us to identify some linguistic characteristics of SZ that are domain-independent, and identify differences in markers of SZ across domains.

This work aims to build on the previous studies that have looked at SZ langauge on social media platforms. Specifially, to our knowledge we present the first complete analysis of LIWC features using Reddit data and compare these results with the previous findings of LIWC features of SZ on social media. Additionally, we analyze all Reddit posts of Reddit users claiming to have received a SZ diagnosis, not just those in forums devoted to discussions of SZ, and compare them to a control group of other Reddit users. We also train a machine learning classifier to automatically identify individuals with SZ, which has not been previously explored using Reddit data. This research will add to the current body of knowledge of linguistic characteristics of individuals with SZ and will hopefully help improve diagnoses and bolster early detection of the disorder.

3 Data

3.1 Reddit Corpus

We used the Python Reddit API Wrapper (PRAW) (Boe, 2016) to collect a corpus of Reddit posts from users who stated that they were diagnosed with SZ and a control group of users. We first compiled a list of users with self-disclosures of SZ by visiting subreddits devoted to discussions about SZ. These included: r/schizophrenia, r/schizophrenic, and r/AskReddit under the topic "Any Redditors With Schizophrenia?". We manually inspected the posts to only include contributors with a clear statement of receiving a formal diagnosis of SZ. For example, a user who referred to "my diagnosis of schizophrenia" would be included in the SZ group.

We also collected a random control group of Reddit users, using the r/random subreddit, which takes you to a random subreddit. To ensure a control sample that is more representative of the overall population, every five usernames that were selected came from a different random subreddit. We collected all public Reddit posts from the SZ and CTL users across all subreddits, and removed any users from the CTL group who mentioned suffering from SZ in any of their posts. We collected data from a total of 159 users for each group (318 total) who had posted at least 10 times on Reddit.

Users in the SZ group made a total of 66,454 comments, and there were 113,570 comments from the CTL users.

We note that this data is not representative of the general population. For example, Reddit users have been found to be predominantly male and young (under 30) (Finlay, 2014). Our findings are limited to this population, and further research is needed to study the effects of gender and age on linguistic markers of SZ. Another limitation of using anonymous social media data for this work is that it is not externally validated; although the users in the SZ group stated that they were diagnosed with SZ, and the CTL users did not, we do not have clinical information to verify this.

3.2 LIWC Features

Having collected this dataset, we analyzed linguistic markers of SZ using Linguistic Inquiry and Word Count (Pennebaker et al., 2015b). LIWC is a text analysis program that computes word counts for semantic classes as well as structural features. LIWC relies on an internal dictionary that maps words to psychologically motivated categories. When analyzing a target text, the program looks up the target words in the dictionary and computes frequencies for each of the dimensions. The categories include standard linguistic dimensions (e.g., percentage of words that are pronouns, articles), markers of psychological processes (e.g., affect, social, cognitive words), punctuation categories (e.g., periods, commas), and formality measures (e.g., fillers, swear words). LIWC dimensions have been used in many studies to predict outcomes including personality (Pennebaker and King, 1999), deception (Newman et al., 2003), and health (Pennebaker et al., 1997). We extracted a total of 93 features using LIWC 2015. A full description of these features is found in (Pennebaker et al., 2015a).

We selected LIWC to analyze linguistic markers of SZ because these features have been widely studied for this purpose in other domains (such as Twitter), which enables a direct comparison of results across domains.

Category	Reddit	Discussion Forums	Twitter		
Paper	Current	Lyons et. al	(A)	(B)	(C)
Linguistic Processes					
Word count	SZ				
Dictionary words	SZ				
Total function words	SZ		SZ	SZ	
Total pronouns	SZ		SZ		
Personal pronouns	SZ	SZ	SZ		
1st person singular	SZ	SZ	SZ		SZ
1st person plural	CTL	CTL			SZ
2nd person	SZ	SZ			
3rd person singular	CTL	SZ			SZ
3rd person plural		SZ	SZ	SZ	SZ
Impersonal pronouns			SZ		SZ
Articles	CTL		SZ	SZ	SZ
Auxiliary verbs	SZ		SZ	SZ	SZ
Common adverbs	SZ				
Conjunctions	SZ		SZ	SZ	
Negations	CTL				SZ
Other Grammar					
Common verbs	SZ				
Numbers	CTL				
Quantifiers				SZ	SZ
Psychological processes					
Affective processes	SZ	SZ			
Positive emotion	SZ	CTL	CTL		SZ
Negative emotion		SZ	SZ		SZ
Anxiety	SZ	SZ		SZ	
Anger	CTL	SZ			
Sadness		SZ			SZ
Social processes	SZ				
Friends					CTL
Male references	CTL				
Cognitive processes	SZ		SZ	SZ	SZ
Insight	SZ		SZ	SZ	SZ
Causation				SZ	SZ
Discrepancy			SZ		SZ
Tentatitve	SZ		SZ	SZ	SZ
Certainty					SZ
Perceptual processes	SZ				SZ
See	CTL		CTL		
Hear	SZ				SZ
Feel	SZ				SZ
Biological Processes	SZ				SZ
Body					SZ
Health	SZ		SZ	SZ	SZ
Sexual					SZ
Drives	SZ				
Achievement					SZ
Power	CTL				

Feature	Current	Lyons	A	B	C
Reward	SZ				
Time orientations	SZ				SZ
Past focus	SZ				SZ
Present focus	SZ				SZ
Future focus			CTL		
Relativity	CTL		CTL	CTL	
Motion	CTL			CTL	
Space			CTL	SZ	
Personal concerns					
Work					SZ
Leisure	CTL		CTL	CTL	
Home			CTL	CTL	SZ
Money	CTL				
Death			SZ	SZ	
Informal language					
Swear words	CTL				SZ
Assent			CTL	CTL	
Punctuation					
Question marks	CTL				
Exclamation marks	SZ		SZ		
Dashes	CTL				
Other punctuation	CTL				

Table 1: LIWC features that were significantly different between SZ and CTL groups, compared across five studies. "Current" indicates the analysis of Reddit posts conducted in this paper, Lyons et al. (2018) studied some LIWC variables in discussion board posts (including Reddit). The three studies that examined Twitter data are: (A) Mitchell et al. (2015); (B): Coppersmith et al. (2015) ; and (C): Birnbaum et al. (2017). Gray cells indicate categories that were not examined in a study (some are due to differences between LIWC 2015 and 2007 versions).

4 Linguistic Characteristics of SZ and CTL Reddit Comments

To identify linguistic markers of SZ, we compared the frequencies of each LIWC dimension in SZ and CTL users. We averaged the frequencies of the LIWC dimensions across all posts per user so that each user was represented once in the dataset. This was done to avoid skewing the data based on a few users who posted a large number of comments. We used an independent samples t-test to determine whether the difference in mean frequency for each LIWC feature between the SZ and CTL groups was statistically significant. All tests for significance correct for family-wise Type I error by controlling the false discovery rate (FDR) at $\alpha = 0.05$ (Benjamini and Hochberg, 1995). The kth smallest p value is considered significant if it is less than $\frac{k*\alpha}{n}$. Table 1 shows the results of this analysis in the "Reddit" column. "SZ" indicates that the feature was significantly more frequent in posts from users with SZ, and "CTL" indicates that the feature was significantly more frequent in posts from the control group of users.

We found significant differences between the SZ group and the CTL group for many of the LIWC features. These differences spanned various linguistic domains including linguistic processes, grammar, psychological processes, and punctuation. In addition to showing the results of our analysis of Reddit posts, Table 1 shows a comparison of our results with four other studies that examined LIWC features and SZ in social media data: one study (Lyons et al., 2018) used data from Reddit and other online discussion forums (but only examined personal pronouns and affective processes), and 3 studies examined Twitter data: (A) Mitchell et al. (2015), (B) Coppersmith et al. (2015), and (C) Birnbaum et al. (2017).

Many of our findings were in line with previous research on other social media platforms, while some of the markers that we identified differed from previous studies. We identified several markers of SZ in our Reddit corpus that have not been previously noted. These include an increased association between users with SZ and the following features: Word count, Dictionary words, Common adverbs, Verbs, Reward, and Drives. Additionally, unlike previous social media studies, we found diminished expression among the following features: 3rd person singular, Articles, Negations, Anger, Male references, Power, Money, Swear

words, Question marks, Dashes, and Other punctuation. It is not surprising that there are discrepancies between this study and others. This type of analysis has not been previously conducted on data taken exclusively from Reddit, and the majority of these features were not analyzed in the discussion forum data by (Lyons et al., 2018). There is a substantial domain mismatch between Reddit and Twitter data, and markers of SZ that have been observed in Twitter data may not generalize to other domains, while other markers that we have observed in the Reddit may not have been observed in previous work with Twitter data due to the character constraints that platform places on users' posts.

On the other hand, some of the findings regarding association between specific LIWC features and SZ are more robust and have been replicated in multiple studies. When comparing results from the five studies that looked at SZ language and social media, at least 3 out of the 5 studies reported increased frequency among users with SZ in the following features: Total function words, Personal Pronouns, 1st person pronouns, 3rd person plural, Articles, Auxiliary verbs, Conjunctions, Negative emotion, Anxiety, Cognitive processes, Insight, Tentative, and Health. Other findings that have been replicated multiple times relate to diminished expression of certain LIWC features among users with SZ in comparison with control users. Three of the five studies found that users with SZ used words associated with the features Relativity and Leisure significantly less than control groups.

4.1 Discussion

The present results are consistent with past studies that have found that users with SZ use words associated with health issues, anxiety, negative emotion and use of 1st person singular pronouns more than control groups. An emphasis on health related matters, expressions of negative emotions, and a focus on one's self are understandable for people suffering from a serious mental illness. It is also somewhat understandable that users with SZ use leisure related words significantly less than controls, since individuals suffering from mental illness appear to be less focused or interested in leisure activities (Thornicroft et al., 2004). However, some of the linguistic features that have been found elevated among users with SZ in multiple studies are not as intuitive, such as usage of 3rd person plural pronouns, Insight words, and Tentative words.

The robust findings of usage of 3rd person plural pronouns may be related to SZ symptomatology. For example, relative excessive use of pronouns such as "they" and "them" may reflect a disaffiliativeness from others that is reflected in symptoms of social anhedonia. Further support for this line of reasoning comes from our finding and findings by Lyons et al. (2018) that members of the SZ group used 1st person plural pronouns such as "we" and "us" less than the CTL group, which may also be an indication of social disaffiliation and withdrawal.

Additionally, the use of 3rd person plural pronouns may reflect positive symptoms common to the disorder (Bentall et al., 2001; APA, 2013). Previous researchers have posited that the increased usage of 3rd person plural pronouns among SZ patients may be a reflection of an externalizing bias, paranoid thinking, and a focus on outside groups (Fineberg et al., 2015; Lyons et al., 2018). The decreased usage of 1st person plural pronouns may also reflect social withdrawal due to paranoid suspicions that result in social anxiety and subsequent isolation.

All of the studies reported here that looked at tentative language in social media data and SZ found that users with SZ used tentative words like "perhaps" and "maybe" significantly more than CTL users. Tausczik and Pennebaker (2010) suggest that tentative language is suggestive of difficulty processing events and forming events into stories and may indicate uncertainty or insecurity about a topic. Use of tentative language may be a manifestation of an impaired sense of agency and diminished self-presence reportedly associated with people with SZ (Jeannerod, 2009; Sass and Parnas, 2003). The increased usage of 1st person pronouns may also be a marker of a hyperreflexivity (exaggerated self-consciousness) experienced by individuals with SZ, as described by Sass and Parnas (2003).

In contrast to earlier social media data we found that the SZ group used punctuation significantly less frequently than the CTL group. The discrepancy between this work and previous work using Twitter data may be due to differences between these two platforms. The character restrictions Twitter places on posts may discourage usage of proper punctuation to preserve space for content words. However, Reddit posts that do not have these restrictions may reflect more natural language of users and allow for additional observations such as differences in punctuation usage. In line with the hypothesis put forth by Fineberg et al. (2015) our finding that users with SZ use punctuation significantly less than CTL users may reflect more disorganized use of language, a prominent symptom of schizophrenia (Covington et al., 2005; APA, 2013).

5 Automatic Identification of Users with Schizophrenia

Having identified many differences in language usage between Reddit users with SZ and the control group, we trained a machine learning classifier to automatically distinguish between the groups, using the LIWC features. We used the scikit-learn (Pedregosa et al., 2011) implementation of a Logistic Regression model using the default parameters. The model was trained and evaluated using stratified 5-fold cross-validation. We averaged the LIWC features across all comments per user and trained the model to determine whether the aggregated LIWC features were from the posts of a user from the SZ group or the CTL group. The random baseline is 50%, since the data is balanced across groups.

The average performance of the classifier across 5 folds was 81.56% accuracy, and the standard deviation was 2.29. The top 10 LIWC dimensions for the SZ and CTL classes, obtained from the logistic regression coefficients, are shown in Table 2. Some of these weighted features were consistent with our statistical analysis of LIWC features. For example, the Health category was highly predictive of SZ, as was the Tentative dimension. Intuitively, Sadness was the strongest (negative) predictor of the control group, and 3rd person singular was also a useful (negative) predictor of the control group.

These findings suggest that linguistic features are useful for automatically identifying social media users with self-described SZ on a large, public, anonymous social media site. The classifier achieved strong performance, 31.56% better than a random baseline. However, although a balanced data set is useful for analyzing linguistic indicators of SZ and for evaluating the machine learning classification results, we note (as do Mitchell et al. (2015)) that this setup is not representative of

Control (CTL)		Schizophrenia (SZ)	
Weight	Feature	Weight	Feature
-1.2748	Sadness	1.6105	Health
-1.1109	Quotation mark	1.0717	Interrogatives
-0.8715	3rd person singular	1.0614	Tentative
-0.7956	Feel	0.9825	Hear
-0.7949	Articles	0.9426	Colon
-0.7302	Nonfluencies	0.9304	Death
-0.6705	Adjectives	0.8021	Biological processes
-0.6329	See	0.7642	1st person singular
-0.6214	Motion	0.6975	Parentheses
-0.6182	Present focus	0.6478	Verbs

Table 2: Top weighted features from the logistic regression classifier for the SZ and CTL groups.

the true distribution of SZ and healthy individuals (only 1% have SZ).

6 Ethical Considerations

Detecting mental health conditions using linguistic features extracted from social media has the potential to enhance detection of disorders for early intervention and improve outcomes for individuals suffering from mental illness. However, there are several important ethical concerns with this line of research, and necessary precautions must be taken. First, is the issue of informed consent. Although social media posts are publicly available, users are typically unaware of the research being conducted and do not explicitly provide consent for their data to be mined for sensitive information. Additionally, individuals with mental illness, and especially young individuals, are a sensitive, at risk population and extra caution must be taken when collecting and analyzing their data to ensure they remain anonymous and unidentifiable.

Submitting to IRB review and obtaining IRB approval or exemption for any study with this population is critical. Extreme caution must be taken to protect this sensitive data, and collected corpora should not be shared without IRB approval. Further, if data is shared with specific parties, the data should be anonymized so that identifying information is not disclosed. As data mining for mental health research becomes more popular and prevalent, it is important to be aware of these ethical considerations and to take the necessary precautions to protect the studied population. For further guidance in this area, Benton et al. (2017) have compiled an excellent review of ethical considerations for social media health research.

7 Conclusion

We collected a corpus of Reddit users claiming to have received a diagnosis of SZ and used natural language processing and statistical techniques to analyze and compare language from their posts and those of a control group comprised of random Reddit users. We identified several linguistic markers of SZ, and compared these findings with previous research on linguistic markers of SZ using data from other social media platforms. This work is useful for identifying markers of SZ that are robust across domains. Finally, we trained a machine learning classifier that identified self-described SZ sufferers on Reddit with over 80% accuracy, using linguistic features. These findings contribute toward the ultimate goal of identifying high risk individuals and providing early intervention to improve overall treatment outcomes.

References

APA. 2013. *Diagnostic and statistical manual of mental disorders (DSM-5®).* American Psychiatric Pub.

Sairam Balani and Munmun De Choudhury. 2015. Detecting and characterizing mental health related self-disclosure in social media. In *Proceedings of the 33rd Annual ACM Conference Extended Abstracts on Human Factors in Computing Systems*, pages 1373–1378. ACM.

Yoav Benjamini and Yosef Hochberg. 1995. Controlling the false discovery rate: a practical and powerful approach to multiple testing. *Journal of the Royal statistical society: series B (Methodological)*, 57(1):289–300.

Richard P Bentall, Rhiannon Corcoran, Robert Howard, Nigel Blackwood, and Peter Kinderman.

2001. Persecutory delusions: a review and theoretical integration. *Clinical psychology review*, 21(8):1143–1192.

Adrian Benton, Glen Coppersmith, and Mark Dredze. 2017. Ethical research protocols for social media health research. In *Proceedings of the First ACL Workshop on Ethics in Natural Language Processing*, pages 94–102.

Natalie Berry, Fiona Lobban, Maksim Belousov, Richard Emsley, Goran Nenadic, and Sandra Bucci. 2017. # whywetweetmh: understanding why people use twitter to discuss mental health problems. *Journal of medical Internet research*, 19(4).

Max Birchwood, Patrick McGorry, and Henry Jackson. 1997. Early intervention in schizophrenia. *The British Journal of Psychiatry*, 170(1):2–5.

Michael L Birnbaum, Sindhu Kiranmai Ernala, Asra F Rizvi, Munmun De Choudhury, and John M Kane. 2017. A collaborative approach to identifying social media markers of schizophrenia by employing machine learning and clinical appraisals. *Journal of medical Internet research*, 19(8).

Bryce Boe. 2016. Python Reddit API Wrapper (PRAW). `https://praw.readthedocs.io/en/v6.1.1/`. Accessed: 2019-03-10.

Andreas Carlborg, Kajsa Winnerbäck, Erik G Jönsson, Jussi Jokinen, and Peter Nordström. 2010. Suicide in schizophrenia. *Expert review of neurotherapeutics*, 10(7):1153–1164.

Glen Coppersmith, Mark Dredze, Craig Harman, and Kristy Hollingshead. 2015. From adhd to sad: Analyzing the language of mental health on twitter through self-reported diagnoses. In *Proceedings of the 2nd Workshop on Computational Linguistics and Clinical Psychology: From Linguistic Signal to Clinical Reality*, pages 1–10.

Michael A Covington, Congzhou He, Cati Brown, Lorina Naçi, Jonathan T McClain, Bess Sirmon Fjordbak, James Semple, and John Brown. 2005. Schizophrenia and the structure of language: the linguist's view. *Schizophrenia research*, 77(1):85–98.

SK Fineberg, S Deutsch-Link, M Ichinose, T McGuinness, AJ Bessette, CK Chung, and PR Corlett. 2015. Word use in first-person accounts of schizophrenia. *The British Journal of Psychiatry*, 206(1):32–38.

S Craig Finlay. 2014. Age and gender in Reddit commenting and success. *Journal of Information Science Theory and Practice*, pages 18–28.

Philip D Harvey and Mark R Serper. 1990. Linguistic and cognitive failures in schizophrenia: A multivariate analysis. *Journal of Nervous and Mental Disease*.

Elizabeth Highton-Williamson, Stefan Priebe, and Domenico Giacco. 2015. Online social networking in people with psychosis: a systematic review. *International Journal of Social Psychiatry*, 61(1):92–101.

Marjolijn Hoekert, René S Kahn, Marieke Pijnenborg, and André Aleman. 2007. Impaired recognition and expression of emotional prosody in schizophrenia: review and meta-analysis. *Schizophrenia research*, 96(1-3):135–145.

Kahyee Hor and Mark Taylor. 2010. Suicide and schizophrenia: a systematic review of rates and risk factors. *Journal of psychopharmacology*, 24(4_suppl):81–90.

Andrew Hutchinson. 2018. Reddit now has as many users as twitter, and far higher engagement rates. `https://www.socialmediatoday.com/news/reddit-now-has-as-many-users-as-twitter-and-far-higher-engagement-rates/521789/`. Accessed: 2019-03-10.

Marc Jeannerod. 2009. The sense of agency and its disturbances in schizophrenia: a reappraisal. *Experimental Brain Research*, 192(3):527.

Simon Kemp. 2014. Social, digital & mobile in 2014. *We Are Social Singapore*, 28.

Gina R Kuperberg. 2010. Language in schizophrenia part 1: an introduction. *Language and linguistics compass*, 4(8):576–589.

Amanda Lenhart, Maeve Duggan, Andrew Perrin, Renee Stepler, Harrison Rainie, Kim Parker, et al. 2015. *Teens, social media & technology overview 2015*. Pew Research Center [Internet & American Life Project].

Minna Lyons, Nazli Deniz Aksayli, and Gayle Brewer. 2018. Mental distress and language use: Linguistic analysis of discussion forum posts. *Computers in Human Behavior*, 87:207–211.

Margaret Mitchell, Kristy Hollingshead, and Glen Coppersmith. 2015. Quantifying the language of schizophrenia in social media. In *Proceedings of the 2nd workshop on Computational linguistics and clinical psychology: From linguistic signal to clinical reality*, pages 11–20.

Matthew L Newman, James W Pennebaker, Diane S Berry, and Jane M Richards. 2003. Lying words: Predicting deception from linguistic styles. *Personality and social psychology bulletin*, 29(5):665–675.

NIMH. 2019. Schizophrenia. `https://www.nimh.nih.gov/health/topics/schizophrenia/index.shtml`. Accessed: 2019-03-10.

Fabian Pedregosa, Gaël Varoquaux, Alexandre Gramfort, Vincent Michel, Bertrand Thirion, Olivier Grisel, Mathieu Blondel, Peter Prettenhofer, Ron Weiss, Vincent Dubourg, et al. 2011. Scikit-learn: Machine learning in python. *Journal of machine learning research*, 12(Oct):2825–2830.

James W Pennebaker, Ryan L Boyd, Kayla Jordan, and Kate Blackburn. 2015a. The development and psychometric properties of liwc2015. Technical report.

James W Pennebaker and Laura A King. 1999. Linguistic styles: language use as an individual difference. *Journal of personality and social psychology*, 77(6):1296.

James W Pennebaker, Tracy J Mayne, and Martha E Francis. 1997. Linguistic predictors of adaptive bereavement. *Journal of personality and social psychology*, 72(4):863.

JW Pennebaker, CK Chung, M Ireland, A Gonzales, and RJ Booth. 2015b. Liwc. *Austin, Texas; 2007. LIWC2007: Linguistic inquiry and word count [software program for text analysis] URL: http://liwc. wpengine. com/[accessed 2017-02-27]*.

Bonnie L Rickelman. 2004. Anosognosia in individuals with schizophrenia: toward recovery of insight. *Issues in Mental Health Nursing*, 25(3):227–242.

S Rochester and JR Martin. 1979. Jr, 1979 crazy talk: A study of the discourse of schizophrenic speakers.

Louis A Sass and Josef Parnas. 2003. Schizophrenia, consciousness, and the self. *Schizophrenia bulletin*, 29(3):427–444.

Yla R Tausczik and James W Pennebaker. 2010. The psychological meaning of words: Liwc and computerized text analysis methods. *Journal of language and social psychology*, 29(1):24–54.

Graham Thornicroft, Michele Tansella, Thomas Becker, Martin Knapp, Morven Leese, Aart Schene, José Luis Vazquez-Barquero, EPSILON Study Group, et al. 2004. The personal impact of schizophrenia in europe. *Schizophrenia research*, 69(2-3):125–132.

Theo Vos, Ryan M Barber, Brad Bell, Amelia Bertozzi-Villa, Stan Biryukov, Ian Bolliger, Fiona Charlson, Adrian Davis, Louisa Degenhardt, Daniel Dicker, et al. 2015. Global, regional, and national incidence, prevalence, and years lived with disability for 301 acute and chronic diseases and injuries in 188 countries, 1990–2013: a systematic analysis for the global burden of disease study 2013. *The Lancet*, 386(9995):743–800.

Semantic Characteristics of Schizophrenic Speech

Kfir Bar[*]
School of Computer Science
College of Management
Academic Studies
Rishon LeZion, Israel
kfirb@colman.ac.il

Vered Zilberstein[*,**]
School of Computer Science
Tel Aviv University
Ramat Aviv, Israel
veredz1@mail.tau.ac.il

Ido Ziv[*]
Department of Psychology
College of Management
Academic Studies
Rishon LeZion, Israel
idoz@colman.ac.il

Heli Baram[*]
Department of Psychology
Ruppin Academic Center
Emek Hefer, Israel
fanta.hchc@gmail.com

Nachum Dershowitz
School of Computer Science
Tel Aviv University
Ramat Aviv, Israel
nachum@tau.ac.il

Samuel Itzikowitz
School of Computer Science
College of Management Academic Studies
Rishon LeZion, Israel
samitz@st.colman.ac.il

Eiran Vadim Harel
Beer Yaakov Mental Heath Center
Beer Yaakov, Israel
eiran.harel@moh.gov.il

Abstract

Natural language processing tools are used to automatically detect disturbances in transcribed speech of schizophrenia inpatients who speak Hebrew. We measure topic mutation over time and show that controls maintain more cohesive speech than inpatients. We also examine differences in how inpatients and controls use adjectives and adverbs to describe content words and show that the ones used by controls are more common than the those of inpatients. We provide experimental results and show their potential for automatically detecting schizophrenia in patients by means only of their speech patterns.

1 Introduction

Thought disorders are described as disturbances in the normal way of thinking. Bleuler (1991) original considered thought disorders to be a speech impairment in schizophrenia patients, but nowadays there is agreement that thought disorders are also relevant to other clinical disorders, including pediatric neurobehavioral disorders like attention deficit hyperactivity disorder and high functioning autism. They can even occur in normal populations, especially in people who have a high level of creativity. Bleuler focused mostly on "loosening of associations", or *derailment*, a thought disorder characterized by the usage of unrelated concepts in a conversation, or in other words, a conversation lacking coherence. The *Diagnostic and Statistical Manual of Mental Disorders (DSM 5)* (Association, 2013) outlines *disorganized speech* as one of the criteria for making a diagnosis of schizophrenia. Morice and Ingram (1982) showed that schizophrenics' speech is built upon a different syntactic structure than normal controls, and that this difference increases over time. Andreasen (1979) suggested several definitions of linguistic and cognitive behaviors frequently observed in patients, and which may be useful for thought-disorder evaluation. Among the definitions presented in that report, one finds the following, which we address in this study:

Incoherence, also known as "word salad", refers to speech that is incomprehensible at times due to multiple grammatical and semantic inaccuracies. In this paper, we focus mostly on the semantic inaccuracies, leaving grammatical issues for future investigation.

Derailment, also known as "loose associations", happens when a speaker shifts among topics that are only remotely related, or are completely unrelated, to the previous ones.

---

[*]Equal contribution.
[**]Supported by the Deutsch Institute.

Proceedings of the Sixth Workshop on Computational Linguistics and Clinical Psychology, pages 84–93
Minneapolis, Minnesota, June 6, 2019. ©2019 Association for Computational Linguistics

Tangentiality occurs when an irrelevant, or just barely relevant, answer is provided for a given question.

We focus here on derailment. But tangentiality has been addressed in some other studies. The two notions are closely related.

One of the main data sources for diagnosing mental disorders is speech, typically collected during a psychiatric interview. Identifying signals that indicate the presence of thought disorders is often challenging and subjective, especially in patients who are not undergoing a psychotic episode at the time of the interview.

In this work, we focus on schizophrenia. We investigate a number of semantic characteristics of transcribed human speech, and propose a way to use them to measure disorganized speech. Natural-language processing software is used to automatically detect those characteristics, and we suggest a way of aggregating them in a meaningful way. We use transcribed interviews, collected from Hebrew-speaking schizophrenia inpatients at a mental health hospital and from a control group. About two thirds of the patients were identified as in schizophrenia remission at the time of the interview.

Following a few previous works (Iter et al., 2018; Bedi et al., 2015), we measure Andreasen's derailment by calculating average semantic similarity between consecutive chunks of a running text to track topical mutations, and show the difference between patients and controls. For incoherence, we look at word modifiers, focusing on adjectives and adverbs, that subjects use to describe the same objects, and then learn the difference between the two groups. As a final step, we use those semantic characteristics in a classification setting and argue for their usability.

This work makes the following contributions:

- We measure derailment in speech using word semantics, similar to (Bedi et al., 2015), this time on Hebrew.

- We explore a novel way of measuring one aspect of speech incoherence, by measuring how similar modifiers (adjectives and adverbs) are to ones used in a reference text to describe the same words.

- Using these measures, we build a classifier for detecting schizophrenia on the basis of

recorded interviews, which achieves 81.5% accuracy.

We proceed as follows: The next section reviews some relevant previous work. In Section 3, we describe how we collected the data. Our main contributions are described in Section 4, followed by some conclusions suggested in the final section.

2   Related Work

There is a large body of work that examines human-generated texts with the aim of learning about the way people who suffer from various mental-health disorders use language in different settings. For example, Al-Mosaiwi and Johnstone (2018) conducted a study in which they analyzed 63 web forums, some related to mental health disorders and others used as control. They ran their analysis with the well-known Linguistic Inquiry and Word Count (Pennebaker et al., 2015) tool to find absolutist words in free text. Overall, they discovered that anxiety, depression, and suicidal-ideation forums contained more absolutist words than control forums.

Recently, social media have become a vital source for learning about how people who suffer from mental-health disorders use language. Several studies collect relevant users from Twitter,[1] by considering users who intentionally write about their diagnosed mental-health disorders. For example, in (De Choudhury et al., 2013; Tsugawa et al., 2015), some language characteristics of Twitter users who claim to suffer from a clinical depression are studied. Similarly, users who suffer from post traumatic stress disorder are addressed in (Coppersmith et al., 2014). Mitchell et al. (2015) analyze tweets posted by schizophrenics, and Coppersmith et al. (2016) investigate the language and emotions that are expressed by users who have previously attempted to commit suicide. Coppersmith et al. (2015) work with users who suffer from a broad range of mental-health conditions and explore language differences between groups. Most of these works found a significant difference in the usage of some linguistic characteristics by the experience group when compared to a control group. Furthermore, different levels of these linguistic characteristics are used as features for training a classifier to detect mental-health disorders prior to the report date.

Reddit[2] has also been identified as a convenient source for collecting data for this goal. Losada and Crestani (2016) outline a methodology for collecting posts and comments of Reddit and Twitter users who suffer from depression. Similarly, a large dataset of Reddit users with depression, manually verified (by lay annotators for an explicit claim of diagnosis), has been released for public use (Yates et al., 2017). In that work, the authors employ a deep neural network on the raw text for detecting clinically depressed people ahead of time, achieving 65% F1 score on an evaluation set.

A few caveats are in order when using social media for analyzing mental health conditions. First, self reporting of a mental health disorder is not a popular course of action. Clearly, then, the experimental group is chosen from a subgroup of the relevant population. Second, the controls, typically collected randomly "from the wild", are not guaranteed to be free of mental-health disorders. Finally, social media posts are considered to be a different form of communication than ordinary speech. For all these reasons, in this work, we use validated experimental and control groups in an interview setting.

Measuring various aspects of incoherence in schizophrenics using computational tools has been previously addressed in (Elvevåg et al., 2007; Bedi et al., 2015; Iter et al., 2018). Elvevåg et al. (2007) analyzed transcribed interviews of inpatients with schizophrenia to measure tangentiality. Moving along the patient's response, they calculated the semantic similarity between text chunks of different sizes and the question that was asked by the interviewer. Semantic similarity was cast by cosine similarity over the latent semantic analysis (LSA) (Deerwester et al., 1990) vectors calculated for each word, and summed across an entire chunk of words. They fitted a linear-regression line to represent the trend of the cosine similarity values, as one moves along the text. The slope of that line was used to measure how quickly the topic diverges from the original question. Overall, they were able to show a significant correlation between those values and a blind human evaluation of the same responses. Furthermore, as chunk size grows larger, the distinction between patients and controls becomes less prominent. One explanation for that could be the large number of mentions of functional and filler words, for which we typically

do not have a good semantic representation. Iter et al. (2018) addressed this suggestion by cleaning the patients' responses of all those words and expressions (e.g. *uh, um, you know*) prior to calculating the semantic scores. This gave a slight improvement, although measured over a relatively small set of participants. Instead of working with chunks of text, they worked with full sentences, and replaced LSA with some modern techniques for sentence embeddings. Likewise, in our work, we use word embeddings instead of LSA.

Bedi et al. (2015) define coherence as an aggregation of the cosine similarity between pairs of consecutive sentences, each represented by the element-wise average vector of the individual words' LSA vectors. They worked with a group of 34 youths at clinical high-risk for psychosis, interviewed them quarterly for 2 1/2 years, and transcribed their answers. Five out of the 34 transitioned to psychosis. They used coherence scores, along with part-of-speech information, to automatically predict transition to psychosis with 100% accuracy.

The goal of all these works, including ours, is to automatically detect disorganized speech in a more objective and reliable way. Inspired by the last three studies described above, we analyzed transcribed responses to 18 open questions given by inpatients with schizophrenia and by controls. Instead of cleaning the text from filler words using a dictionary – as proposed by (Iter et al., 2018), we take a deeper look into the syntactic roles the words play, and calculate semantic similarity over a filtered version of the text, every time using different sets of part-of-speech categories. We report on the results of two sets of experiments: (1) We measure derailment by calculating the semantic similarity of adjacent words of various part-of-speech categories. (2) We measure semantic coherence by looking at the choices of modifiers (adjectives, adverbs) used in responses by inpatients and controls, as compared to those used in ordinary discourse.

Generally speaking, not too much is known about the role played by adjectives and adverbs in thought disorders. Modifiers are often not included in language tests, as they usually need to be presented together with the noun or verb they modify. Some previous works (Obrębska and Obrębski, 2007) have reported a significantly smaller number of adjectives used by schizophren-

---

	Control	**Patients**
N	27	24
Age, Mean (SD)	30.3 (8.26)	38.3 (10.43)
Edu., HS	68%	75%
Edu., Post HS	20%	4%
Loc., South	40%	20%
Loc., Center	44%	33%
M.S., Single	80%	95%
Income, Avg/low	84%	83%

Table 1: Demographics by group. Edu. = Education (HS = High School); Loc. = Location in Israel; M.S. = Marital Status.

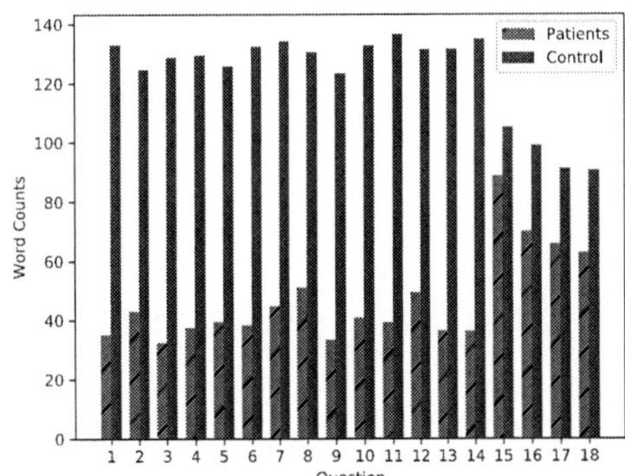

Figure 1: Word counts per question.

ics. In the current study, we use computational tools to investigate the semantic relation between modifiers and objects, and its attribution to speech incoherence.

3 Data Collection

We interviewed 51 men, aged 19–63, divided into control and patient groups, all speaking Hebrew as their mother tongue. The patient group comprised 24 inpatients at Beer Yaakov Mental Health Center in Israel who were officially diagnosed with schizophrenia. The control group includes 27 people, mainly recruited via an advertisement that we placed on social media. Most of the participants are single, with average-to-lower monthly income. Demographics for the two groups are presented in Table 1.

Ethics statement: The institutional review board of the College of Management Academic Studies of Rishon LeZion, as well as of the Beer Yaakov–Ness Ziona Mental Health Center, approved these experiments, and informed consent was obtained for all subjects.

3.1 Interviews

Overall, the participants were asked 18 questions, out of which 14 were thematic-apperception-test (TAT) pictures that participants were requested to describe, followed by 4 questions that require the participant to share some personal thoughts and emotions. Both the control and patient groups completed a demographic questionnaire. To monitor the mental-health condition of the control group, they were requested to complete Beck's Depression Inventory-II (BDI-II) and the State and Trait Anxiety Inventory (STAI). The patient group also completed BDI-II, as well as a Hebrew translation (Katz et al., 2012) of the Positive and Neg-

ative Syndrome Scale–6 (PANSS-6, a shorter version of PANSS-30) questionnaire, in order to assess symptoms of psychosis (Østergaard et al., 2016). Scores for the two questionnaires were found to be highly correlated. Out of the patient group, 66.7% were assigned a score below 14, a recommended preliminary threshold indicating schizophrenia remission.

The interviews were recorded and then manually transcribed by Hebrew-speaking students from our lab. The TAT pictures presented to participants during the interview were: 1, 2, 3BM, 4, 5, 6BM, 7GF, 8BM, 9BM, 12M, 13MF, 13B, 14, 3GF. Table 2 lists the questions that were presented to the participants during the interview. All the transcripts are written in Hebrew. Figure 1 shows average word counts by question, per group. Clearly, the patients spoke fewer words than the controls. The difference becomes less significant for the open-ended questions.

3.2 Preprocessing

Hebrew being a highly-inflected language, we preprocessed the texts with the Ben-Gurion University Morphological Tagger (Adler, 2007), a context-sensitive morphological analyzer for Modern Hebrew. Given a running text, the tagger breaks the text into words and provides morphological information for every word, including the disambiguated part-of-speech tag and lemma. There were no specific instructions given to the transcribers for how to punctuate, which led to an inconsistency in the way punctuation was used in the transcriptions. We used the tags to clean up all punctuation marks by removing all tokens tagged as such.

ID	Question
1	Tell me as much as you can about your bar mitzvah.
2	What do you like to do, mostly?
3	What are the things that annoy you the most?
4	What would you like to do in the future?

Table 2: Four open questions asked during the interview.

4 Tools and Method

We report on two sets of experiments. In the first, we measure derailment by calculating the semantic similarity between adjacent words in running text. In the second set of experiments, we investigate the modifiers that the two groups use to describe specific nouns and verbs. As a final step, we measure the contribution of the semantic characteristics that we compute in the experiments, for automatic classification of schizophrenia.

4.1 Experiment 1: Measuring Derailment

We calculate a derailment score for each response and use it to measure derailment.

Tools: To measure derailment, we calculate the semantic similarity of adjacent words in the answers provided by the participants during the interview. We use word embeddings to represent each word by means of a mathematical vector that captures its meaning. These vectors were created automatically by characterizing words by the surrounding contexts in which they are mentioned in a large corpus of documents. Specifically, we used Hebrew pretrained vectors provided by `fastText` (Grave et al., 2018), which were created from Wikipedia,[3] as well as from other content extracted from the web with Common Crawl.[4] Overall, 97% of the words in our corpus exist in `fastText`. Hebrew words are inflected for person, number and gender; prefixes and suffixes are added to indicate definiteness, conjunction, prepositions, and possessive forms. On the other hand, `fastText` was trained for surface forms. Therefore, we work on the surface-form level. To measure semantic similarity between two words, we use the common cosine-similarity function that calculates the cosine of the angle between the two corresponding vectors. The score ranges from -1 to $+1$, with $+1$ representing maximal similarity.

Method: (1) For each sufficiently long response,

R, we retrieve the `fastText` vector v_i for every word R_i, $i = 0 \ldots n$, in the response. (2) For each word, we calculate the average pairwise cosine similarity between this word and the k following words. The integer k is a parameter; we experimented with different values. (3) We take the average of all the individual cosine similarity scores and form a single score for each response.

In this experiment, we consider only responses that are long enough to allow topic mutation to develop. Therefore, we use only the four questions from Table 2 for which the participants provided a relatively long response. Accordingly, we drop responses of fewer than 50 words. As mentioned above, we consider that the existence of some word types, like fillers and functional words, might introduce some noise, which might harm the calculation process. We would rather focus on words that convey real content. Therefore, we calculate scores separately using all words and using only *content words*, which we take to be nouns, verbs, adjectives, and adverbs. We detected a few types of text repetitions, which may bias the derailment score. One type is when a word is said twice or more for emphasis; for example, "quickly, quickly" (מהר מהר) (i.e. very quickly). To mitigate this bias, we keep only one word out of a pair of consecutive identical words. Another type is when a whole phrase is repeated; for example, "She's in a big hurry; she's in a big hurry" (היא ממהרת מאוד, היא ממהרת מאוד). Handling this problem is left for future work.

We calculate derailment scores for the responses provided by all participants and compare the means of the two groups.

Results: When using all words, we could not detect a significant difference between patients and controls. However, when using content words only, patients scored lower on derailment than the controls, for all window widths k, suggesting that focusing only on content words is the more robust approach for calculating derailment. This finding is consistent with previous work (Iter et al., 2018).

---

[3] `https://www.wikipedia.org`
[4] `http://commoncrawl.org`

Overall, coherence decreases as k increases. Table 3 summarizes the results. To confirm the significance we are seeing in the results, is due to the diagnosis and not due to other characteristics of the participants, we aggregated the same scores for the different age groups and education levels, regardless of the diagnosis status; all these results did not appear to be significant. Figure 2 shows the trend of the average derailment score from Table 3, running with different values of k. The left plot was produced for all word types, and the right plot using only content words. We clearly observe a slight increase of the entire control curve and a slight decrease of the patients curve, when restricting to content words.

4.2 Experiment 2: Incoherence

In this experiment, we examine the way patients use adjectives and adverbs (hereafter, *modifiers*) to describe specific nouns and verbs, respectively. Our goal is to measure the difference between modifiers used by patients and the ones used by controls, when describing the same nouns and verbs. We suggest this as a tool for measuring incoherence in speech. For example, inspecting the responses for the first TAT image, we learn that patients typically use the adjectives "new" (חדש) and "good" (טוב) to modify the noun "violin" (כינור), while controls use the adjectives "old" (ישן), "sad" (עצוב), and "significant" (משמעותי).

Tools: To detect all noun-adjective and verb-adverb pairs in the responses, we use a dependency parser, which analyzes the grammatical structure of a sentence and builds links between "head" words and their modifiers. Specifically, we use YAP (More and Tsarfaty, 2016), a dependency parser for Modern Hebrew, and process each sentence individually. Among other things, YAP provides a word-dependency list, shaped as a list of tuples, each includes a head word, a dependent word, and the kind of dependency. We use the relevant types (e.g. *advmod*, *amod*) for finding all noun-adjective and verb-adverb pairs. For example, Figure 3 shows the dependencies returned by YAP for the input sentence: "I ate a tasty candy" (אכלתי סוכריה טעימה). From this sentence we extract the noun "candy" (סוכריה), which is modified by the adjective "tasty" (טעימה).

Method: To measure the difference between the modifiers that are used by patients and controls, we compare them to the modifiers that are com-

monly used to describe the same nouns and verbs. For example, given an answer with only one noun "violin" (כינור) that is modified by the adjective "sad" (עצוב), we calculate a score that reflects how similar the adjective "sad" is to adjectives that are typically used to describe a violin.

We take the following steps:

(1) We convert each sentence into a list of noun-adjective and verb-adverb pairs using YAP.

(2) To compare each modifier with the modifiers that are typically used to describe the same noun or verb, we use external corpora as reference. These were taken from various sources reflecting the health domain we are working in.[9] Table 4 lists the sources and the corresponding number of documents and words that they contain. Each document in these sources was processed in exactly the same way to find all noun-adjective and verb-adverb pairs.

(3) Given a list of noun-adjective and verb-adverb pairs of one response, we calculate the similarity score of every modifier that describes a specific noun or verb with the set of modifiers describing exactly the same noun or verb in the reference corpus. Looking at our example above, we would want to calculate a similarity score between the adjective "old" (ישן) and all the adjectives that are used to describe "violin" (כינור) in the reference corpus. Searching for instances of the same Hebrew word is challenging due to Hebrew's rich morphology. Hebrew words are inflected for person, number, and gender; prefixes and suffixes are added to indicate definiteness, conjunction, various prepositions, and possessive forms. Therefore, we work on the lemma (base-form) level. Most vowels in Hebrew are not indicated in standard writing; therefore, Hebrew words tend to be ambiguous, and determining the correct lemma for a word is nontrivial. We use the lemmas provided by YAP.

Another challenge is how to compare a single modifier with a group of modifiers that were taken from the reference corpus. We take the fastText vectors of the modifiers that were extracted from the reference corpus and aggregate them into a single vector. Then, we take cosine similarity between the modifier from the response and the aggregated vector of the modifiers from the reference corpus. As an aggregation function, we use element-wise weighted average of the individual modifiers' fastText vectors, and define

	All words			**Content words**		
k	Control	Patients	t	Control	Patients	t
1	0.270 (0.014)	0.257 (0.025)	2.004*	0.265 (0.019)	0.240 (0.020)	2.968*
2	0.246 (0.017)	0.239 (0.025)	1.173	0.256 (0.018)	0.231 (0.025)	2.687*
3	0.237 (0.017)	0.233 (0.025)	0.476	0.250 (0.018)	0.225 (0.026)	2.614*
4	0.233 (0.018)	0.229 (0.025)	0.471	0.245 (0.018)	0.221 (0.026)	2.539*
5	0.230 (0.017)	0.226 (0.026)	0.528	0.241 (0.018)	0.218 (0.023)	2.598*

Table 3: Results for Experiment 1. Comparing average derailment scores of patients and controls. The numbers are provided as average across patients and controls, with standard deviation in parentheses, $*p < 0.05$.

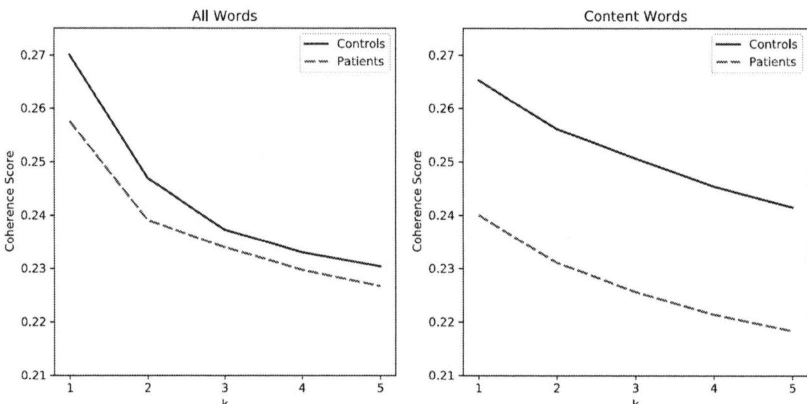

Figure 2: Derailment scores for different values of k. The left plot shows the results for all word types, and the right plot shows the results for content words only.

Corpus	Description	# Documents	# Words
Doctors[5]	Articles from the Doctors medical website	239	187,938
Infomed[6]	Question-and-answer discussions from the Infomed website's medical forum, January 2006 – September 2007	749	128,090
To Be Healthy[7]	Articles and forum discussions from the To Be Healthy (L'Hiyot Bari, 2b-bari) medical website	137	112,839
HaAretz[8]	News and articles from the HaAretz news website, 1991	4,920	250,399

Table 4: The external Hebrew corpora used to collect modifiers of nouns and verbs that are typically used.

the weights to be the inverse-document-frequency (IDF) score to account more for modifiers that describe the noun or verb more uniquely. We calculate IDF scores using the reference corpora. For this purpose, a "qualified" word is a noun or verb that has an IDF score and that has at least one modifier linked to it in either the control or patient corpus. Most of the nouns and verbs are non-qualified; we only consider qualified words in this investigation.

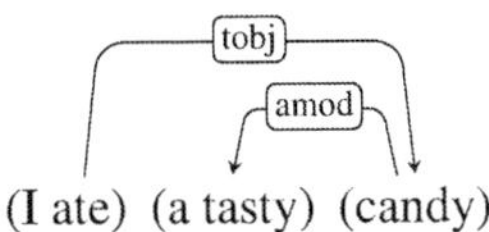

Figure 3: The dependencies returned by YAP for the sentence "(I ate) (a tasty) candy". The parentheses delimit the translations for each of the three Hebrew words in the sentence.

	Control	Patients	t
Adj	0.5891 (0.0301)	0.5498 (0.0284)	4.7765***
Adv	0.6880 (0.0251)	0.6254 (0.0709)	4.2961***

Table 5: Results for Experiment 2. The numbers are average coherence scores across patients and controls (with standard deviations); ***$p < 0.001$.

	Control		Patients	
	Total	Qual.	Total	Qual.
Nouns	934	226	242	90
Adjectives	573	371	204	127
Verbs	699	60	204	34
Adverbs	166	104	86	50

Table 6: Experiment 2: Counts of nouns, verbs, and their modifiers, across the two groups. Qual. = Qualified.

Classifier	Acc.	Prec.	Recall
Random Forest	81.5%	91.3%	71.8%
XGBoost	80.5%	86.8%	73.1%
SVM	70.4%	72.1%	47.3%

Table 7: Classification results for each classifier.

(4) For each response, we calculate two scores, individually. The adjective-similarity score is the IDF-weighted average of the individual adjective scores we calculate in the previous step. Similarly, the adverb-similarity score is the IDF-weighted average of the individual adverb scores we calculate in the previous step.

(5) To calculate a score on the participant level, we average the scores of all the individual responses provided by the participant.

The output of this process is a pair of scores, one for adjectives and one for adverbs, calculated for each participant. The higher a score is, the more similar the modifiers are to ones that are typically used to describe the same noun or verb.

Results: Table 5 summarizes the results. Overall, controls have significantly higher scores for both modifier types, indicating a higher agreement on modifiers by the controls and external writers.

There are more nouns and adjectives than verbs and adverbs, as summarized in Table 6. On average, participants use more adjectives to describe nouns than adverbs to describe verbs. Controls use about 0.61 adjectives per noun, while patients use 0.84 adjectives on average. Similarly, patients use more adverbs to describe a verb on average than

controls do. While patients use about 0.42 adverbs per verb, controls use only 0.23. However, these differences are not significant.

4.3 Classification

As a final step, we train several classifiers to distinguish between controls and patients. We represent participants with the characteristics we compute in the two experiments. Specifically, each subject is represented by the following: (1) noun and verb derailment scores; (2) coherence scores for 5 windows, using all words; and (3) coherence scores for 5 windows, using only content words. In total, we use 12 scores per subject. Each classifier was trained using a 10-fold cross-validation evaluation of prediction quality over the 51 participants. For each classifier, we report on the overall prediction accuracy, as well as precision and recall for the prediction of the patients group. The classification algorithms we tried are Random Forest (Breiman, 2001) and XGBoost (Chen and Guestrin, 2016), both based on decision trees, and, in addition, linear support vector machines (SVM) (Cortes and Vapnik, 1995). Table 7 summarizes the results per classifier with respect to the different metrics.

We used the decision-tree based classifiers to calculate the most important features, that is, the ones that have the greatest impact on prediction decisions. The most important features were found to be the two derailment scores, as expected.

---

[5]https://www.doctors.co.il

[6]https://www.infomed.co.il

[7]https://tobehealthy.co.il

[8]https://www.haaretz.co.il

[9]All were downloaded from MILA Knowledge Center for Processing Hebrew: http://mila.cs.technion.ac.il/resources_corpora.html.

5   Conclusions

With the aim of detecting speech disturbances, we have analyzed transcribed Hebrew speech, produced by schizophrenia inpatients and compared it with those of controls. We believe that speech produced during a psychiatric interview is a more reliable data source for detecting disturbances than are social media posts.

Generally speaking, we find that patients talk significantly less in interviews than controls do.

In one experiment, we use word embeddings to detect derailment, that is, when a speaker shifts to a topic that is not strongly related to previously discussed ones. The results show that controls have higher scores, indicating that they keep the topic more cohesive than patients do. These results are in line with previous studies on English (Bedi et al., 2015), which showed that schizophrenics have a lower score, calculated by a similar mathematical procedure.

In a second experiment, we examine the difference in how patients and controls use adjectives and adverbs to describe nouns and verbs, respectively. Our results show that the adjectives and adverbs that are used by the controls are more similar to the typical ones used to describe the same nouns and verbs. For now, we consider this difference as related to speech incoherence; however, we plan to continue investigating this direction in the near future, when more data become available.

Analyzing Hebrew is more challenging than analyzing English due to Hebrew's rich morphology, as well as the absence of written vowels. In the first experiment, we work with `fastText`, which provides word embeddings on the surface-form level. In the second, we used lemmata rather than the word surface forms, so we can find multiple instances of the same lexeme.

As we did not measure the IQ of participants, some of the results might, to a certain extent, be attributable to differences in intellect. Moreover, as can be seen in Table 1, about 20% of the control participants have some sort of post high-school education, while most of the inpatients did not continue beyond high-school. We plan to address these questions in followup work. Another limitation that we are aware of is related to the classification results, as the number of participants we use for training the classifiers might be considered relatively small.

Overall, we found the semantic characteristics that we compute in this study to be beneficial for the task of detecting thought disorders in Hebrew speech. We plan to collect speech samples from more subjects, and to continue to explore additional semantic – as well as grammatical – textual characteristics to support the automatic detection of various mental disorders.

References

Meni Adler. 2007. *Hebrew Morphological Disambiguation: An Unsupervised Stochastic Word-based Approach.* Ph.D. thesis, Ben-Gurion University of the Negev, Beer-Sheva, Israel.

Mohammed Al-Mosaiwi and Tom Johnstone. 2018. In an absolute state: Elevated use of absolutist words is a marker specific to anxiety, depression, and suicidal ideation. *Clinical Psychological Science*, 6(4):529–542.

Nancy C. Andreasen. 1979. Thought, language, and communication disorders: I. clinical assessment, definition of terms, and evaluation of their reliability. *Archives of General Psychiatry*, 36(12):1315–1321.

American Psychological Association. 2013. *Diagnostic and Statistical Manual of Mental Disorders (5th ed.).* Washington, DC.

Gillinder Bedi, Facundo Carrillo, Guillermo A. Cecchi, Diego Fernández Slezak, Mariano Sigman, Natália B. Mota, Sidarta Ribeiro, Daniel C. Javitt, Mauro Copelli, and Cheryl M. Corcoran. 2015. Automated analysis of free speech predicts psychosis onset in high-risk youths. *npj Schizophrenia*, 1:15030.

Eugen Bleuler. 1991. Dementia praecox oder Gruppe der Schizophrenien. In G. Aschaffenburg, editor, *Handbuch der Psychiatrie*, volume Spezieller Teil. 4. Abteilung, 1. Hälfte. Franz Deuticke, Leipzig.

Leo Breiman. 2001. Random forests. *Mach. Learn.*, 45(1):5–32.

Tianqi Chen and Carlos Guestrin. 2016. XGBoost: A scalable tree boosting system. In *Proceedings of the 22nd ACM SIGKDD International Conference on Knowledge Discovery and Data Mining*, KDD '16, pages 785–794, New York, NY. ACM.

Glen Coppersmith, Mark Dredze, Craig Harman, and Kristy Hollingshead. 2015. From ADHD to SAD: Analyzing the language of mental health on Twitter through self-reported diagnoses. In *Proceedings of the 2nd Workshop on Computational Linguistics and Clinical Psychology: From Linguistic Signal to Clinical Reality*, pages 1–10.

Glen Coppersmith, Craig Harman, and Mark Dredze. 2014. Measuring post traumatic stress disorder in Twitter. In *ICWSM 2014*.

Glen Coppersmith, Kim Ngo, Ryan Leary, and Anthony Wood. 2016. Exploratory analysis of social media prior to a suicide attempt. In *Proceedings of the Third Workshop on Computational Lingusitics and Clinical Psychology*, pages 106–117.

Corinna Cortes and Vladimir Vapnik. 1995. Support-vector networks. *Mach. Learn.*, 20(3):273–297.

Munmun De Choudhury, Michael Gamon, Scott Counts, and Eric Horvitz. 2013. Predicting depression via social media. *ICWSM*, 13:1–10.

Scott Deerwester, Susan T. Dumais, George W. Furnas, Thomas K. Landauer, and Richard Harshman. 1990. Indexing by latent semantic analysis. *Journal of the American Society for Information Science*, 41(6):391–407.

Brita Elvevåg, Peter W. Foltz, Daniel R. Weinberger, and Terry E. Goldberg. 2007. Quantifying incoherence in speech: an automated methodology and novel application to schizophrenia. *Schizophrenia research*, 93(1–3):304–316.

Edouard Grave, Piotr Bojanowski, Prakhar Gupta, Armand Joulin, and Tomas Mikolov. 2018. Learning word vectors for 157 languages. In *Proceedings of the International Conference on Language Resources and Evaluation (LREC 2018)*.

Dan Iter, Jong Yoon, and Dan Jurafsky. 2018. Automatic detection of incoherent speech for diagnosing schizophrenia. In *Proceedings of the Fifth Workshop on Computational Linguistics and Clinical Psychology: From Keyboard to Clinic*, pages 136–146. Association for Computational Linguistics.

Gregory Katz, Leon Grunhaus, Shukrallah Deeb, Emi Shufman, Rachel Bar-Hamburger, and Rimona Durst. 2012. A comparative study of Arab and Jewish patients admitted for psychiatric hospitalization in Jerusalem: the demographic, psychopathologic aspects, and the drug abuse comorbidity. *Comprehensive Psychiatry*, 53(6):850–853.

David E. Losada and Fabio Crestani. 2016. A test collection for research on depression and language use. In *International Conference of the Cross-Language Evaluation Forum for European Languages*, pages 28–39. Springer.

Margaret Mitchell, Kristy Hollingshead, and Glen Coppersmith. 2015. Quantifying the language of schizophrenia in social media. In *CLPsych@HLT-NAACL*.

Amir More and Reut Tsarfaty. 2016. Data-driven morphological analysis and disambiguation for morphologically rich languages and universal dependencies. In *Proceedings of COLING 2016*.

Rodney D. Morice and John C. L. Ingram. 1982. Language analysis in schizophrenia: Diagnostic implications. *Australian & New Zealand Journal of Psychiatry*, 16(2):11–21. PMID: 6957177.

M Obrębska and T Obrębski. 2007. Lexical and grammatical analysis of schizophrenic patients' language: A preliminary report. *Psychology of Language and Communication*, 11(1):63–72.

Soren Dinesen Østergaard, Ole Michael Lemming, Ole Mors, Christoph U. Correll, and Per Bech. 2016. PANSS-6: A brief rating scale for the measurement of severity in schizophrenia. *Acta Psychiatrica Scandinavica*, 133(6):436–444.

James W. Pennebaker, Ryan L. Boyd, Kayla Jordan, and Kate Blackburn. 2015. The development and psychometric properties of LIWC2015. Technical report, University of Texas at Austin, Austin, TX.

Sho Tsugawa, Yusuke Kikuchi, Fumio Kishino, Kosuke Nakajima, Yuichi Itoh, and Hiroyuki Ohsaki. 2015. Recognizing depression from Twitter activity. In *Proceedings of the 33rd Annual ACM Conference on Human Factors in Computing Systems*, pages 3187–3196. ACM.

Andrew Yates, Arman Cohan, and Nazli Goharian. 2017. Depression and self-harm risk assessment in online forums. *CoRR*, abs/1709.01848.

Computational Linguistics for Enhancing Scientific Reproducibility and Reducing Healthcare Inequities

Julia Parish-Morris, PhD[1,2,3]

[1] Center for Autism Research, Children's Hospital of Philadelphia (CHOP)
[2] Departments of Biomedical and Health Informatics and Child and Adolescent Psychiatry, CHOP
[3] Department of Psychiatry, Perelman School of Medicine of the University of Pennsylvania

Abstract

Computational linguistics holds promise for improving scientific integrity in clinical psychology, and for reducing longstanding inequities in healthcare access and quality. This paper describes how computational linguistics approaches could address the "reproducibility crisis" facing social science, particularly with regards to reliable diagnosis of neurodevelopmental and psychiatric conditions including autism spectrum disorder (ASD). It is argued that these improvements in scientific integrity are poised to naturally reduce persistent healthcare inequities in neglected subpopulations, such as verbally fluent girls and women with ASD, but that concerted attention to this issue is necessary to avoid reproducing biases built into training data. Finally, it is suggested that computational linguistics is just *one* component of an emergent digital phenotyping toolkit that could ultimately be used for clinical decision support, to improve clinical care via precision medicine (i.e., personalized intervention planning), granular treatment response monitoring (including remotely), and for gene-brain-behavior studies aiming to pinpoint the underlying biological etiology of otherwise behaviorally-defined conditions like ASD.

1 Introduction

Humans are complex social beings, and the intricacies of language manifest this richness. Although language emanates from the brain, it has not yet been fully leveraged in the service of understanding brain-based psychiatric variation (e.g., disorders such as schizophrenia, bipolar disorder, and autism). Efforts to incorporate computational linguistics approaches into the mental health system have primarily focused on mining electronic medical records (Doshi-Velez, Ge, & Kohane, 2014; Lingren et al., 2016). While valuable, these efforts are often limited to analyzing text generated by doctors or other programs (Tran et al., 2014), rather than directly assessing specific psychiatric issues in patients themselves. This paper discusses ways in which analyzing spoken language in psychiatric contexts can move the needle on two persistent challenges: reproducibility in human social sciences (Section 2), and inequities in mental health care (Section 3).

2 Reproducibility

In 2015, an article appeared in the journal *Science,* which suggested that the majority of published experiments in psychology are not reproducible (Open Science Collaboration, 2015). Out of 100 experiments, only 39 replicated in a new sample, despite careful methods and communication with original authors (see (Gilbert, King, Pettigrew, & Wilson, 2016) for a comment, and (Anderson et al., 2016) for a response). In this and subsequent analyses, lack of scientific reproducibility has been argued to be due to a number of factors, including p-hacking, selective reporting of results, over-emphasis on innovation and novelty over stability, poor experimental training for scientists, lack of power (small sample sizes), and inadequate measurement (Button et al., 2013; National Science Foundation, 2015). The first part of this short paper focuses on reproducibility challenges that result from traditional methods of psychiatric diagnosis and symptom measurement, and proposes that computational linguistics is a promising tool for improving reliability and enhancing fine-grained characterization efforts.

2.1 Psychiatric Diagnosis

Reproducible methods in the field of clinical psychology and psychiatry require, first and foremost, accurate characterization of the

Proceedings of the Sixth Workshop on Computational Linguistics and Clinical Psychology, pages 94–102
Minneapolis, Minnesota, June 6, 2019. ©2019 Association for Computational Linguistics

condition under study. However, potential error is inherent in how psychiatric diagnoses are traditionally made. Although significant resources have been devoted to identifying biological causes of psychiatric conditions like schizophrenia, and some non-diagnostic brain-based (Ecker, Bookheimer, & Murphy, 2015; McDonald et al., 2005; Zalesky, Fornito, & Bullmore, 2010) and genetic (Geschwind et al., 2001) differences have been identified, the majority of mental health disorders are <u>still diagnosed using behavior alone</u> (American Psychiatric Association, 2013).

Whether or not a person has a psychiatric condition may seem obvious, but a number of factors complicate reliable diagnosis. First, in the absence of biological ground truth (e.g., a blood test or a brain scan), clinicians must grapple with wide behavioral heterogeneity that can cause two people with the same disorder to appear very different from one another. For example, ASD symptoms often manifest differently from one person to the next. Within a single subject, behavioral profiles may vary from week-to-week or even day-to-day. An individual may appear very typical in one context (e.g., familiar, low-stress environments), but their autistic behaviors could become very obvious in others (e.g., novel, high-stress environments). The consequences of this variability are measurable, such that a large, multi-site study of ASD found relatively low diagnostic agreement between expert clinicians at different sites (Catherine Lord, 2012).

Low diagnostic agreement has significant implications for the reliability of human scientific research. For example, in order to test whether ASD *causes* differences in executive function, a study should control every other variable except diagnosis. That is, two groups are assembled: individuals with ASD and neurotypical controls. Groups are matched on important variables like sex ratio, race/ethnicity, chronological age, full-scale IQ, verbal IQ, nonverbal IQ, maternal education (a strong predictor of offspring language ability, which has associations with executive function), etc. An executive function task is administered, and if the groups differ, it may be inferred that the difference is due to ASD. However, if the diagnostic category of ASD is in any way unreliable, another researcher following the exact same procedure with a new sample may not produce the same result due to differences in the ASD group.

Poor diagnostic reliability is a long-standing problem in psychiatric research. Some have suggested that larger sample sizes could reduce the impact of the problem, but the low incidence of ASD [current estimates suggest that approximately 1.5% of the population has ASD (Christensen, 2016)], in combination with long and expensive diagnostic processes, make it challenging to assemble high-powered samples. Recent research suggests that computational linguistics could provide objective diagnostic decision support (through direct measurement) in ways that might speed the process and make it more reliable.

2.2 Objective Measurement for Clinical Characterization

The process of making a mental health diagnosis is often mediated by language; primary diagnostic tools for many psychiatric conditions include structured or semi-structured interviews, wherein a clinical psychologist or psychiatrist asks patients about their thoughts, feelings, and experiences (Kaufman et al., 1997; Lord et al., 1989), comparing patterns of responding to diagnostic symptom checklists or scoring algorithms. After incorporating other relevant information (e.g., family/medical history, current stressors), clinicians use their best judgment to determine diagnostic category. When individuals are nonverbal or minimally verbal, these interviews may be conducted with family members who know the person well (Rutter, LeCouteur, & Lord, 2008). Characteristics of patient speech and language are often noted in the course of clinical evaluations, but they are often only minimally quantified; that is, presence or absence of atypical speech-language characteristics are noted, but highly detailed information is often not systematically gathered. Thus, one valuable application for computational linguistics within clinical psychology and psychiatry is to enhance existing phenotypic characterization methods by adding fine-grained measures of patient speech and language produced during diagnostic evaluations.

In recent years, linguists and computer scientists have begun to analyze clinical evaluations using computational approaches (Black et al., 2011; Kiss, Santen, Prud'Hommeaux, & Black, 2012; Kumar et al., 2016). For example, it has been shown that not only do children with ASD speak differently than neurotypical peers during diagnostic assessments (Parish-Morris et al.,

2016), but characteristics of the interviewer's language predict children's symptom severity as well (Bone, Bishop, Gupta, Lee, & Narayanan, 2016).

Beyond applying computational linguistics approaches to audio recordings of clinical assessments (which remain expensive and complicated to collect, and are not very ecologically valid), researchers have begun to explore whether computational linguistics could be used to characterize psychiatric disorders using everyday language samples (Parish-Morris et al., 2018). Naturalistic samples are challenging to study for a variety of reasons, including the myriad uncontrolled (and perhaps uncontrollable) variables inherent in dynamic human interaction. Consider two people meeting each other for the first time. Each person's behavior is influenced not only by their genetically-linked dispositions, but also a lifetime of experiences, and immediate factors (e.g., did they eat breakfast that day?). When the two people begin to converse, their behavior becomes bi-directionally influential (e.g., each person dynamically reacts to the other in real time, which affects the next moment, and so on). When one or more participants brings extreme psychiatric variation (e.g., active psychosis) to the conversation – the interaction itself changes, and the course of the interaction will likely also fall outside the norm. Despite the challenges associated with measuring two people in an uncontrolled context instead of one person in a controlled context (as in a clinical evaluation), basing future research on naturalistic samples is key; the generalizability gap between research and the real world will shrink as we increase the ecological validity of our research samples.

Importantly, tools from computational linguistics might also be used to directly influence diagnostic decision making in ways that make it more reproducible. Rather than replacing clinicians, the current promise of computational linguistics is to develop objective and granular metrics for use as clinical decision *support* tools. For example, objective linguistic analysis could be used to flag subtle atypical patterns that are not perceptible to the naked ear [e.g., slightly elevated disfluency rates, or reduced lexical diversity; (Parish-Morris et al., 2017, 2018)]. Clinicians provided with this type of evidence could use it, in combination with other information like family history, as part of the diagnostic decision process.

In summary, using computational linguistics to more accurately specify behavioral phenotypes in psychiatry will not only improve our ability to quickly and objectively diagnose patients, but will also improve our efforts to understand the biological underpinnings of these disorders, by helping us identify diagnostic groups that can be carved along objective joints. Improved characterization of psychiatric conditions will allow researchers to assemble experimental groups that are more homogeneous than broad "ASD" vs. "neurotypical" designations. Reducing sample heterogeneity (noise) through improved characterization could increase the likelihood of identifying true signal in scientific studies, thus improving reproducibility. Finally, objective computational linguistics tools that do not require human intervention could be used by clinicians for clinical decision support, ultimately improving diagnostic reliability.

3 Healthcare Inequities

Computational linguistics has the potential improve *human behavioral science* by addressing problems with reproducibility, but it can also improve the state of *mental health care* by reducing inequities related to access and provider biases.

Persistent race-, sex-, and income-related inequities in health outcomes have been extensively documented across a wide variety of domains. These have been attributed, in part, to reduced access in some cases (Ahmed, Lemkau, Nealeigh, & Mann, 2001) and deep-seated provider biases in others (Burgess, van Ryn, Dovidio, & Saha, 2007; Chapman, Kaatz, & Carnes, 2013). This is especially problematic in psychiatry and clinical psychology, given recent estimates suggesting that nearly 1 in 5 people lives with a mental health condition (Hedden et al., 2015). Below, it is argued that some inequities could be addressed using tools developed jointly by computational linguists and clinicians.

3.1 Sources of Inequity: Access

Inter-related barriers to healthcare access include geographic distance, mental health provider shortages, and socio-economic disadvantages (expensive care). High-quality mental health care availability varies widely by region in the United States. Geographically remote individuals – those living far from a population center – currently have limited access to psychiatric screening and services (New American Economy, 2017). Even in population centers, a significant shortage of mental health providers leads to long wait lists for care (National Council for Behavioral Health, 2017). Given this shortage and lower reimbursement rates for mental vs. physical care (Melek, Perlman, & Davenport, 2017), many mental health providers choose not to accept insurance. Thus, if a patient does not have the economic resources to pay privately, they may not be able to receive care in their area, or may need to wait months to begin the intake and assessment process, much less engage in treatment.

3.2 Improving Access

Computational linguistics approaches, particularly when integrated into web- and phone-based telemedicine, could address some of these barriers to access. For example, long wait lists for screening or assessment of ASD could be shortened by the introduction of home- or school-based audio/video algorithms that measure how severely a person is impacted (and thus, help short-handed clinicians triage potential patients). Although this is not a complete fix (it addresses only one part of a larger problem), it could help overburdened clinicians organize their time and effort more efficiently to help those most immediately in need of assessment and services. Similarly, telemedicine approaches to depression monitoring could use vocal features (Yang, Fairbairn, & Cohn, 2013) alone or in combination with facial markers (Williamson, Quatieri, Helfer, Ciccarelli, & Mehta, 2014) to track change over time and signal the need for urgent intervention; moving people to the top of the waitlist. While expensive to initially build, these kinds of algorithms could reduce costs over time, as more people access health services through supportive automation.

3.3 Sources of Inequity: Biases

A growing body of research delineates deep and enduring biases within the medical and mental health treatment communities that negatively impact care for patients from racial/ethnic minority backgrounds, individuals born into poverty, immigrants/refugees/non-Western peoples, people with disabilities, gender minorities, and women (Conner et al., 2010; Fiscella, Franks, Doescher, & Saver, 2002; McCann & Sharek, 2016; Nadeem et

al., 2007; Ojeda & Bergstresser, 2008; Puhl & Brownell, 2001; Sentell, Shumway, & Snowden, 2007; Winter et al., 2016). One potential source of bias is baked into mental health assessment tools: often, the tools used to assess, intervene, and monitor treatment response were not developed on the populations to whom they are currently being applied, and may therefore be inappropriate for entire segments of people. For example, when "depression inventories" were developed in the 1950s and 60s, who was included in the norming sample?

Depression was once thought to be much more common in women than men, and thus "depression" was conceptualized using women as prototypical exemplars. However, research suggests that the stereotypical conceptualization of depression as feelings of extreme sadness, while true for many women, does not hold true for many men. For men, depression may be more likely to manifest as irritability and aggression (Martin, Neighbors, & Griffith, 2013), leading many men to live their lives undiagnosed and untreated.

On the flip side of the coin, autism was originally described in predominantly male samples (Asperger, 1944; Kanner, 1943). Subsequently, most established assessment tools are male-referenced. Unfortunately, failure to understand the female autistic phenotype has led to systematic *under-diagnosis* of girls and women with ASD, who are either missed entirely or misdiagnosed with other disorders instead (Loomes, Hull, & Mandy, 2017). Incorrect or missed diagnoses are a serious concern in ASD, as early intervention has been shown to improve later outcomes (Howlin, Magiati, & Charman, 2009). Although some researchers have developed sex-referenced norms for social characterization (Constantino, 2012), the primary diagnostic tools for ASD still do not acknowledge the ways in which the disorder may manifest differently in girls vs. boys (American Psychiatric Association, 2013; Lord, Risi, & Bishop, 2012; Rutter et al., 2008).

These two examples spark further questions: how might depression and autism look different in cultural subgroups, such as recent immigrants from various parts of the world? Questions about whether historical norming and development samples are truly representative of the diverse set of people now seeking help for mental health issues in the U.S. have significant implications for accurately identifying the needs of a diverse patient population, and for providing effective services.

3.4 Reducing Biases

Language is one of the primary mediums through which behavioral diagnoses like autism, ADHD, depression, and anxiety are made, so it is important to recognize that language is also one of the mediums through which biases operate most efficiently. Accents, grammar, prosody, and word choice are all features that may be associated with unconscious biases (e.g., negative stereotypes could be activated by accents typical of rural populations in the U.S., slang used in inner cities, upspeak/vocal fry, accents of individuals learning English as a second language, etc.).

The challenge that computational linguistics can address, at least in part, is to provide objective metrics for quantifying language in a way that could reduce the effects of these linguistic biases. Much like orchestral auditions that, when conducted behind a curtain, result in significantly more women being hired than when the judge sees the person performing (Goldin & Rouse, 2000), biases that affect clinician judgements could be significantly reduced – or perhaps even eliminated – through the application of more objective measurement approaches developed by computational linguists.

The goal of objective measurement is to circumvent identified problems with bias that affect the likelihood of understudied subgroups getting referred, evaluated, diagnosed, and treated appropriately (e.g., men with depression, girls and women with ASD). However, the promise of comprehensive digital phenotyping (to include audio, video, web- and phone-based methods, and wearables) is not that measurement in the social sciences will suddenly be perfect. Rather, it is hoped that the quest to develop objective metrics for use in mental health research and practice will shed light on biases that operate in assessment and treatment contexts, and will allow those biases to be purposefully counteracted. This effort has significant implications for how we detect and treat mental health conditions in diverse patient populations.

4 Limitations

Like humans, computerized algorithms and "objective" computational approaches for addressing mental health conditions are not without their weaknesses. For example, well-intentioned efforts to use machine learning in support of policing has led to unjust racial profiling; this profiling was largely due to racially-biased training data (Chander, 2017). If training data is biased, the algorithm will be biased too. In the case of ASD, labeled language training data is subject to the problems associated with systematic, long-term under-diagnosis of girls. This begs the question: How can we use computational linguistics or digital phenotyping to support clinician decision-making when available training data is biased against females, or racial/ethnic minorities, or economically disadvantaged individuals? It is critical to grapple with these questions while simultaneously forging ahead to collect new (less biased) data, and develop tools that purposefully counteract these biases while eliminating barriers to access for underserved populations.

5 Conclusion

Objective phenotyping approaches based in computational linguistics will likely prove useful for scientific reasons like reproducibility and measurement granularity. Importantly, these methods also hold promise as tools to improve healthcare access and equity. Groups that have been historically understudied, subject to bias, and otherwise disenfranchised from getting early accurate mental health screening and personalized treatment, with negative impacts on long-term outcomes, stand to benefit from carefully implemented digital phenotyping efforts that identify/correct deeply problematic biases and barriers to equitable research and care.

Acknowledgments

This work was supported by an Autism Science Foundation postdoctoral fellowship to J.P.M., and generous gifts from the Eagles Charitable Foundation and the Allerton Foundation to R.T. Schultz at the Center for Autism Research, CHOP.

References

Ahmed, S. M., Lemkau, J. P., Nealeigh, N., & Mann, B. (2001). Barriers to healthcare access in a non-elderly urban poor American population. *Health & Social Care in the Community, 9*(6), 445–453.

https://doi.org/10.1046/j.1365-2524.2001.00318.x

American Psychiatric Association. (2013). *Diagnostic and Statistical Manual of Mental Disorders, 5th Edition: DSM-5* (5 edition). Washington, D.C: American Psychiatric Publishing.

Anderson, C. J., Bahník, Š., Barnett-Cowan, M., Bosco, F. A., Chandler, J., Chartier, C. R., … Zuni, K. (2016). Response to Comment on "Estimating the reproducibility of psychological science." *Science (New York, N.Y.)*, *351*(6277), 1037. https://doi.org/10.1126/science.aad9163

Asperger, H. (1944). Die "Autistischen Psychopathen" im Kindesalter. *Archiv für Psychiatrie und Nervenkrankheiten*, *117*(1), 76–136. https://doi.org/10.1007/BF01837709

Black, M., Bone, D., Williams, M. E., Gorrindo, P., Levitt, P., & Narayanan, S. S. (2011). The USC CARE Corpus: Child-Psychologist Interactions of Children with Autism Spectrum Disorders. *INTERSPEECH*, 1497–1500. Retrieved from http://www.researchgate.net/profile/Daniel_Bone/publication/221485841_The_USC_CARE_Corpus_Child-Psychologist_Interactions_of_Children_with_Autism_Spectrum_Disorders/links/09e4150c125896017d000000.pdf

Bone, D., Bishop, S., Gupta, R., Lee, S., & Narayanan, S. (2016). Acoustic-prosodic and turn-taking features in interactions with children with neurodevelopmental disorders. *Interspeech 2016*, 1185–1189.

Burgess, D., van Ryn, M., Dovidio, J., & Saha, S. (2007). Reducing Racial Bias Among Health Care Providers: Lessons from Social-Cognitive Psychology. *Journal of General Internal Medicine*, *22*(6), 882–887. https://doi.org/10.1007/s11606-007-0160-1

Button, K. S., Ioannidis, J. P. A., Mokrysz, C., Nosek, B. A., Flint, J., Robinson, E. S. J., & Munafò, M. R. (2013). Power failure: why small sample size undermines the reliability of neuroscience. *Nature Reviews Neuroscience*, *14*(5), 365. https://doi.org/10.1038/nrn3475

Chander, A. (2017). The Racist Algorithm? *Michigan Law Review*, *115*, 24.

Chapman, E. N., Kaatz, A., & Carnes, M. (2013). Physicians and Implicit Bias: How Doctors May Unwittingly Perpetuate Health Care Disparities. *Journal of General Internal Medicine*, *28*(11), 1504–1510. https://doi.org/10.1007/s11606-013-2441-1

Christensen, D. L. (2016). Prevalence and Characteristics of Autism Spectrum Disorder Among Children Aged 8 Years—Autism and Developmental Disabilities Monitoring Network, 11 Sites, United States, 2012. *MMWR. Surveillance Summaries*, *65*. Retrieved from http://www.cdc.gov/mmwr/volumes/65/ss/ss6503a1.htm

Conner, K. O., Copeland, V. C., Grote, N. K., Koeske, G., Rosen, D., Reynolds, C. F., & Brown, C. (2010). Mental Health Treatment Seeking Among Older Adults with Depression: The Impact of Stigma and Race. *The American Journal of Geriatric Psychiatry : Official Journal of the American Association for Geriatric Psychiatry*, *18*(6), 531–543. https://doi.org/10.1097/JGP.0b013e3181cc0366

Constantino, J. N. (2012). *SRS-2 (Social Responsiveness Scale, Second Edition)*. Retrieved from http://www4.parinc.com/Products/Product.aspx?ProductID=SRS-2

Doshi-Velez, F., Ge, Y., & Kohane, I. (2014). Comorbidity Clusters in Autism Spectrum Disorders: An Electronic Health Record Time-Series Analysis. *Pediatrics*, *133*(1), e54–e63. https://doi.org/10.1542/peds.2013-0819

Ecker, C., Bookheimer, S. Y., & Murphy, D. G. M. (2015). Neuroimaging in autism spectrum disorder: brain structure and function across the lifespan. *The Lancet Neurology*, *14*(11), 1121–1134. https://doi.org/10.1016/S1474-4422(15)00050-2

Fiscella, K., Franks, P., Doescher, M. P., & Saver, B. G. (2002). Disparities in Health Care by Race, Ethnicity, and Language among the Insured: Findings from a National Sample. *Medical Care*, *40*(1), 52–59. Retrieved from JSTOR.

Geschwind, D. H., Sowinski, J., Lord, C., Iversen, P., Shestack, J., Jones, P., … Spence, S.

J. (2001). The Autism Genetic Resource Exchange: A Resource for the Study of Autism and Related Neuropsychiatric Conditions. *The American Journal of Human Genetics, 69*(2), 463–466. https://doi.org/10.1086/321292

Gilbert, D. T., King, G., Pettigrew, S., & Wilson, T. D. (2016). Comment on "Estimating the reproducibility of psychological science." *Science (New York, N.Y.), 351*(6277), 1037. https://doi.org/10.1126/science.aad7243

Goldin, C., & Rouse, C. (2000). Orchestrating Impartiality: The Impact of "Blind" Auditions on Female Musicians. *American Economic Review, 90*(4), 715–741. https://doi.org/10.1257/aer.90.4.715

Hedden, S. L., Kennet, J., Lipari, R., Medley, G., Tice, P., Copello, E. A. P., & Kroutil, L. A. (2015). *Key Substance Use and Mental Health Indicators in the United States: Results from the 2015 National Survey on Drug Use and Health.* 74.

Howlin, P., Magiati, I., & Charman, T. (2009). Systematic Review of Early Intensive Behavioral Interventions for Children With Autism. *American Journal on Intellectual and Developmental Disabilities, 114*(1), 23–41. https://doi.org/10.1352/2009.114:23-41

Kanner, L. (1943). Autistic disturbances of affective contact. *Nervous Child, 2*(3), 217–250.

Kaufman, J., Birmaher, B., Brent, D., Rao, U., Flynn, C., Moreci, P., … Ryan, N. (1997). Schedule for Affective Disorders and Schizophrenia for School-Age Children-Present and Lifetime Version (K-SADS-PL): Initial Reliability and Validity Data. *Journal of the American Academy of Child & Adolescent Psychiatry, 36*(7), 980–988. https://doi.org/10.1097/00004583-199707000-00021

Kiss, G., Santen, J. P. van, Prud'Hommeaux, E., & Black, L. M. (2012). Quantitative analysis of pitch in speech of children with neurodevelopmental disorders. *Thirteenth Annual Conference of the International Speech Communication Association.* Retrieved from http://people.rit.edu/emilypx/papers/Inte rspeech12-GK.pdf

Kumar, M., Gupta, R., Bone, D., Malandrakis, N., Bishop, S., & Narayanan, S. S. (2016, September 8). *Objective Language Feature Analysis in Children with Neurodevelopmental Disorders During Autism Assessment.* 2721–2725. https://doi.org/10.21437/Interspeech.201 6-563

Lingren, T., Chen, P., Bochenek, J., Doshi-Velez, F., Manning-Courtney, P., Bickel, J., … Savova, G. (2016). Electronic Health Record Based Algorithm to Identify Patients with Autism Spectrum Disorder. *PLOS ONE, 11*(7), e0159621. https://doi.org/10.1371/journal.pone.015 9621

Loomes, R., Hull, L., & Mandy, W. P. L. (2017). What Is the Male-to-Female Ratio in Autism Spectrum Disorder? A Systematic Review and Meta-Analysis. *Journal of the American Academy of Child & Adolescent Psychiatry, 56*(6), 466–474. https://doi.org/10.1016/j.jaac.2017.03.01 3

Lord, C., Risi, S., & Bishop, S. L. (2012). *Autism diagnostic observation schedule, second edition (ADOS-2).* Torrance, CA: Western Psychological Services.

Lord, C., Rutter, M., Goode, S., Heemsbergen, J., Jordan, H., Mawhood, L., & Schopler, E. (1989). Autism diagnostic observation schedule: a standardized observation of communicative and social behavior. *Journal of Autism and Developmental Disorders, 19*(2), 185–212.

Lord, Catherine. (2012). A Multisite Study of the Clinical Diagnosis of Different Autism Spectrum Disorders. *Archives of General Psychiatry, 69*(3), 306. https://doi.org/10.1001/archgenpsychiatr y.2011.148

Martin, L. A., Neighbors, H. W., & Griffith, D. M. (2013). The Experience of Symptoms of Depression in Men vs Women: Analysis of the National Comorbidity Survey Replication. *JAMA Psychiatry, 70*(10), 1100–1106. https://doi.org/10.1001/jamapsychiatry.2 013.1985

McCann, E., & Sharek, D. (2016). Mental Health Needs of People Who Identify as Transgender: A Review of the Literature. *Archives of Psychiatric Nursing, 30*(2), 280–285. https://doi.org/10.1016/j.apnu.2015.07.0 03

McDonald, C., Bullmore, E., Sham, P., Chitnis, X., Suckling, J., Maccabe, J., … Murray, R. M. (2005). Regional volume deviations of brain structure in schizophrenia and psychotic bipolar disorder. *British Journal of Psychiatry*, *186*(05), 369–377. https://doi.org/10.1192/bjp.186.5.369

Melek, S. P., Perlman, D., & Davenport, S. (2017). *Addiction and mental health vs. physical health: Analyzing disparities in network use and provider reimbursement rates* (pp. 1–56) [Milliman Research Report].

Nadeem, E., Lange, J. M., Edge, D., Fongwa, M., Belin, T., & Miranda, J. (2007). Does Stigma Keep Poor Young Immigrant and U.S.-Born Black and Latina Women From Seeking Mental Health Care? *Psychiatric Services*, *58*(12), 1547–1554. https://doi.org/10.1176/ps.2007.58.12.1547

National Council for Behavioral Health. (2017). *The Psychiatric Shortage: Causes and Solutions.* Retrieved from https://www.thenationalcouncil.org/wp-content/uploads/2017/03/Psychiatric-Shortage_National-Council-.pdf

National Science Foundation. (2015). *Social, Behavioral, and Economic Sciences Perspectives on Robust and Reliable Science.* Retrieved from https://www.nsf.gov/sbe/AC_Materials/SBE_Robust_and_Reliable_Research_Report.pdf

New American Economy. (2017). *The Silent Shortage: How Immigration Can Help Address the Large and Growing Psychiatrist Shortage in the United States* (pp. 1–31) [Health]. Retrieved from http://www.newamericaneconomy.org/wp-content/uploads/2017/10/NAE_PsychiatristShortage_V6-1.pdf

Ojeda, V. D., & Bergstresser, S. M. (2008). Gender, Race-Ethnicity, and Psychosocial Barriers to Mental Health Care: An Examination of Perceptions and Attitudes among Adults Reporting Unmet Need. *Journal of Health and Social Behavior*, *49*(3), 317–334. Retrieved from JSTOR.

Open Science Collaboration. (2015). Estimating the reproducibility of psychological science. *Science*, *349*(6251), aac4716. https://doi.org/10.1126/science.aac4716

Parish-Morris, J., Liberman, M., Ryant, N., Cieri, C., Bateman, L., Ferguson, E., & Schultz, R. T. (2016). Exploring Autism Spectrum Disorders Using HLT. *Proceedings of the 3rd Workshop on Computational Linguistics and Clinical Psychology: From Linguistic Signal to Clinical Reality*, *3*, 74–84. Retrieved from http://languagelog.ldc.upenn.edu/myl/CLPsych2016_FINAL1.pdf

Parish-Morris, J., Liberman, M. Y., Cieri, C., Herrington, J. D., Yerys, B. E., Bateman, L., … Schultz, R. T. (2017). Linguistic camouflage in girls with autism spectrum disorder. *Molecular Autism*, *8*(1). https://doi.org/10.1186/s13229-017-0164-6

Parish-Morris, J., Sariyanidi, E., Zampella, C., Bartley, G. K., Ferguson, E., Pallathra, A. A., … Tunc, B. (2018). Oral-Motor and Lexical Diversity During Naturalistic Conversations in Adults with Autism Spectrum Disorder. *Proceedings of the Fifth Workshop on Computational Linguistics and Clinical Psychology: From Keyboard to Clinic*, 147–157. https://doi.org/10.18653/v1/W18-0616

Puhl, R., & Brownell, K. D. (2001). Bias, Discrimination, and Obesity. *Obesity Research*, *9*(12), 788–805. https://doi.org/10.1038/oby.2001.108

Rutter, M., LeCouteur, A., & Lord, C. (2008). *Autism Diagnostic Interview - Revised (ADI-R).* Los Angeles: Western Psychological Services.

Sentell, T., Shumway, M., & Snowden, L. (2007). Access to Mental Health Treatment by English Language Proficiency and Race/Ethnicity. *Journal of General Internal Medicine*, *22*(S2), 289–293. https://doi.org/10.1007/s11606-007-0345-7

Tran, T., Luo, W., Phung, D., Harvey, R., Berk, M., Kennedy, R. L., & Venkatesh, S. (2014). Risk stratification using data from electronic medical records better predicts suicide risks than clinician assessments. *BMC Psychiatry*, *14*(1), 76. https://doi.org/10.1186/1471-244X-14-76

Williamson, J. R., Quatieri, T. F., Helfer, B. S., Ciccarelli, G., & Mehta, D. D. (2014). Vocal and Facial Biomarkers of

Depression based on Motor Incoordination and Timing. *Proceedings of the 4th International Workshop on Audio/Visual Emotion Challenge - AVEC '14*, 65–72. https://doi.org/10.1145/2661806.266180 9

Winter, S., Diamond, M., Green, J., Karasic, D., Reed, T., Whittle, S., & Wylie, K. (2016). Transgender people: health at the margins of society. *The Lancet, 388*(10042), 390–400. https://doi.org/10.1016/S0140-6736(16)00683-8

Yang, Y., Fairbairn, C., & Cohn, J. F. (2013). Detecting Depression Severity from Vocal Prosody. *IEEE Transactions on Affective Computing, 4*(2), 142–150. https://doi.org/10.1109/T-AFFC.2012.38

Zalesky, A., Fornito, A., & Bullmore, E. T. (2010). Network-based statistic: Identifying differences in brain networks. *NeuroImage, 53*(4), 1197–1207. https://doi.org/10.1016/j.neuroimage.201 0.06.041

Temporal Analysis of the Semantic Verbal Fluency Task in Persons with Subjective and Mild Cognitive Impairment

Nicklas Linz[1], Kristina Lundholm Fors[2], Hali Lindsay[1],
Marie Eckerström[2], Jan Alexandersson[1], Dimitrios Kokkinakis[2]
[1]German Research Center for Artificial Intelligence (DFKI), Saarbrücken, Germany
[2]University of Gothenburg, Gothenburg, Sweden
nicklas.linz@dfki.de, kristina.lundholmfors@gu.se, hali.lindsay@dfki.de
marie.eckerstrom@neuro.gu.se, jan.alexandersson@dfki.de, dimitrios.kokkinakis@gu.se

Abstract

The Semantic Verbal Fluency (SVF) task is a classical neuropsychological assessment where persons are asked to produce words belonging to a semantic category (e.g., animals) in a given time. This paper introduces a novel method of temporal analysis for SVF tasks utilizing time intervals and applies it to a corpus of elderly Swedish subjects (mild cognitive impairment, subjective cognitive impairment and healthy controls). A general decline in word count and lexical frequency over the course of the task is revealed, as well as an increase in word transition times. Persons with subjective cognitive impairment had a higher word count during the last intervals, but produced words of the same lexical frequencies. Persons with MCI had a steeper decline in both word count and lexical frequencies during the third interval. Additional correlations with neuropsychological scores suggest these findings are linked to a person's overall vocabulary size and processing speed, respectively. Classification results improved when adding the novel features ($AUC = 0.72$), supporting their diagnostic value.

1 Introduction

Verbal fluency is a widely adapted neuropsychological test. Historically, Schiller (1947) used the "spontaneous naming by free association"-test for the assessment of brain injuries, becoming one of the first recorded instances of what would later be referred to as "category fluency". Category fluency, or semantic verbal fluency (SVF), requires the verbal production of as many different items from a given category, e.g., animals, as possible within a given timeframe. A large body of evidence substantiates the discriminative power of semantic verbal fluency for dementia due to Alzheimers Disease (AD) and its precursor Mild Cognitive Impairment (MCI) (Henry et al., 2004; Auriacombe et al., 2006; Gomez and White, 2006; Raoux et al., 2008; Linz et al., 2017).

As there is currently no cure for AD, preventive medication labeled to delay the onset or worsening of symptoms is the primary course of action, with an emphasis on early intervention being a beneficial factor for effective treatment. Early identification of subtle symptoms is also valuable for drug trial screening programs and supports early behavioral interventions that can delay the onset of the disease (Ashford et al., 2007; Zucchella et al., 2018).

SVF has been used to identify the early stages of dementia through traditional crude measures, such as the total number of unique words produced. This may overlook persons with very subtle cognitive impairment because they lack statistically significant differences from healthy controls. Thus, additional sensitive measures of performance are needed. Further analysis of SVF has often looked at the production as a series of *clusters* and *switches*, where a cluster is a group of semantically similar words (e.g. pets such as 'cat', 'dog' and 'hamster') and a switch is the task of changing semantic focus from one group of animals to another (e.g. switching from enumerating pets to producing animals that live in Africa) (Troyer et al., 1997). Authors have also suggested approaches to clustering and switching that solely rely on temporal information (Troeger et al., 2019).

SVF has been shown to be a valid measure of executive function and verbal ability, specifically vocabulary size and lexical access speed (Shao et al., 2014). It has been sug-

103

Proceedings of the Sixth Workshop on Computational Linguistics and Clinical Psychology, pages 103–113
Minneapolis, Minnesota, June 6, 2019. ©2019 Association for Computational Linguistics

gested that word production in SVF is moderated by different cognitive processes over time, where the initial process is a semi-automatic retrieval of commonly used and readily available words, whereas later stages demand more effortful processing (Demetriou and Holtzer, 2017).

In this paper, we examine SVF results of three groups of Swedish participants; those with Subjective Cognitive Impairment (SCI), with MCI and healthy controls (HC). By analysing the data temporally, we are able to reveal differences that are not evident when looking at the SVF as a whole. This paper is structured in the following way: An overview of related work is given, with a focus on performance on the SVF by persons with MCI and SCI. Then the dataset and methodology are described as well as the features that were extracted. Finally, the results of our analyses and machine learning experiments are presented and discussed in tandem with other relevant neuropsychological metrics.

2 Related work

Performance of SVF tasks in healthy older adults tends to decline with age, and is partially attributed to a decrease in processing speed, rather than a diminished verbal knowledge (Elgamal et al., 2011). In line with this reasoning, Tallberg et al. (2008) found that the performance of Swedish speakers on SVF is negatively correlated with age and positively correlated with years of education. Healthy participants in the age range 65-89 with ≤12 years of education produced a mean of 14.9±6.4 animals, whereas those in the same age range but with an education of >12 years produced 19.4±5.6 animals in the same task.

The deterioration of cognition in MCI, with impairment both in processing speed and switching attention (Ashendorf et al., 2008), results in persons with amnestic MCI (aMCI) producing smaller clusters and fewer switches than healthy controls (Peter et al., 2016). This reduction across strategy generalises to persons with aMCI producing significantly less categorical words (Price et al., 2012; Mueller et al., 2015).

Nikolai et al. (2018) found categorical differences between naming animals and vegetables when comparing participants with SCI and HC on the SVF test. While the animal category revealed no differences, persons with SCI generated significantly fewer vegetables, specifically in the later 30 seconds. Participants with SCI produced smaller clusters and made more switches in the animal category. The groups did not differ significantly on any demographic variables (age, education, gender) or on the Mini-Mental State Examination (MMSE; Folstein et al. (1975)).

Throughout the SVF, word production rate decreases regardless of the presence of cognitive impairment. To further explore the performance of persons with MCI and healthy controls, Demetriou and Holtzer (2017) divided and analyzed the task into three 20-second sections with two substantial findings; both groups declined over time and generated more words in the first time span. However, persons with MCI performing within normal limits produced fewer words in the first time interval. Slow initiation of lexical search process suggests that MCI inhibits early semi-automatic word retrieval processes. This is in line with previous research showing that the last 30 seconds of the verbal fluency task does not differ between participants, whereas the first 30 seconds contain discriminating information (Fernaeus et al., 2008).

When performing an even finer-grained temporal analysis based on ten second intervals, Fernaeus et al. (2008) found that intervals 1 and 2 were useful in distinguishing persons with AD and MCI, and interval 3 made it possible to differentiate between persons with MCI and SCI, and MCI and AD respectively.

3 Methods

3.1 Recruitment and Data Acquisition

All the participants in the current study on "Linguistic and extra-linguistic parameters for early detection of cognitive impairment" were recruited from the Gothenburg MCI study (Wallin et al., 2016). All participants were speakers of Swedish, selected according to detailed inclusion and exclusion criteria (Kokkinakis et al., 2017). Data collection took place in a quiet lab environment where participants were fitted with a lapel microphone (AudioTechnica ATR3350) and digitally recorded with a Zoom H4n Pro recorder (44.1 kHz sampling rate; 16bit resolution). The following instruction was given in Swedish: "Your task is to think of words. I want you to tell me all the different *animals* you can think of. You have 60 seconds. Do you have any questions? Are you ready? Go ahead and start." If the participant seemed unsure, they were told "any animals are okay: big ones, little ones, etc.". At the end of the 60 seconds, a timer would go off and the test leader would let the participant know that 60 seconds had passed. The resulting audio files were manually transcribed and manually time aligned in Praat (Boersma and Weenink, 2018). All animals named were transcribed on a separate tier.

A future follow-up visit at the memory clinic in 2019, after a second round of language tests, will include a renewed GDS (Global Deterioration Scale) classification and neuropsychological tests. The study was approved by local ethical committee (ref. number: 206-16, 2016 and T021-18, 2018).

3.2 Clinical Assessments

Participants in the Gothenburg MCI study were classified as having SCI, MCI, or dementia, and the controls were recruited separately and evaluated to ascertain that they were cognitively healthy. The classification is based on the Global Deterioration Scale (GDS), where level 1 codes for cognitively healthy, level 2 SCI, level 3 MCI and level 4 and above dementia (Auer and Reisberg, 1997; Wallin et al., 2016). Participants were further evaluated with neuropsychological tests, magnetic resonance imaging (MRI), blood samples, and spinal fluid samples (Wallin et al., 2016).

Compared to the other study participants, the persons with SCI were relatively young, had higher levels of education, higher prevalence of stress conditions and depressive symptoms as well as a family history of dementia (Eckerström et al., 2016).

3.3 Features

3.3.1 Traditional measures

From the manual transcripts, traditional SVF performance metrics were automatically extracted. The word count was determined as the number of unique, correctly named

animals. Clusters and switches were determined based on a temporal metric proposed by Troeger et al. (2019). In this approach, the cluster structure is solely determined by the temporal position of words in the recording. Consecutive words are clustered if the transition time between them is shorter than then average transition time over the sample. This threshold is furthermore scaled over the process of the task to account for the decline in production speed. The mean number of clusters and the number of switches between them is extracted.

3.3.2 Temporally resolved measures

To explore different cognitive processes engaged over the course of the one minute task, SVF performance is examined in 10 second steps. Words in the transcript were assigned to a temporal interval based on their onset. Word count is determined for each interval, disregarding repetitions from earlier intervals. Lexical frequency of words were determined using the KORP collection of Swedish corpora (Borin et al., 2012). Transition times between consecutive words were defined as the difference between the end of the current word and the onset of the next. Word frequency and transition times are reported as the average over each interval.

3.4 Statistical analysis

Statistical analysis was performed using R (software version 3.4.0). For group comparisons of traditional measures, linear models with the measure as a function of diagnostic group were examined. Temporally resolved measures were examined with separate linear mixed effects analysis, one for each response variable –word count, lexical frequency and transition time– using the *lme4* (Bates et al., 2014) package. Each time interval is modelled as a single data point and with age and education level, as well as the interaction between the time interval (T) and diagnosis, as fixed effects. The participant identifier was modelled as a random intercept. Spearman correlations between the interval word count and neuropsychological scores were examined. Age and education were chosen as demographic variables. As neuropsychological correlates, the following scores were used: the Trail Making Test

Part A (TMT-A), as an indicator for processing speed; the Boston Naming Test (BNT; Kaplan et al. (1983)), which assess language ability with a spectrum of high to low frequency words as a proxy of vocabulary size; and the Wechsler Adult Intelligence Scale Similarities (WAIS-Similarities), which measures abstract thinking, concept formation and verbal reasoning (Wechsler, 1999).

3.5 Machine Learning

The predictive power of the proposed temporal and semantic features were validated with machine learning experiments for the HC and MCI populations. For each transcribed speech sample, the features described in Section 3.3.1 and 3.3.2 were extracted and label in accordance to their diagnostic category. Logistic Regression (LR) and Support Vector Machine (SVM) models, as implemented by the scikit-learn (Pedregosa et al., 2011) framework, were trained as binary classifiers to separate the groups. First, models were trained with only word count, to establish a baseline, and then, on the complete feature set, utilizing univariate feature selection.

Area under the Receiver-Operator curve (AUC) is reported as the evaluation parameter. Due to the small size of the dataset, we used leave-pair-out cross validation (LPO-CV), which has been shown to produce an unbiased estimate for AUC on small datasets (Airola et al., 2009). We also computed the standard deviation in AUC as described by Roark et al. (2011).

Feature scaling and hyper-parameter optimisation were done on the training set in each fold. Features were scaled using min-max scaling between 0 and 1. For both SVMs and LR, C was optimised between $C \in [10^{-4}, ..., 10^{4}]$ using a grid search. LR models were trained with both L1 and L2 loss; for SVM a linear and an rbf kernel were used.

For the extended feature set, feature selection based on χ^2-tests was applied to the training set in each fold. The number of selected features was scaled between 1 and the maximum of 30.

	HC	SCI	MCI
N	32	19	24
Sex (M/F)	12/20	8/11	11/13
Age (years)	68.1 (7.2)	66.0 (6.7)	70.8 (5.6)
Education (years)	13.2 (3.5)	16.0 (2.3)	13.8 (3.5)
MMSE (max 30)	29.7 (0.5)	29.6 (0.8)	28.5 (1.4)

Table 1: Demographic information; the MMSE (Mini Mental State Exam) is a general screening test of cognitive status and has a maximum score of 30.

4 Results

4.1 Demographic information

Demographic information by diagnostic group is reported in Table 1. The SCI group is slightly younger and has a higher education level than the other two groups. The MMSE, a general index of cognitive status with a maximum score of 30, is lower in the MCI group. With an average MMSE of 28.5, this MCI population is still quite functional in comparison to other MCI populations (mean MMSE score can vary between 23 and 29 in the MCI group) (Lonie et al., 2009). Note that cut-off points for MMSE may vary slightly: for Swedish, a cut-off value between 25 and 27 indicates possible cognitive impairment which should be further evaluated (Palmqvist et al., 2013) while other studies consider an "abnormal" MMSE score to be lower or equal to 25 (Zadikoff et al., 2008).

4.2 Traditional measures

A linear model of word count as a function of diagnosis revealed a significant main effect ($F(2, 72) = 8.57, p < 0.01$). Compared to the control group ($WC = 24.06 \pm 6.37$), the SCI group ($WC = 27.84 \pm 5.6$) had a significantly increased word count ($3.78 \pm 1.8, p < 0.5$); the MCI group ($WC = 20.12 \pm 6.08$) a significantly lowered one ($-3.94 \pm 1.6, p < 0.5$). No significant effects for the size of temporal clusters ($F(2, 72) = 2.59, p = 0.08$) or the number of temporal switches ($F(2, 72) = 1.64, p = 0.2$) as a function of diagnosis are found.

4.3 Temporally resolved measures

Word count, lexical word frequency and transition times by 10 second intervals is visualized in Figure 1 and the results of linear mixed random effects models are presented in Table 2.

Variable	Estimate	t	95% CI	p-Value
$\mathbf{WC_{T_1-T_2}}$	-0.456	-6.196	[-0.529, -0.382]	$< .01$
$\mathbf{WC_{T_1-T_3}}$	-0.698	-7.898	[-0.787, -0.61]	$< .01$
$\mathbf{WC_{T_1-T_4}}$	-0.937	-8.681	[-1.046, -0.83]	$< .01$
$\mathbf{WC_{T_1-T_5}}$	-1.301	-8.675	[-1.452, -1.152]	$< .01$
$\mathbf{WC_{T_1-T_6}}$	-1.290	-8.690	[-1.439, -1.142]	$< .01$
Age	-0.011	-3.294	[-0.014, -0.008]	$< .01$
Education	-0.003	-0.411	[-0.010, 0.004]	.68
SCI	-0.086	-1.128	[-0.164, -0.010]	.26
SCI x T				
$\mathbf{SCI \times WC_{T_1-T_2}}$	0.247	2.161	[0.133, 0.361]	$< .03$
SCI x $WC_{T_1-T_3}$	0.155	1.102	[0.014, 0.296]	.27
SCI x $WC_{T_1-T_4}$	0.180	1.068	[0.012, 0.349]	.29
$\mathbf{SCI \times WC_{T_1-T_5}}$	0.543	2.738	[0.345, 0.742]	$< .01$
$\mathbf{SCI \times WC_{T_1-T_6}}$	0.575	2.959	[0.381, 0.770]	$< .01$
MCI	-0.041	-0.602	[-0.111, 0.028]	.55
MCI x T				
MCI x $WC_{T_1-T_2}$	-0.088	-0.724	[-0.210, 0.034]	.47
$\mathbf{MCI \times WC_{T_1-T_3}}$	-0.383	-2.176	[-0.559, -0.207]	$< .05$
MCI x $WC_{T_1-T_4}$	-0.015	-0.089	[-0.189, 0.158]	.93
MCI x $WC_{T_1-T_5}$	-0.101	-0.396	[-0.354, 0.153]	.69
MCI x $WC_{T_1-T_6}$	-0.299	-1.046	[-0.585, -0.013]	.30

(a) Word Count

Variable	Estimate	t	95% CI	p-Value
$\mathbf{WF_{T_1-T_2}}$	-0.774	-2.558	[-1.077, -0.472]	$< .05$
$\mathbf{WF_{T_1-T_3}}$	-0.696	-2.298	[-0.999, -0.393]	$< .05$
$\mathbf{WF_{T_1-T_4}}$	-1.274	-4.208	[-1.577, -0.971]	$< .01$
$\mathbf{WF_{T_1-T_5}}$	-1.386	-4.578	[-1.689, -1.083]	$< .01$
$\mathbf{WF_{T_1-T_6}}$	-1.514	-5.000	[-1.816, -1.211]	$< .01$
Age	0.023	2.600	[0.014, 0.032]	$< .05$
Education	0.000	0.003	[-0.018, 0.018]	0.99
SCI	0.228	0.642	[-0.127, 0.582]	.52
SCI x T				
SCI x $WF_{T_1-T_2}$	-0.549	-1.108	[-1.045, -0.053]	.27
SCI x $WF_{T_1-T_3}$	-0.763	-1.539	[-1.259, -0.267]	.12
SCI x $WF_{T_1-T_4}$	-0.123	-0.248	[-0.619, 0.373]	.80
SCI x $WF_{T_1-T_5}$	-0.138	-0.279	[-0.634, 0.358]	.78
SCI x $WF_{T_1-T_6}$	-0.575	-1.159	[-1.071, -0.079]	.25
MCI	0.193	0.588	[-0.135, 0.521]	.56
MCI x T				
MCI x $WF_{T_1-T_2}$	-0.261	-0.564	[-0.723, 0.202]	.57
$\mathbf{MCI \times WF_{T_1-T_3}}$	-0.936	-2.025	[-1.399, -0.474]	$< .05$
MCI x $WF_{T_1-T_4}$	-0.356	-0.769	[-0.818, 0.107]	.44
MCI x $WF_{T_1-T_5}$	-0.256	-0.554	[-0.719, 0.206]	.58
MCI x $WF_{T_1-T_6}$	-0.282	-0.610	[-0.745, 0.180]	.54

(b) Word frequency

Variable	Estimate	t	95% CI	p-Value
$L_{T_1-T_2}$	0.986	1.460	[0.311, 1.662]	.15
$L_{T_1-T_3}$	2.557	3.786	[1.882, 3.233]	< .01
$L_{T_1-T_4}$	2.641	3.911	[1.966, 3.317]	< .01
$L_{T_1-T_5}$	5.245	7.766	[4.570, 5.921]	< .01
$L_{T_1-T_6}$	5.641	8.352	[4.965, 6.316]	< .01
Age	0.028	1.029	[0.001, 0.055]	.31
Education	-0.074	-1.355	[-0.129, -0.019]	.18
SCI	0.311	0.365	[-0.541, 1.163]	.72
SCI x T				
SCI x $L_{T_1-T_2}$	-0.703	-0.635	[-1.81, 0.404]	.53
SCI x $L_{T_1-T_3}$	-1.429	-1.291	[-2.536, -0.322]	.20
SCI x $L_{T_1-T_4}$	-0.803	-0.726	[-1.910, 0.303]	.47
SCI x $L_{T_1-T_5}$	-2.528	-2.284	[-3.634, -1.421]	< .05
SCI x $L_{T_1-T_6}$	-2.384	-2.154	[-3.490, -1.277]	< .05
MCI	0.22	0.281	[-0.564, 1.004]	.78
MCI x T				
MCI x $L_{T_1-T_2}$	0.167	0.162	[-0.865, 1.198]	.87
MCI x $L_{T_1-T_3}$	0.510	0.494	[-0.522, 1.542]	.62
MCI x $L_{T_1-T_4}$	0.724	0.702	[-0.308, 1.756]	.48
MCI x $L_{T_1-T_5}$	-1.212	-1.175	[-2.244, -0.18]	.24
MCI x $L_{T_1-T_6}$	0.41	0.397	[-0.622, 1.441]	.69

(c) Transition Length

Table 2: Linear Mixed Random Effects model examining the effects of time interval, diagnosis, age and education on one of three variables, while controlling random effects per subject. Significant values ($p < .05$) are indicated in bold.

A general decline in the word count for each time interval is visible and reflected in the model, regardless of diagnostic group. A significant effect for age is present, implicating that higher age leads to a reduced word count. For the SCI group, there is a significant interaction between the diagnostic group and the decline in WC_{T2}, WC_{T5} and WC_{T6}. In these intervals, the decline of the SCI group is less severe. The MCI diagnostic group shows a significant interaction with the decline in WC_{T3}, with a stronger decline in word count than the other groups.

For lexical word frequency, again, a significant decline over time is visible, regardless of diagnostic group, which means that participants produce more common words at the start of the task, and less common words towards the end. Older participants produce words that are significantly more frequent. The MCI group has a significant interaction with WF_{T3}, indicating this group uses lower frequency words in this time interval.

Starting from the third interval, a significant increase in word transition times is visible. A significant interaction between the SCI group and the fifth and sixth interval, indicates the SCI group shows significantly lower transition times in these intervals.

4.4 Correlation analysis

Spearman correlations between the word count by time interval, neuropsychological scores and demographic information is displayed in Figure 2. Only significant correlations are displayed.

Significant positive correlations between the BNT score and the word count in the last three time intervals are observed. The WAIS Similarity score shows positive correlations with the word count of the last two intervals. Negative correlations are observed between TMT A and the second and third interval, as well as between age and these two intervals (for the TMT A a lower score indicates a better per-

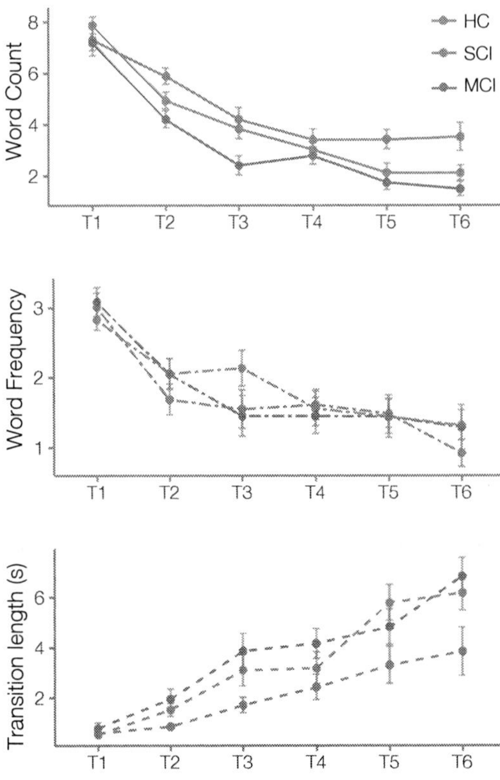

Figure 1: Word Count, Word Frequency and Transition length by time interval and for each group separately. Error bars display standard error.

formance).

4.5 Machine Learning

Figure 3 displays the results of the machine learning experiments. AUC is plotted, while varying the number of features chosen in feature selection, using different classifiers.

The baseline performances of models using just the word count is $AUC = 0.64$ for LR, both with $L1$ and $L2$ loss, and the linear SVM. The SVM with an rbf kernel only achieves $AUC = 0.62$ with the word count feature. Generally, the models using all features outperform the baseline. The best performance of $AUC = 0.72$ is observed for a linear SVM with 20 features. Generally, the linear and rbf SVM and the LR with $L1$ loss show similar performance patterns, across all number of features. The LR with $L2$ shows steadily increasing performance. The SVM with rbf kernel outperforms the other models with a lower number of features.

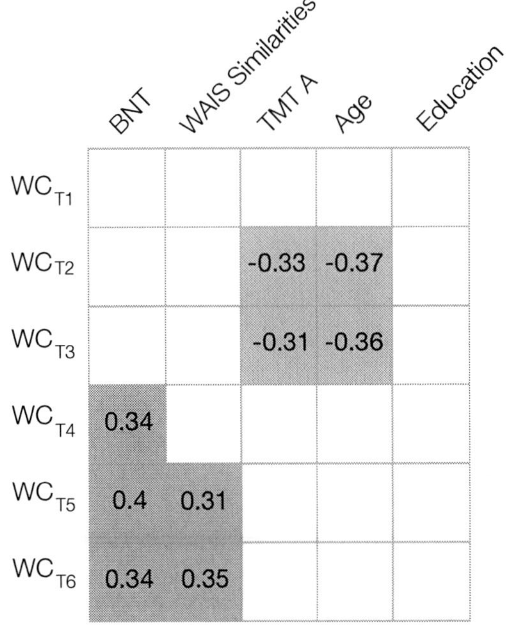

Figure 2: Spearman correlation between 10 second word count (WC) intervals and neuropsychological test scores. Only significant correlations are shown. Positive correlations in blue, negative ones in red.

5 Discussion

Reviewing the overall performance on the SVF, a significant difference in word count was found between the groups, but no differences in cluster size or number of temporal clusters. The temporally resolved measures showed that the MCI, SCI and HC group follow similar trends with regard to word count, word frequency and transition length: word count and word frequency generally decrease over time, while average transition times increase. Significant differences between the MCI group and the other two groups were found mainly for the third interval, where the participants in the MCI group produce fewer and less frequent words. For the word count, this is in line with previous findings from Fernaeus et al. (2008), and the lower word frequency in the third interval indicates that persons with MCI have to resort to low frequency words earlier in the task, switching from semi-automatic retrieval of more common words to effortful retrieval at an earlier point than the other groups.

The persons with SCI showed an increased

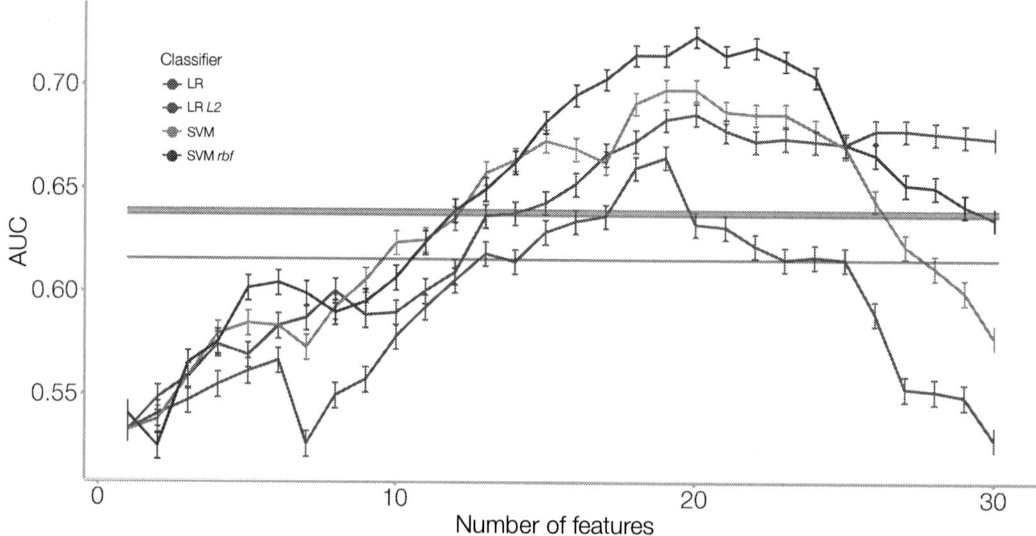

Figure 3: Area under the curve (AUC) of different classification models separating HC and MCI, plotted against number of features selected through univariate feature selection. Horizontal lines show the performance of models solely trained on the word count. Error bars indicate standard deviation of performance.

word count in the second, fifth and sixth interval, and reduced transition times in the fifth and the sixth interval. This suggests that they were able to sustain a continuous production for longer. The words they produced in the last intervals did not differ in frequency from the other groups, but the persons with SCI seemed to have access a larger store of words. Participants in the SCI group had a longer education than the general population, and one possibility is that the participants with SCI in the Gothenburg MCI study perform better because of higher premorbid functioning (Eckerström et al., 2016).

Correlation analysis with additional psychometric data lends a deeper understanding of the results, and significant correlations showed that higher BNT and WAIS similarities scores were associated with a higher word count in the latter part of the SVF. This suggests that having a broader vocabulary, as measured by the BNT, predicts a higher word count in the second half of the SVF. When reviewing the word count graph in Figure 1 and comparing the groups, it is evident that the ability of participants with SCI to sustain performance in the later time intervals can be explained by the access to a larger vocabulary as measured by the BNT. Age and TMT-A both show significant negative correlation with the second and third time intervals of the SVF. TMT-A is a measure of processing speed, and it decreases with increasing age. A decrease in processing speed seems to specifically inhibit production in the second and third interval. Demetriou and Holtzer (2017) suggested a semi-automatic retrieval phase at the beginning and a more effortful retrieval at the end of the task. Our findings support the notion of these phases occurring over the course of task, where the first phase is more influenced by processing speed and the later benefits more strongly from a larger vocabulary.

The benefits of temporal analysis were apparent in the increase of the ability to correctly classify participants as HC or MCI, compared to a classification based solely on word count. In the best case, the performance of the SVM with *rbf* kernel improved from $AUC = 0.62$ to $AUC = 0.72$ with temporal analysis. While this study was based on manually transcribed data, previous research shows that this type of analysis can be done fully automatically including ASR, which allows for easy scaling of the task (König et al., 2018).

6 Conclusion

This paper introduced a novel, interval-based temporal analysis method for SVF tasks. The resulting outcome revealed distinct patterns that differentiated the groups: persons with SCI had a higher word count and sustained lexical frequency level during the last intervals, while persons with MCI had a steeper decline in both word count and lexical frequencies during the third interval. Correlations with neuropsychological scores suggested that the superior performance of the SCI group could be attributed to vocabulary size. Classification results improved when adding the novel features ($AUC = 0.72$), supporting their diagnostic value. This increase over the baseline performance underlines the value of using novel methods in addition to clinical standards.

The results of group comparisons and correlations are in line with previous findings about phases of production in SVF. The special role of the third time interval in discriminating MCI patients is also supported by previous research. Future research should strive to validate these findings on larger data sets, for other languages and other semantic categories.

Based on our findings, we suggest that temporal analysis of the SVF may be useful as a screening tool, when assessing persons with self-perceived memory problem, as this type of analysis seems to highlight the subtle differences between the groups. We see it as a strength that instead of adding new tasks, we are using an already clinically validated tool in an innovative and new manner.

7 Acknowledgements

This research has been funded by Riksbankens Jubileumsfond - The Swedish Foundation for Humanities and Social Sciences, through the grant agreement no: NHS 14-1761:1. We would also like to thank the Gothenburg University Centre for Ageing and Health (AgeCap) and Sahlgrenska University Hospital for supporting this study.

References

Antti Airola, Tapio Pahikkala, Willem Waegeman, Bernard De Baets, and Tapio Salakoski. 2009. A comparison of AUC estimators in small-sample studies. In *Machine Learning in Systems Biology*, pages 3–13.

Lee Ashendorf, Angela L Jefferson, Maureen K O'Connor, Christine Chaisson, Robert C Green, and Robert A Stern. 2008. Trail Making Test errors in normal aging, mild cognitive impairment, and dementia. *Archives of Clinical Neuropsychology*, 23:129–137.

J Wesson Ashford, Soo Borson, Ruth O'Hara, Paul Dash, Lori Frank, Philippe Robert, William R Shankle, Mary C Tierney, Henry Brodaty, Frederick A Schmitt, Helena C Kraemer, Herman Buschke, and Howard Fillit. 2007. Should older adults be screened for dementia? It is important to screen for evidence of dementia! *Alzheimer's & dementia : the journal of the Alzheimer's Association*, 3(2):75–80.

Stefanie Auer and Barry Reisberg. 1997. The GDS/FAST staging system. *International Psychogeriatrics*, 9(SUPPL. 1):167–171.

Sophie Auriacombe, Nathalie Lechevallier, Hélène Amieva, Sandrine Harston, Nadine Raoux, and J-F Dartigues. 2006. A Longitudinal Study of Quantitative and Qualitative Features of Category Verbal Fluency in Incident Alzheimer's Disease Subjects: Results from the PAQUID Study. *Dementia and geriatric cognitive disorders*, 21(4):260–266.

Douglas Bates, Martin Mächler, Ben Bolker, and Steve Walker. 2014. Fitting linear mixed-effects models using lme4. *arXiv preprint arXiv:1406.5823*.

Paul Boersma and David Weenink. 2018. Praat: doing phonetics by computer. version 6.0.40. Computer program.

Lars Borin, Markus Forsberg, and Johan Roxendal. 2012. Korp the corpus infrastructure of Språkbanken. In *The 8th international conference on Language Resources and Evaluation (LREC)*, pages 474–478, Istanbul, Turkey.

Eleni Demetriou and Roee Holtzer. 2017. Mild Cognitive Impairments Moderate the Effect of Time on Verbal Fluency Performance. *Journal of the International Neuropsychological Society*, 23:44–55.

Marie Eckerström, Anne Ingeborg Berg, Arto Nordlund, Sindre Rolstad, Simona Sacuiu, and Anders Wallin. 2016. High Prevalence of Stress and Low Prevalence of Alzheimer Disease CSF Biomarkers in a Clinical Sample with Subjective Cognitive Impairment. *Dementia and Geriatric Cognitive Disorders*, 42(1-2):93–105.

Safa A. Elgamal, Eric A. Roy, and Michael T. Sharratt. 2011. Age and Verbal Fluency: The Mediating Effect of Speed of Processing. *Canadian Geriatrics Journal*, 14(3).

Sven Erik Fernaeus, Per Östberg, Åke Hellström, and Lars Olof Wahlund. 2008. Cut the coda: Early fluency intervals predict diagnoses. *Cortex*, 44(2):161–169.

Marshal F Folstein, Susan E Folstein, and Paul R. McHugh. 1975. Mini-mental status. a practical method for grading the cognitive state of patients for the clinician. *Journal of Psychiatric Research*, 12(3):189–198.

Rowena G. Gomez and Desire A. White. 2006. Using verbal fluency to detect very mild dementia of the Alzheimer type. *Archives of Clinical Neuropsychology*, 21(8):771 – 775.

Julie D Henry, John R Crawford, and Louise H Phillips. 2004. Verbal fluency performance in dementia of the alzheimer's type: a meta-analysis. *Neuropsychologia*, 42(9):1212–1222.

Edith Kaplan, Harold Goodglass, Sandra Weintraub, and Osa Segal. 1983. Boston naming test. In *Psychological Corporation*, Philadelphia: Lea & Febiger.

Dimitrios Kokkinakis, Kristina Lundholm Fors, Eva Björkner, and Arto Nordlund. 2017. Data Collection from Persons with Mild Forms of Cognitive Impairment and Healthy Controls - Infrastructure for Classification and Prediction of Dementia. In *Proceedings of the 21st Nordic Conference of Computational Linguistics*, volume 75, pages 172–182. Linköping University Electronic Press.

Alexandra König, Nicklas Linz, Johannes Tröger, Maria Wolters, Jan Alexandersson, and Phillipe Robert. 2018. Fully automatic speech-based analysis of the semantic verbal fluency task. *Dementia and geriatric cognitive disorders*, 45(3-4):198–209.

Nicklas Linz, Johannes Tröger, Jan Alexandersson, and Alexandra Konig. 2017. Using Neural Word Embeddings in the Analysis of the Clinical Semantic Verbal Fluency Task. In *IWCS 2017 - 12th International Conference on Computational Semantics*, pages 1–7, Montpellier, France.

Jane A Lonie, Kevin M Tierney, and Klaus P Ebmeier. 2009. Screening for mild cognitive impairment: A systematic review. *International Journal of Geriatric Psychiatry*, 24(9):902–915.

Kimberly Diggle Mueller, Rebecca L. Koscik, Asenath LaRue, Lindsay R. Clark, Bruce Hermann, Sterling C. Johnson, and Mark A. Sager. 2015. Verbal fluency and early memory decline: Results from the wisconsin registry for alzheimer's prevention. *Archives of Clinical Neuropsychology*, 30(5):448–457.

Tomas Nikolai, Ondrej Bezdicek, Hana Markova, Hana Stepankova, Jiri Michalec, Miloslav Kopecek, Monika Dokoupilova, Jakub Hort, and Martin Vyhnalek. 2018. Semantic verbal fluency impairment is detectable in patients with subjective cognitive decline. *Applied Neuropsychology:Adult*, 25(5):448–457.

Sebastian Palmqvist, B Terzis, C Strobel, and Anders Wallin. 2013. Mmse-sr: Mini mental state examination - swedish revision, version 2.

F. Pedregosa, G. Varoquaux, A. Gramfort, V. Michel, B. Thirion, O. Grisel, M. Blondel, P. Prettenhofer, R. Weiss, V. Dubourg, J. Vanderplas, A. Passos, D. Cournapeau, M. Brucher, M. Perrot, and E. Duchesnay. 2011. Scikit-learn: Machine Learning in Python. *Journal of Machine Learning Research*, 12:2825–2830.

Jessica Peter, Jannis Kaiser, Verena Landerer, Lena Köstering, Christoph P Kaller, Bernhard Heimbach, Michael Hüll, Tobias Bormann, and Stefan Klöppel. 2016. Category and design fluency in mild cognitive impairment: Performance, strategy use, and neural correlates. *Neuropsychologia*, 93:21–29.

Sarah E. Price, Glynda J. Kinsella, Ben Ong, Elsdon Storey, Elizabeth Mullaly, Margaret Phillips, Lanki Pangnadasa-Fox, and Diana Perre. 2012. Semantic verbal fluency strategies in amnestic mild cognitive impairment. *Neuropsychology*, 26(4):490–497.

Nadine Raoux, Hélène Amieva, Mélanie Le Goff, Sophie Auriacombe, Laure Carcaillon, Luc Letenneur, and Jean-François Dartigues. 2008. Clustering and switching processes in semantic verbal fluency in the course of Alzheimer's disease subjects: Results from the PAQUID longitudinal study. *Cortex*, 44(9):1188–1196.

Brian Roark, Margaret Mitchell, John-Paul Hosom, Kristy Hollingshead, and Jeffery Kaye. 2011. Spoken language derived measures for detecting mild cognitive impairment. *IEEE Transactions on Audio, Speech, and Language Processing*, 19(7):2081–2090.

F Schiller. 1947. Aphasia studied in patients with missile wounds. *J Neurol Neurosurg Psychiatry*, 10(4):183–197.

Zeshu Shao, Esther Janse, Karina Visser, and Antje S. Meyer. 2014. What do verbal fluency tasks measure? Predictors of verbal fluency performance in older adults. *Frontiers in Psychology*, 5(JUL):1–10.

Ing Mari Tallberg, E. Ivachova, K. Jones Tinghag, and Per Östberg. 2008. Swedish norms for word fluency tests: FAS, animals and verbs. *Scandinavian Journal of Psychology*, 49(5):479–485.

Johannes Troeger, Nicklas Linz, Alexandra Koenig, Jessica Peter, Philippe Robert, and Jan Alexandersson. 2019. Exploitation vs. ExplorationComputational Temporal and Semantic Analysis Explains Semantic Verbal Fluency Impairment in Alzheimers Disease. *Neuropsychologia*. Submitted.

Angela K Troyer, Morris Moscovitch, and Gordon Winocur. 1997. Clustering and Switching as Two Components of Verbal Fluency: Evidence From Younger and Older Healthy Adults. *Neuropsychology*, 11(1):138–146.

Anders Wallin, Arto Nordlund, Michael Jonsson, Karin Lind, Åke Edman, Mattias Göthlin, Jacob Stålhammar, Marie Eckerström, Silke Kern, Anne Börjesson-Hanson, Mårten Carlsson, Erik Olsson, Henrik Zetterberg, Kaj Blennow, Johan Svensson, Annika Öhrfelt, Maria Bjerke, Sindre Rolstad, and Carl Eckerström. 2016. The Gothenburg MCI study: Design and distribution of Alzheimer's disease and subcortical vascular disease diagnoses from baseline to 6-year follow-up. *Journal of cerebral blood flow and metabolism: official journal of the International Society of Cerebral Blood Flow and Metabolism*, 36(1):114–31.

D Wechsler. 1999. Wechsler abbreviated intelligence scale. In *Psychological Corporation*, San Antonio, TX, USA.

Cindy Zadikoff, Susan H. Fox, David F. TangWai, Teri Thomsen, Rob M.A. de Bie, Pettarusup Wadia, Janis Miyasaki, Sarah DuffCanning, Anthony E. Lang, and Connie Marras. 2008. A comparison of the mini mental state exam to the montreal cognitive assessment in identifying cognitive deficits in parkinsons disease. *Movement disorders*, 23(2):297–299.

Chiara Zucchella, Elena Sinforiani, Stefano Tamburin, Angela Federico, Elisa Mantovani, Sara Bernini, Casale Roberto, and Michelangelo Bartolo. 2018. The multidisciplinary approach to alzheimer's disease and dementia. a narrative review of non-pharmacological treatment. *Front. Neurol.*, 9(1058).

Mental Health Surveillance over Social Media
with Digital Cohorts

Silvio Amir, Mark Dredze and **John W. Ayers** [†]

Center for Language & Speech Processing, Johns Hopkins University, Baltimore, MD
[†]Division of Infectious Diseases & Global Public Health, University of California, La Jolla, CA
`samir@jhu.edu, mdredze@cs.jhu.edu, ayers.john.w@gmail.com`

Abstract

The ability to track mental health conditions via social media opened the doors for large-scale, automated, mental health surveillance. However, inferring accurate population-level trends requires representative samples of the underlying population, which can be challenging given the biases inherent in social media data. While previous work has adjusted samples based on demographic estimates, the populations were selected based on specific outcomes, e.g. specific mental health conditions. We depart from these methods, by conducting analyses over demographically representative *digital cohorts* of social media users. To validated this approach, we constructed a cohort of US based Twitter users to measure the prevalence of depression and PTSD, and investigate how these illnesses manifest across demographic subpopulations. The analysis demonstrates that cohort-based studies can help control for sampling biases, contextualize outcomes, and provide deeper insights into the data.

1   Introduction

The ability of social media analysis to support computational epidemiology and improve public health practices is well established (Culotta, 2010; Paul and Dredze, 2011; Salathe et al., 2012; Paul and Dredze, 2017). The field has seen particular success around the diagnosis, quantification and tracking of mental illnesses (Hao et al., 2013; Schwartz et al., 2014; Coppersmith et al., 2014a, 2015a,c; Amir et al., 2017). These methods have utilized social media (Coppersmith et al., 2014b; Kumar et al., 2015; De Choudhury et al., 2016), as well as other online data sources (Ayers et al., 2017, 2013, 2012; Arora et al., 2016), to obtain population level estimates and trends around mental health topics.

Accurately estimating population-level trends requires obtaining representative samples of the general population. However, social media has many well know biases, e.g. young adults tend be over-represented (demographic bias). Yet, most social media analyses tend ignore these issues, either by assuming that all the data is equally relevant, or by selecting data for specific outcome. For example, studying depression from users who talk about depression instead of first selecting a population and then measuring outcomes. Outcome based data selection can also introduce biases, such as over-representing individuals vocal about the topic of interest (*self-selection* bias). Consequently, trends or insights gleaned from these analyses might not be generalizable to the broader population.

Fortunately, these problems are well understood in traditional health studies, and well-established techniques from polling and survey-based research are routinely used to correct for these biases. For example, medical studies frequently utilize a cohort based approach in which a group is pre-selected to study disease causes or to identify connections between risk factors and health outcomes (Prentice, 1986). We can replicate these universally accepted approaches by conducting analyses over *digital cohorts* of social media users, characterized with respect to key demographic attributes. In this work, we propose to use such a social media based cohort for the purposes of mental health surveillance. We developed a digital cohort by sampling a large number of Twitter users at random (not based on outcomes), and then using demographic inference techniques to infer key demographics for the users namely, the age, gender, location and race/ethnicity. Then, we used the cohort to measure relative rates of both depression and PTSD, using supervised classifiers for each mental health condition. The inferred de-

114

Proceedings of the Sixth Workshop on Computational Linguistics and Clinical Psychology, pages 114–120
Minneapolis, Minnesota, June 6, 2019. ©2019 Association for Computational Linguistics

mographic information allowed us to observe clear differences in how these illnesses manifest in the population. Moreover, the analysis demonstrates how social media based cohort studies can help to control for sampling biases and contextualize the outcomes.

2 Methodology

We now briefly describe our approach for cohort-based studies over social media. A more detailed description of the proposed methodology will appear in a forthcoming publication. Most works on social media analysis estimate trends by aggregating document-level signals inferred from arbitrary (and biased) data samples selected to match a predefined outcome. While some recent work has begun incorporating demographic information to contextualize analyses (Mandel et al., 2012; Mitchell et al., 2013; Huang et al., 2017, 2019) and to improve representativeness of the data (Coppersmith et al., 2015b; Dos Reis and Culotta, 2015), these studies still select on specific outcomes.

We depart from these works by constructing a demographically representative digital cohort of social media users *prior* to the analyses, and then conducting cohort-based studies over this pre-selected population. While a significant undertaking in most medical studies, the vast quantities of available social media data make assembling social media cohorts feasible. Such cohorts can be used to support longitudinal and cross-sectional studies, allowing experts to contextualize the outcomes, produce externally valid trends from inherently biased samples and extrapolate those trends to a broader population. Similar strategies have been utilized in online surveys, which can have comparable validity to other survey modalities simply by controlling for basic demographic features such as the location, age, ethnicity and gender (Duffy et al., 2005).

2.1 Building Digital Cohorts

Our cohort construction process entails two key steps: first, randomly selecting a large sample of Twitter users; and second, annotating those users with key demographic attributes. While such attributes are not provided by the API, automated methods can be used to infer such traits from data (Cesare et al., 2017). Following this approach, we develop a demographic inference pipeline to automatically infer **age**, **gender**, **race/ethnicity** and **location** for each cohort candidate.

Age Identifying age based on the content of a user can be challenging, and exact age often cannot be determined based on language use alone. Therefore, we use discrete categories that provide a more accurate estimate of age: *Teenager* (below 19), *20s*, *30s*, *40s*, *50s* (50 years or older).

Gender The gender was inferred using *Demographer*, a supervised model that predicts the (binary) gender of Twitter users with features based on the *name* field on the user profile (Knowles et al., 2016).

Race/Ethnicity The standard formulation of race and ethnicity is not well understood by the general public, so categorizing social media users along these two axes may not be reasonable. Therefore, we use a single measure of multi-cultural expression that includes five categories: *White* (W), *Asian* (A), *Black* (B), *Hispanic* (H), and *Other*.

Location The location was inferred using *Carmen*, an open-source library for geolocating tweets that uses a series of rules to lookup location strings in a location knowledge-base (Dredze et al., 2013). We use the inferred location to select users that live in the United States.

The age and race/ethnicity attributes were inferred with custom supervised classifiers based on Amir et al. (2017)'s user-level model. The classifiers were trained and evaluated on a dataset of 5K annotated users, attaining performances of 0.28 and 0.41 Average F_1, respectively. See the supplemental notes for additional details on these experiments[1].

2.2 Mental Health Classifiers

We build on prior work on supervised models for mental health inference over social media data. We focus on two mental health conditions — depression and PTSD — and develop classifiers with the *self-reported* datasets created for CLPysch 2015 (Mitchell et al., 2015; Coppersmith et al., 2015b). These labeled datasets derive from users that have publicly disclosed on Twitter a diagnosis of depression (327 users) or PTSD (246 users), with an equal number of randomly selected demographically-matched (with respect to age and gender) users as *controls*. For each user, the asso-

---

[1] `https://samiroid.github.io/assets/demos.pdf`

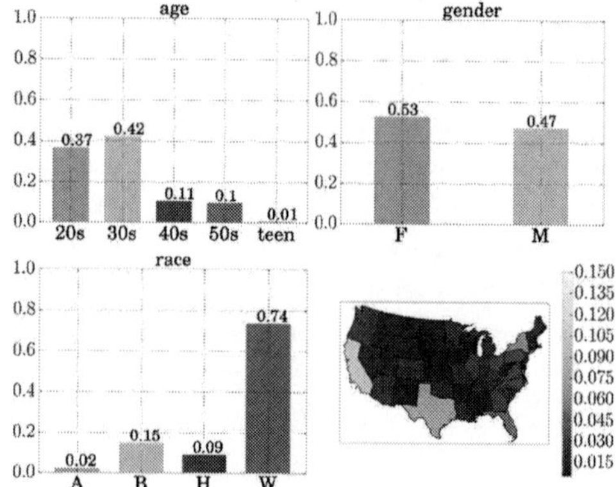

Figure 1: Demographics of the digital cohort.

ciated metadata and posting history was also collected — up to the 3000 most recent *tweets*, per limitations of the Twitter API.

The participants of the task proposed a host of methods ranging from rule-based systems to various supervised models (Pedersen, 2015; Preotiuc-Pietro et al., 2015; Coppersmith et al., 2015b). More recently, the neural user-level classifier proposed by Amir et al. (2017) showed not only good performance on this task, but also the ability to capture implicit similarities between users affected by the same diseases, thus opening the door to more interpretable analyses[2]. Hence, we adopt their model for this analysis.

3 Analysis

We constructed a cohort for our analysis by randomly selecting a sample of Twitter users and processing it with the aforementioned demographic inference pipeline. After discarding accounts from users located outside the United States, we obtained a cohort of 48K Twitter users with the demographic composition shown in Figure 1. Some demographic groups are over-represented (e.g. young adults) while others are grossly under-represented (e.g. teenagers) which illustrates the need for methodologies that can take these disparities into account.

We then processed the cohort through the mental-health classifiers to estimate the prevalence of depression and PTSD, and examine how these illnesses manifest across the population. The analysis revealed that 30.2% of the cohort members are likely to suffer from depression, 30.8% from

PTSD, and 20% from both. We observe a significant overlap between people affected by depression **and** PTSD, which is not surprising given that the comorbidity of these disorders is well-known, with approximately half of people with PTSD also having a diagnosis of major depressive disorder (Flory and Yehuda, 2015).

How do these conditions affect different parts of the population? To answer this question, we looked at the affected users and measured how the demographics of individual sub-populations differ from those of the cohort as a whole. Figures 2 and 3 show the estimates for depression, PTSD and both, controlled for the cohort demographics. We observe large generational differences — PTSD seems to be more prevalent among older people whereas depression affects predominantly younger people. We also observe that in all cases Women are more susceptible than Men, and Blacks and Hispanics are more likely to be affected than Whites. This may represent a bias in the underlying data used to construct the classifiers, or a difference in how social media is used by different demographic groups. For example, models that were trained with a majority of data from White users maybe oversensitive to specific dialects used by other communities.

3.1 Discussion

Comparing our estimates with the current statistics provided by the NIH — a prevalence of 6.7% for depression[3] and 3.6% for PTSD[4] —, we can see that ours are much higher. It should be noted however, that the NIH reports refers to Major Depression episodes whereas our classifiers maybe also be sensitive to mild depressions which may never be diagnosed as such. Moreover, these estimates are not directly comparable since the NIH statistics are outdated (the estimates are from 2003 and 2015 for PTSD and depression, respectively) and our cohort was not adjusted to match the demographics of the US population. Nevertheless, it is worth noting that the relative prevalence rates, per demographic group, we obtained correlate with the NIH reports. For example, we observe similar distributions in terms of age and gen-

---

[2] a similar finding to Benton et al. (2017)

[3] https://www.nimh.nih.gov/health/statistics/major-depression.shtml

[4] https://www.nimh.nih.gov/health/statistics/post-traumatic-stress-disorder-ptsd.shtml

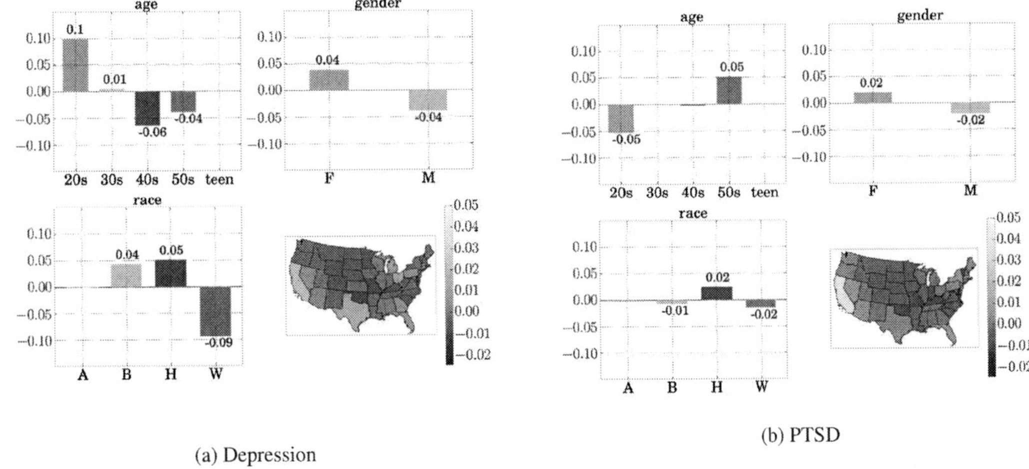

(a) Depression   (b) PTSD

Figure 2: The prevalence of two mental health conditions in the cohort.

der. However, we found that Blacks and Hispanics are more likely to be affected by mental illnesses, whereas the NIH reports a higher prevalence among Whites.

One possible reason for these disparities is that racial minorities are more likely to come from communities with lower education rates and socioeconomic status (SES), and to be in a position where they lack proper health coverage and mental-health care. Reports from the NIH and other US governmental agencies show that 46.3% of Whites suffering from a mental-illness were subjected to some form treatment, but this was case for only 29.8% of Blacks and 27.3% of Hispanics[5]. There may also be a bias in reporting within different racial and ethnic groups, as prevalence estimates can be biased by access to mental health care and social stigma. Recent studies show that factors such as discrimination and perceived inequality have a stronger influence on mental-health than it was previously supposed, even when controlling for the SES (Budhwani et al., 2015). Others have found that acute and chronic discrimination causes racial disparities in health to be even more pronounced at the upper ends of the socioeconomic spectrum. One of the reasons being that for Whites, improvements in SES result in improved health and significantly less exposure to discrimination, whereas for Blacks and Hispanics upwards mobility significantly increases the likelihood of discrimination and unfair treatment,

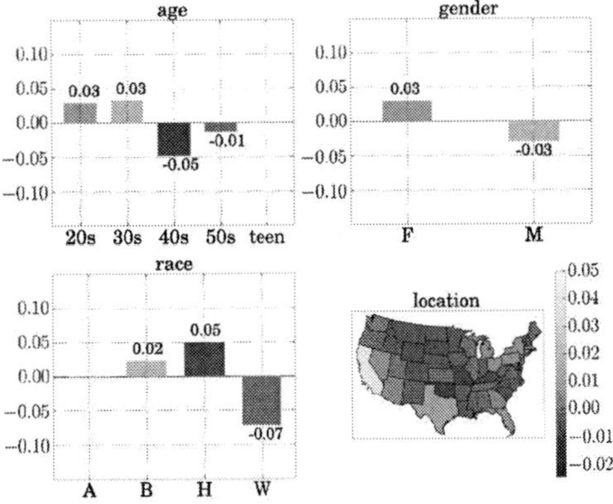

Figure 3: Depression and PTSD

as they move into predominantly White neighborhoods and work environments (Colen et al., 2017).

While an in-depth analysis of this issue is beyond the scope of this work, these results suggest that it deserves further investigation. A follow-up study to investigate the role of discrimination in mental-health could be conducted by adding a model to identify users who reported instances of discrimination and compare the prevalence of mental-illness with a control group.

4 Conclusions

We have presented the first cohort based study of mental health trends on Twitter. Instead of conducting the analysis over arbitrary data samples selected to match a given outcome, we first developed a digital cohort of social media users char-

---

[5] https://www.integration.samhsa.gov/MHServicesUseAmongAdults.pdf

acterized with respect to key demographic traits. We used this cohort to measure relative rates of depression and PTSD, and examine how these illnesses affect different demographic strata. The ability to disaggregate the estimates per demographic group allowed us to observe clear differences in how these illnesses manifest across different parts of the population — something that would not be possible with typical social media analysis methodologies. This brings social media analysis methodologies closer to universally accepted practices in surveillance based research.

Information about how different subpopulations perceive or are affected by certain health issues, could also improve public health policies and inform intervention campaigns targeted for different demographics. Moreover, the fact that some of our estimates correlate with statistics obtained through traditional methodologies suggests that this might be a promising approach to complement current epidemiology practices. Indeed, this opens the door to more responsive and deliberate public health interventions, and allow experts to track the progress or the effects of targeted interventions, in near real-time.

4.1 Privacy and Ethical Considerations

The majority of social media analysis approaches try to extract signals from individual posts and thus do not need to record any personal information. However, as we start moving towards user-level analyses, we are collecting and storing complete records of social media users communications. Even though this information is publicly available, people might not be consciously aware of the implications of sharing all their data and certainly have not given explicit consent for their data to be analyzed in aggregate. This is even more pertinent for analyses involving sensitive information (e.g. health related issues). As it has been demonstrated by the recent incidents involving companies inadvertently sharing or failing to protect users personal data, there is a serious danger of abuse and exploitation for systems that collect and store large amounts of personal data.

Even though this is in large part an ethical question, there are technical solutions that can be used to partially address this issue. One is to use anonymization techniques to obfuscate any details that allow third parties (even analysts) to identify the individuals that are involved in the study. Another is to store only abstract representations — which can still be updated and consumed by predictive models — , and discard the actual content. In regards to consent, there are initiatives to support voluntary data donation for research purposes, e.g. the *Our Data Helps* program[6].

References

Silvio Amir, Glen Coppersmith, Paula Carvalho, Mario J. Silva, and Bryon C. Wallace. 2017. Quantifying mental health from social media with neural user embeddings. In *Proceedings of the 2nd Machine Learning for Healthcare Conference*, volume 68 of *Proceedings of Machine Learning Research*, pages 306–321, Boston, Massachusetts. PMLR.

Vischal S Arora, David Stuckler, and Martin Mckee. 2016. Tracking search engine queries for suicide in the united kingdom, 2004–2013. *Public health*, 137:147–153.

John W Ayers, Benjamin M Althouse, Jon-Patrick Allem, Matthew A Childers, Waleed Zafar, Carl Latkin, Kurt M Ribisl, and John S Brownstein. 2012. Novel surveillance of psychological distress during the great recession. *Journal of affective disorders*, 142(1-3):323–330.

John W Ayers, Benjamin M Althouse, Jon-Patrick Allem, J Niels Rosenquist, and Daniel E Ford. 2013. Seasonality in seeking mental health information on google. *American journal of preventive medicine*, 44(5):520–525.

John W Ayers, Benjamin M Althouse, Eric C Leas, Mark Dredze, and Jon-Patrick Allem. 2017. Internet searches for suicide following the release of 13 reasons why. *JAMA internal medicine*, 177(10):1527–1529.

Adrian Benton, Margaret Mitchell, and Dirk Hovy. 2017. Multitask learning for mental health conditions with limited social media data. In *Proceedings of the 15th Conference of the European Chapter of the Association for Computational Linguistics: Volume 1, Long Papers*, volume 1, pages 152–162.

Henna Budhwani, Kristine Ria Hearld, and Daniel Chavez-Yenter. 2015. Depression in racial and ethnic minorities: the impact of nativity and discrimination. *Journal of racial and ethnic health disparities*, 2(1):34–42.

Nina Cesare, Christan Grant, and Elaine O Nsoesie. 2017. Detection of user demographics on social media: A review of methods and recommendations for best practices. *arXiv preprint arXiv:1702.01807*, 0.

---

[6]https://ourdatahelps.org/

Cynthia G Colen, David M Ramey, Elizabeth C Cooksey, and David R Williams. 2017. Racial disparities in health among nonpoor african americans and hispanics: the role of acute and chronic discrimination. *Social Science & Medicine*, 0.

Glen Coppersmith, Mark Dredze, and Craig Harman. 2014a. Quantifying mental health signals in twitter. In *Proceedings of the Workshop on Computational Linguistics and Clinical Psychology: From Linguistic Signal to Clinical Reality*, pages 51–60. Association for Computational Linguistics.

Glen Coppersmith, Mark Dredze, Craig Harman, and Kristy Hollingshead. 2015a. From ADHD to SAD: Analyzing the language of mental health on twitter through self-reported diagnoses. In *Proceedings of the 2nd Workshop on Computational Linguistics and Clinical Psychology: From Linguistic Signal to Clinical Reality*, pages 1–10. Association for Computational Linguistics.

Glen Coppersmith, Mark Dredze, Craig Harman, Kristy Hollingshead, and Margaret Mitchell. 2015b. Clpsych 2015 shared task: Depression and ptsd on twitter. In *Proceedings of the 2nd Workshop on Computational Linguistics and Clinical Psychology: From Linguistic Signal to Clinical Reality*, pages 31–39. Association for Computational Linguistics.

Glen A Coppersmith, Mark Dredze, Craig Harman, and Kristy Hollingshead. 2015c. From adhd to sad: analyzing the language of mental health on twitter through self-reported diagnoses. In *NAACL Workshop on Computational Linguistics and Clinical Psychology*, pages 1–10.

Glen A Coppersmith, Craig Harman, and Mark Dredze. 2014b. Measuring post traumatic stress disorder in twitter. In *International Conference on Weblogs and Social Media (ICWSM)*, pages 579–582.

Aron Culotta. 2010. Towards detecting influenza epidemics by analyzing Twitter messages. In *Proceedings of the First Workshop on Social Media Analytics*, SOMA '10, pages 115–122, New York, NY, USA. ACM.

Munmun De Choudhury, Emre Kiciman, Mark Dredze, Glen Coppersmith, and Mrinal Kumar. 2016. Discovering shifts to suicidal ideation from mental health content in social media. In *Proceedings of the 2016 CHI Conference on Human Factors in Computing Systems*, CHI '16, pages 2098–2110, New York, NY, USA. ACM.

Virgile Landeiro Dos Reis and Aron Culotta. 2015. Using matched samples to estimate the effects of exercise on mental health from twitter. In *Proceedings of the Twenty-Ninth AAAI Conference on Artificial Intelligence*, AAAI'15, pages 182–188. AAAI Press.

Mark Dredze, Michael J Paul, Shane Bergsma, and Hieu Tran. 2013. Carmen: A Twitter geolocation system with applications to public health. In *AAAI Workshop on Expanding the Boundaries of Health Informatics Using AI (HIAI)*.

Bobby Duffy, Kate Smith, George Terhanian, and John Bremer. 2005. Comparing data from online and face-to-face surveys. *International Journal of Market Research*, 47(6):615.

Janine D Flory and Rachel Yehuda. 2015. Comorbidity between post-traumatic stress disorder and major depressive disorder: alternative explanations and treatment considerations. *Dialogues in clinical neuroscience*, 17(2):141.

Bibo Hao, Lin Li, Ang Li, and Tingshao Zhu. 2013. Predicting mental health status on social media. In *Cross-Cultural Design. Cultural Differences in Everyday Life.CCD 2013*, Lecture Notes in Computer Science, pages 101–110. Springer, Berlin, Heidelberg.

Xiaolei Huang, Michael Smith, Michael Paul, Dmytro Ryzhkov, Sandra Quinn, David Broniatowski, and Mark Dredze. 2017. Examining patterns of influenza vaccination in social media. In *AAAI Workshops*.

Xiaolei Huang, Michael C Smith, Amelia M Jamison, David A Broniatowski, Mark Dredze, Sandra Crouse Quinn, Justin Cai, and Michael J Paul. 2019. Can online self-reports assist in real-time identification of influenza vaccination uptake? a cross-sectional study of influenza vaccine-related tweets in the usa, 2013–2017. *BMJ open*, 9(1):e024018.

Rebecca Knowles, Josh Carroll, and Mark Dredze. 2016. Demographer: Extremely simple name demographics. In *Proceedings of the First Workshop on NLP and Computational Social Science*, pages 108–113.

Mrinal Kumar, Mark Dredze, Glen Coppersmith, and Munmun De Choudhury. 2015. Detecting changes in suicide content manifested in social media following celebrity suicides. In *Proceedings of the 26th ACM conference on Hypertext and hypermedia*. ACM.

Benjamin Mandel, Aron Culotta, John Boulahanis, Danielle Stark, Bonnie Lewis, and Jeremy Rodrigue. 2012. A demographic analysis of online sentiment during hurricane irene. In *Proceedings of the 2nd Workshop on Language in Social Media*, pages 27–36. Association for Computational Linguistics.

Lewis Mitchell, Kameron Decker Harris, Morgan R Frank, Peter Sheridan Dodds, and Christopher M Danforth. 2013. The geography of happiness: connecting Twitter sentiment and expression, demographics, and objective characteristics of place. *PLOS ONE*, 8(5).

Margaret Mitchell, Glen Coppersmith, and Kristy Hollingshead, editors. 2015. *Proceedings of the*

119

Workshop on Computational Linguistics and Clinical Psychology: From Linguistic Signal to Clinical Reality. North American Association for Computational Linguistics, Denver, Colorado, USA.

Michael J Paul and Mark Dredze. 2011. You are what you tweet: Analyzing Twitter for public health. In *Proceedings of the 5th International AAAI Conference on Weblogs and Social Media*, ICWSM11. Association for the Advancement of Artificial Intelligence.

Michael J. Paul and Mark Dredze. 2017. *Social Monitoring for Public Health*. Morgan & Claypool Publishers.

Ted Pedersen. 2015. Screening Twitter users for depression and PTSD with lexical decision lists. In *Proceedings of the Workshop on Computational Linguistics and Clinical Psychology: From Linguistic Signal to Clinical Reality*, pages 46–53.

Ross L Prentice. 1986. A case-cohort design for epidemiologic cohort studies and disease prevention trials. *Biometrika*, 73(1):1–11.

Daniel Preotiuc-Pietro, Maarten Sap, H Andrew Schwartz, and LH Ungar. 2015. Mental illness detection at the world well-being project for the clpsych 2015 shared task. *NAACL HLT 2015*, page 40.

Marcel Salathe, Linus Bengtsson, Todd J Bodnar, Devon D Brewer, John S Brownstein, Caroline Buckee, Ellsworth M Campbell, Ciro Cattuto, Shashank Khandelwal, Patricia L Mabry, et al. 2012. Digital epidemiology. *PLoS computational biology*, 8(7):e1002616.

H. Andrew Schwartz, Johannes Eichstaedt, Margaret L. Kern, Gregory Park, Maarten Sap, David Stillwell, Michal Kosinski, and Lyle Ungar. 2014. Towards assessing changes in degree of depression through Facebook. In *Proceedings of the Workshop on Computational Linguistics and Clinical Psychology: From Linguistic Signal to Clinical Reality*, pages 118–125. Association for Computational Linguistics.

Reviving a psychometric measure: Classification and prediction of the Operant Motive Test

Dirk Johannßen
MIN Faculty,
Dept. of Computer Science
Universität Hamburg
& Nordakademie

Chris Biemann
MIN Faculty,
Dept. of Computer Science
Universität Hamburg
22527 Hamnburg, Germany

David Scheffer
Dept. of Economics
Nordakademie
25337 Elmshorn, Germany

http://lt.informatik.uni-hamburg.de/
{biemann, johannssen}@informatik.uni-hamburg.de
{david.scheffer, dirk.johannssen}@nordakademie.de

Abstract

Implicit motives allow for the characterization of behavior, subsequent success and long-term development. While this has been operationalized in the operant motive test, research on motives has declined mainly due to labor-intensive and costly human annotation. In this study, we analyze over 200,000 labeled data items from 40,000 participants and utilize them for engineering features for training a logistic model tree machine learning model. It captures manually assigned motives well with an F-score of 80%, coming close to the pairwise annotator intraclass correlation coefficient of r = .85. In addition, we found a significant correlation of r = .2 between subsequent academic success and data automatically labeled with our model in an extrinsic evaluation.

1 Introduction

In psychology, texts have been analyzed for so-called motives since the 1930s Schultheiss and Brunstein (2010a). Implicit motives are unconscious motives, which are measurable by operant methods. Operant methods, in turn, are psychometrics, which are captured by having participants write free texts, i.e. participants are asked ambiguous questions or are shown faint images, which they describe or interpret. Classically, motives are labeled manually in these descriptions for further analysis (Schultheiss, 2008). Knowledge of operant motives facilitate clinical research on e.g. traumas, as conducted by Weindl and Lueger-Schuster (2016). According to Schultheiss (2008), there are three main motives of the operant system: i) affiliation (hereafter referred to as A), which is a desire for establishing positive relationships, ii) achievement (hereafter referred to as L), described as the capacity of mastering challenges and gaining satisfaction

from such and iii) power (hereafter referred to as M), which is the desire to have an impact on one's fellows. Originally, psychological motives were measured with projective techniques, such as the thematic apperception test (TAT, (Murray, 1943)) or with questionnaires (Schüler et al., 2015). During the TAT, participants were shown between 8 and 30 colorless images in two sessions and were asked to tell stories for each of the 10 images per sessions, which took about 20-30 minutes. Besides this time consumption, the TAT showed variable objectivity, thus an acceptable inter-rater agreement could not be achieved. Motives can be also measured by questionnaires, which helps to achieve objectivity but measure something different, i.e. explicit motives. The hypothesis of those independent motivational systems (explicit, implicit) was proposed and shown by McClelland et al. (1989). Implicit motives are aroused by affective incentives that promise direct emotional rewards, whilst explicit motives are aroused by rational incentives, which include social expectations (Schüler et al., 2015).

Even though it is possible to predict the hierarchical development of managers, subsequent academic success and preferred clothing brands (as reviewed in Section 3), research on motives has declined mainly due to labor-intensive and costly human annotation by well-trained psychologists. In this work, we examine how far processing with natural language processing (NLP) techniques can automatize the assignment of operant motives. We evaluate our approach intrinsically as well as extrinsically for the prediction of subsequent academic success as reflected in grades of final student's bachelor's theses.

As far as we are aware, this is the first work that uses the OMT for training a machine learning algorithm in order to classify yet unlabeled data and investigate measurable connections between oper-

121

Proceedings of the Sixth Workshop on Computational Linguistics and Clinical Psychology, pages 121–125
Minneapolis, Minnesota, June 6, 2019. ©2019 Association for Computational Linguistics

ant motives and subsequent academic success.

2   The OMT and MIX

The operant motive test (OMT) was originally developed by Kuhl and Scheffer (1999). Different to the TAT by Murray (1943), for measuring motives with the OMT, participants are shown sketched scenarios with multiple persons in underspecified situations, such as displayed in Figure 1.

The OMT has the two main advantages, that participants are asked to state very short answers in contrast to whole stories of the TAT and that the OMT introduces additional *levels* of affective valence to the three main motives ranging from 1 to 5, allowing psychologists to differentiate affects of participants even further. Level 1 stands for self-regulating, 2 for incentive-driven, 3 for self-driven, 4 for active avoidance and 5 for passive avoidance.

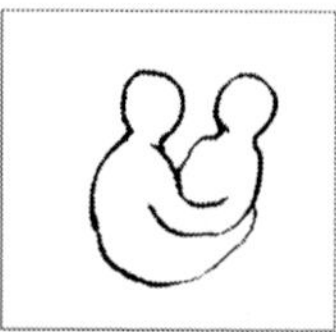

Figure 1: Sketched scenery for participant to answer four (OMT) questions on the narration and involved emotions (Kuhl and Scheffer, 1999)

A so-called zero-motive or zero-level (annotated as 0 for both, the motives and levels) are labeled if no clear motive or level can be identified, resulting in 4 X 6 possible target classes (0, A, M, L with levels 0 to 5). Even though cases are rare, it is possible to assign a level other than 0 with a 0 motive, i.e. no motive could be identified since motives and levels are orthogonal classifications.

A closely related psychometric test is the so-called Motive Index (MIX), developed by Scheffer and Kuhl (2006). The MIX is measured similarly to the OMT with slightly altered questions for an even faster assessment, making the MIX suitable for shortened aptitude diagnostics.

3   Related Work

McClelland and Boyatzis (1982) showed during an assessment center study that managers with a highly developed power motive were significantly more likely to reach higher hierarchy levels within 18 years. Weindl and Lueger-Schuster

(2016) utilized the OMT for clinically investigating survivors of childhood abuse in foster care settings, finding connections between certain motive level constellations and symptoms of abuse. Schmidt and Frieze (1997) utilized the motive model of McClelland and Boyatzis (1982) on 142 college students and concluded that a stronger power motive occurrence mediated product involvement such as expensive cars or interview clothing, whilst affiliation was associated with purchasing gift cards. Schultheiss and Brunstein (2010b) analyzed CEO speeches and were able to predict individual and collective behavior of company members or companies. Schüler et al. (2015) compared and related three different motive measures, namely the picture story exercise (PSE, (Schultheiss and Pang, 2007)), the OMT and the multi-motive grid (Sokolowski et al., 2000), and showed that the measures differ in their scoring system and thus show little overlap, indicating them being unexchangeable. It is controversial whether the achievement motive is connected with academic success: Scheffer (2004) was able to predict grades with a significant correlation of r = .2, attributed to the intrinsic desire for excellence, whilst McClelland (1988) found that the power motive is rather correlated with academic success if grades are exposed to peers due to the desire to impress fellows.

Those studies show the validity and promising predictive power of the OMT, which can be utilized for aptitude diagnostics of different fields. In terms of the bachelor thesis grades, which are perceptible by peers, the predictability by the power motive can be hypothesized.

4   Data

Data has been collected by having 40,000 anonymized participants textually associate images in German such as the one in Figure 1 on the two questions i) Who is the main person and what is important for that person? ii) How does that person feel? The participants gave 220,859 answers on 15 different images. After filtering (cf. Section 5.1), we retain 209,716 text instances.

Each answer was labeled manually with the motives 0, A, L or M and a level ranging from 0 to 5. The annotators were psychologists, trained by the OMT manual by Kuhl and Scheffer (1999). The inter-annotator agreement with previously coded motives using the Winter scale (Winter, 1994)

reached as high as 97% and 95% for the two annotators after the manual training. The pairwise intraclass correlation coefficient is an often utilized agreement measure, developed by Shrout and Fleiss (1979). This coefficient was measured to be .85 on average for the three motives (Schüler et al., 2015), thus showing the difficulty to standardize the labeling process.

The class distributions of motives and levels displayed in Table 1 show that the power motive (M) is with 59% nearly three times as frequent as the second largest class of achievement (L) with 19%. Furthermore, levels 4 and 5 together represent more than half of all level-labeled instances.

In addition to the roughly 220,000 labeled OMT text data instances, a small dataset of related but unlabeled MIX texts from 105 participants is available, which come with the additional information of the bachelor thesis grades of the anonymized participants. We will use this dataset for the extrinsic evaluation below.

5 Methodology

The main goal of this work is the automatization of the motive classification by training a machine learning model. Another goal will be the first and basic validation of the trained model by classifying the yet unlabeled 105 additional texts and hypothesizing a correlation between the achievement or the power motive with the bachelor thesis grades.

5.1 Pre-processing

We pre-processed the data by first removing spam, which mostly contained the same letters repeated, empty answers or a random variation of symbols. Also, we removed entries in different languages other than German. Lastly, texts with encoding problems were either resolved or removed. After this pre-processing, the whole dataset consisted of 209,716 texts. The distribution of filtered questions is uneven.

	0	1	2	3	4	5	Σ
0	7,921	0	2	1	2	6	7,932
A	11	2,888	9,581	1,361	7,617	6,822	28,280
L	6	2,455	12,697	6,405	7,542	3,742	32,847
M	25	11,338	12,353	15,248	36,103	23,610	98,677
Σ	7,963	16,681	34,633	23,015	51,264	34,180	167,736

Table 1: The OMT's training classes distribution after filtering and removing a held-out test and development set (10% each).

5.2 Feature engineering

For engineering features, the texts mostly were tokenized and processed per token. Engineered features were the type-token-ratio, the ratio of spelling mistakes and frequencies between 3 and 10 appearances.

Further features are LIWC and language model perplexities. The psychometric dictionary and software *language inquiry and word count* (LIWC) was developed by Pennebaker et al. (1999) and later transferred to German by Wolf et al. (2008). The German LIWC allowed for 96 categories to be assigned to each token, ranging from rather syntactic features such as personal pronouns to rather psychometric values such as familiarity, negativity or fear.

Part-of-speech (POS) tags were assigned to each token and thereafter counted and normalized to form a token ratio. We trained a POS tagger via the natural language toolkit (NLTK) on the TIGER corpus, assembled by Brants et al. (2004) and utilizing the STTS tagset, containing 54 individual POS tags.

We trained a bigram language model for each class and incorporated Good-Turing smoothing for calculating the perplexity. During training, we tuned parameters (e.g. which smoothing to use) via development set and tested the model with a held-out test set of 20,990 instances. The perplexity of a model q is:

$$2^{-\frac{1}{N} \sum_{i=1}^{N} \log_2 q(x_i)}$$

with p being an unknown probability distribution, $x_1, x_2, \ldots x_N$ being the sequence (i.e. the sentence) drawn from p and q being the probability model.

5.3 Model training

Even though deep learning has shown to be powerful, it often comes with a cost of losing transparency, which is crucial for our task, in which we seek to better understand the connection between psychology and language. Therefore we utilized different classical machine learning algorithms such as Naïve Bayes, LMT or regression and found the logistic model tree (LMT) implementation of Landwehr et al. (2005) to be the best-performing one amongst the tested. A LMT is a decision tree, which performs logistic regressions at its leaves. The root differentiates the language model's perplexities (A, M, and L) and thereafter performs the logistic regressions based on further features.

A qualitative post-hoc analysis by psychologists has resulted in an agreement with the model's predictions, except for too many assigned 0 labels and motives.

6 Results

Based on the correlation-based *Feature Subset Selection* by Hall (2000), the most influential features are the LIWC categories *I, Anger, Communication, Friends, Down, Motion, Occup, Achieve* and *TV*, as well as the perplexities of the language models affiliation (A), performance (L) and power (M) and attributive possessive pronoun (PPOSAT) POS tag frequency.

When classifying unlabeled OMT related texts of 105 anonymized participants, counting the motive predictions and analyzing a possible connection with the bachelor thesis grade and said counts, a weak but significant Pearson correlation coefficient of r = .2 could be found between the power motive and the thesis grade value (shown in Figure 2), whilst the achievement motive did not show any correlation. A wordlist-based model, which consists of 415 affiliation, 512 achievement, and 572 power words showed an insignificant correlation of r = .07 with an F-score of 61.07%.

		0	**A**	**L**	**M**	**Σ**
	0	338	92	163	427	1,020
	A	51	2,667	105	708	3,531
Actual	**L**	115	66	3,151	804	4,136
	M	209	573	556	10,965	12,303
	Σ	713	3,398	3,975	12,904	20,990

Predicted

Table 2: The confusion matrix of the motive classification task (without the levels) on the test set (10% of available data) with filtered values.

The confusion matrices in Table 2 illustrate the model's performance for each class. The model scores an F1 score of 65.4% for classifying the levels and 80.1% for classifying the motives.

An error analysis revealed that misclassified instances contain more words on average (24.2 versus 21.04). Also, misclassifications contain four times the amount of fillers (e.g. you know, like, i mean, Pennebaker et al. (1999)). Those instances are focused on plural personal pronouns twice as often and show a higher amount of answer particle. Moreover, misclassified instances contain 50% more often religious expressions, metaphors,

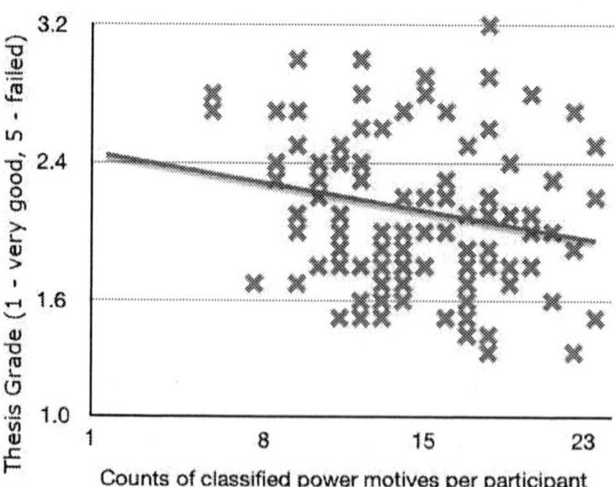

Figure 2: Correlation of r = -0.20 between LMT classifier predicted counts of power motive answers and the bachelor thesis grades. The German grading system ranges from 1.0 (very good) to 5.0 (failed).

and topics of sadness. Most of the misclassified instances show high perplexity scores of either one motive, are written in all caps and contain one-word sentences. When referring to the OMT manual Kuhl and Scheffer (1999) used for training psychologists on that labeling task, it is controversial whether all caps words should be viewed as a feature in itself and whether single word sentences qualify for being labeled different than 0, hence the OMT asks participants for stories rather than keywords. The annotators seem to have developed an intuition besides the OMT manual, as reflected in their high intraclass correlation coefficients.

7 Conclusion

The psychometric OMT is hampered by costly and labor-intensive manual annotation. Automatization is possible by utilizing the proposed model for motive and level classification. The annotators have had an average intraclass correlation coefficient of .85, whilst the overall F-score has reached 80.1%, clearly exceeding F = 61.07% of the wordlist-based model. Even though both measures are not directly comparable, the respectable F-scores suggest that the feature-engineered machine learning model is approaching human-like performance. Interestingly, the most influential features relate to the OMT theory. Lastly, a first theory validation has resulted in a significant r = .2 correlation between the predicted power motive and bachelor thesis grades. Furthermore, often better performing neural approaches should be considered for future work.

References

Sabine Brants, Stephaie Dipper, Silvia Hansen, Wolfgang Lezius, and George Smith. 2004. The tiger treebank. *Journal of Language and Computation*, 2:597–620.

Mark Andrew Hall. 2000. *Correlation-Based Feature Selection for Machine Learning.* dissertation, University of Auckland, New Zealand.

Julius Kuhl and David Scheffer. 1999. *Der operante Multi-Motiv-Test (OMT): Manual [The operant multi-motive-test (OMT): Manual].* Impart, Osnabrück, Germany: University of Osnabrück.

Niels Landwehr, Mark Andrew Hall, and Eibe Frank. 2005. Logistic Model Trees. *Machine Learning*, 59(1):161–205.

David Clarance McClelland. 1988. *Human Motivation.* Cambridge University Press.

David Clarance McClelland and Richard Boyatzis. 1982. Leadership Motive Pattern and Long-Term Success in Management. *Journal of Applied Psychology*, 67:737–743.

David Clarance McClelland, Richard Koestner, and Joel Weinberger. 1989. How do self-attributed and implicit motives differ? *Psychological Review*, 96(4):690–702.

Henry Alexander Murray. 1943. *Thematic apperception test.* Thematic apperception test. Harvard University Press, Cambridge, MA, US.

James Pennebaker, Martha Eileen Francis, and Roger John Booth. 1999. Linguistic inquiry and word count (LIWC). *Software manual.* http://liwc.wpengine.com (visited: 2019-01-17).

David Scheffer. 2004. *Implizite Motive: Entwicklung, Struktur und Messung [Implicit Motives: Development, Structure and Measurement]*, 1st edition. Hogrefe Verlag, Göttingen, Germany.

David Scheffer and Julius Kuhl. 2006. *Erfolgreich motivieren: Mitarbeiterpersönlichkeit und Motivationstechniken [Motivate Successfully: Employer Personality and Motivational Techniques]*, 1st edition. Hogrefe Verlag, Göttingen, Germany.

Laura Schmidt and Irene Hanson Frieze. 1997. A mediational model of power, affiliation and achievement motives and product involvement. *Journal of Business and Psychology*, 11(4):425–446.

Oliver Schultheiss. 2008. Implicit motives. *Handbook of personality: Theory and research*, pages 603–633.

Oliver Schultheiss and Joachim Brunstein. 2010a. *Implicit Motives.* Oxford University Press, Oxford, New York.

Oliver Schultheiss and Joachim Brunstein. 2010b. *Implicit Motives.* Oxford University Press, Oxford, New York.

Oliver Schultheiss and Joyce Pang. 2007. Measuring implicit motives. In *Handbook of research methods in personality psychology*, pages 322–344, New York, NY, US. Guilford Press.

Julia Schüler, Veronika Brandstätter, Mirko Wegner, and Nicola Baumann. 2015. Testing the convergent and discriminant validity of three implicit motive measures: PSE, OMT, and MMG. *Motivation and Emotion*, 39(6):839–857.

Patrick Shrout and Joseph Fleiss. 1979. Intraclass correlations: Uses in assessing rater reliability. *Psychological Bulletin*, 86(2):420–428.

Kurt Sokolowski, Heinz-Dieter Schmalt, Thomas Langens, and Rosa Maria Puca. 2000. Assessing Achievement, Affiliation, and Power Motives All at Once: The Multi-Motive Grid (MMG). *Journal of Personality Assessment*, 74(1):126–145.

Dina Weindl and Brigitte Lueger-Schuster. 2016. Institutional Abuse (IA) and Implicit Motives of Power, Affiliation, and Achievement - an Alternative Perspective on Trauma-Related Psychological Responses. In *ISTSS International Society for Traumatic Stress Studies 32nd Annual Meeting*, Dallas, Texas, USA.

David Winter. 1994. *Manual for scoring motive imagery in running text.* Dept. of Psychology, University of Michigan (unpublished).

Markus Wolf, Andrea Horn, Matthias Mehl, Severin Haug, James Pennebaker, and Hans Kordy. 2008. Computergestützte quantitative Textanalyse: Äquivalenz und Robustheit der deutschen Version des Linguistic Inquiry and Word Count. *Diagnostica*, 54:85–98.

Coherence models in schizophrenia

Sandra Just[1], Erik Haegert[2], Nora Kořánová[2], Anna-Lena Bröcker[1], Ivan Nenchev[1], Jakob Funcke[1], Christiane Montag[1] and Manfred Stede[2]

[1] Department of Psychiatry and Psychotherapy, Campus Charité Mitte, (Psychiatric University Clinic at St. Hedwig Hospital), Charité – Universitätsmedizin Berlin, corporate member of Freie Universität Berlin, Humboldt-Universität zu Berlin, and Berlin Institute of Health
[2]Applied Computational Linguistics, UFS Cognitive Science, University of Potsdam
sandra-anna.just@charite.de

Abstract

Incoherent discourse in schizophrenia has long been recognized as a dominant symptom of the mental disorder (Bleuler, 1911/1950). Recent studies have used modern sentence and word embeddings to compute coherence metrics for spontaneous speech in schizophrenia. While clinical ratings always have a subjective element, computational linguistic methodology allows quantification of speech abnormalities. Clinical and empirical knowledge from psychiatry provide the theoretical and conceptual basis for modelling. Our study is an interdisciplinary attempt at improving coherence models in schizophrenia. Speech samples were obtained from healthy controls and patients with a diagnosis of schizophrenia or schizoaffective disorder and different severity of positive formal thought disorder. Interviews were transcribed and coherence metrics derived from different embeddings. One model found higher coherence metrics for controls than patients. All other models remained non-significant. More detailed analysis of the data motivates different approaches to improving coherence models in schizophrenia, e.g. by assessing referential abnormalities.

1 Introduction

Language impairments in schizophrenia are frequent (Kuperberg, 2010), can impede communication and social integration, and are usually a predictor for poorer outcome (Roche et al., 2015). They include difficulties with structural aspects and pragmatic use of language as well as deficits in cohesion (Abu-Akel, 1997; Bartolucci and Fine, 1987; Chaika and Lambe, 1989) and semantic coherence (Bedi et al., 2015; Ditman and Kuperberg, 2010; Elvevag et al., 2007; Iter et al., 2018). Although incoherent speech is a prominent symptom of schizophrenia (American Psychiatric Association, 2013; Andreasen, 1979a; Ditman and Kuperberg, 2010), there have been few collaborations of psychiatry and linguistics to analyze the symptom with linguistic quantitative methodology.

In psychopathological terms, incoherent speech is usually not described as a *language* disorder but as one possible manifestation of formal *thought* disorder (FTD) – a symptom occurring in a wide range of disorders, albeit predominantly in psychosis (Andreasen and Grove, 1986; Mercado et al., 2011; Roche et al., 2015). FTD comprises diverse abnormalities of speech and thought, such as neologisms, flight of ideas, rumination and perseveration, and negative symptoms like alogia (Broome et al., 2017; Roche et al., 2015) – all of which are not necessarily related to incoherent speech. For example, neologisms might impair coherence (Lecours and Vanier-Clément, 1976), but can also facilitate expressing ideas (Bleuler, 1911/1975; Covington et al., 2005). Another example is perseveration, where constant repetitions indicate speech abnormality but do not have to impede coherence (Liddle et al., 2002). Still, especially ratings of

Proceedings of the Sixth Workshop on Computational Linguistics and Clinical Psychology, pages 126–136
Minneapolis, Minnesota, June 6, 2019. ©2019 Association for Computational Linguistics

positive FTD bear relevance to assessing incoherent speech. For example, in the Scale for the Assessment of Positive Symptoms (SAPS), incoherent speech is defined as loss of associations within sentences which can result in incomprehensible "schizophasia" or "word salad" (Andreasen, 1979a). It is linked to other forms of positive FTD such as tangentiality (i.e. irrelevant responses to questions), derailment (i.e. loss of associations between larger units of speech), illogical, and indirect speech (Andreasen, 1984). It should be noted that incoherent speech varies across patients depending on the phase of illness and the presence and severity of other symptoms (Allen et al., 1993; Chaika, 1974; Roche et al., 2015).

In linguistics, incoherence refers to the deeper semantic sense of speech transcending the meaning of individual sentences. It is present locally, within and between sentences, as well as globally, as the overall topic or function of speech (Stede, 2007, p. 24f.). Rhetorical Structure Theory, for example, defines coherence through establishing relations between minimal discourse units and thereby building a structure which is reflective of the internal organization of discourse (Mann and Thompson, 1987).

A linguistic, valid, reliable and objective measure of incoherent speech could serve to find a common language between psychiatry and linguistics and specify the definition of incoherence as part of FTD. This could be useful for further examining the concept and underlying mechanisms such as neurological correlates as well as for assessing prognosis and treatment responsiveness.

1.1 Automated speech and coherence analysis

Ditman and Kuperberg (2010) suggest that incoherent speech in schizophrenia appears to be connected to abnormal use of referential markers (see also Docherty et al. (1998), Rochester (2013) or Hinzen and Rosselló (2015)) and problems in "integrating meaning across clauses" (p. 7) which can lead to a lower similarity between sentences in schizophrenia. This latter observation invites for automated coherence analysis that models coherence as lexical cohesion or concept overlap. Latent Semantic Analysis (LSA) (Landauer and Dumais, 1997) is such a measure and has been tested in schizophrenia research (Bedi et al., 2015;

Elvevag et al., 2007). In a recent study, Iter et al. (2018) could not distinguish between schizophrenia patients and healthy controls replicating the LSA-based models used in Bedi et al. (2015) and Elvevag et al. (2007). They point out three major shortcomings of the models: (1) the misinterpretation of verbal fillers as incoherent speech, (2) a bias to judging longer sentences as more coherent than short ones, as well as (3) a bias to judging repetitions as more coherent. Iter et al. (2018) were able to improve coherence models by Elvevag et al. (2007) and Bedi et al. (2015) by preprocessing their dataset and using modern word and sentence embedding techniques which have been shown to outperform LSA (Fang et al., 2016; Levy et al., 2015). Moreover, they credit the mentioned observations of referential problems in schizophrenia and propose a referential coherence model based on classifying ambiguous pronoun use to further improve the predictive value of their results.

Our study aims to (1) assess whether the models used by Iter et al. (2018) can be transferred to the German language, and (2) to apply them to a larger sample of patients of varying stability. Specifically, we aim to examine (1) whether schizophrenia patients and controls can be differentiated based on automated coherence analysis, and (2) whether schizophrenia patients of varying stability can be differentiated not only based on clinical rating scales but also based on automated coherence analysis. (3) We aim to extend attempts by Iter et al. (2018) to further improve coherence models by quantifying idiosyncrasies of speech in schizophrenia.

2 Method

2.1 Participants

$N = 30$ participants took part in this study (see Table 1 for characteristics of the sample). $n = 10$ were patients from the Psychiatric University Clinic at St. Hedwig Hospital Berlin and $n = 10$ patients were recruited from the pool of participants in the MPP-S study (clinical trials ID: NCT02576613). Participants were: (1) inpatients ($n = 5$) or outpatients ($n = 15$) with a diagnosis of schizophrenia ($n = 15$) or schizoaffective disorder ($n = 5$) according to Diagnostic and Statistical Manual of Mental Disorders, Fourth Edition, Text Revision (DSM-IV-TR), confirmed by trained

	Patients $n = 20$		Controls $n = 10$	Statistics
	with positive FTD $n = 10$	without positive FTD $n = 10$		
Age (years)	48.1 (12.17) [†]	45.7 (11.7)	44.5 (13.79)	$F^{\,a} = .21$
Sex (male)	$n = 8$	$n = 5$	$n = 5$	$\chi^{2\,c} = 2.5$
Verbal IQ	104.5 (15.39)	106.6 (14.17)	106.6 (9.28)	$F = .08$
Inpatients	$n = 5$	$n = 0$	-	$\chi^2 = 6.67**$
F20.0	$n = 7$	$n = 7$	-	
F25.0	$n = 3$	$n = 3$	-	
Antipsychotic medication	$n = 9$	$n = 10$	-	$\chi^2 = 1.05$
CGI	5.9 (.88)	4.2 (1.48)	-	$t = -3.13**$
Duration of illness (years)	21.5 (13.7)	15.2 (11.74)	-	$t = -1.12$
SAPS			-	
positive FTD	2.9 (.74)	.4 (.52)		$t = -8.78**$
Incoherence	1.9 (1.45)	.1 (.32)		$t = -3.84**$
Tangentiality	2.4 (.7)	.1 (.32)		$t = -9.48**$
Derailment	2.4 (1.51)	.0		$t = -5.04**$
Illogicality	1.9 (1.45)	.0		$t = -3.48**$
Circumstantiality	1.5 (1.65)	.7 (.95)		$t = -1.33$
Pressured speech	2.1 (1.45)	.2 (.63)		$t = -3.8**$
Distractibility	1.8 (1.4)	.0		$t = -4.07**$
Clanging	1.2 (1.14)	.0		$t = -3.34**$
Hallucinations	1.9 (1.91)	1.3 (1.77)		$t = -.73$
Delusions	3.2 (.79)	.9 (1.2)		$t = -5.07**$
Bizarre Behavior	1.6 (1.35)	.1 (.32)		$t = -3.42**$
Inappropriate Affect	1.1 (1.37)	.0		$t = -2.54**$
SANS			-	
Flat Affect	1.9 (1.66)	1.7 (1.16)		$t = -.31$
Alogia	1.2 (1.32)	1.1 (1.29)		$t = -.17$
Avolition/Apathy	2.3 (1.49)	2.1 (1.37)		$t = -.31$
Anhedonia/Asociality	2.6 (1.43)	2.5 (1.35)		$t = -.16$
Attention	1.2 (1.32)	.3 (.95)		$t = -1.75$

[†] Mean (SD); [a] ANOVA; [b] t-test independent samples; [c] χ^2-test; **$p < .05$

Table 1: Characteristics of sample.

clinicians; (2) showed native proficiency in German language; (3) had no organic mental disorder or relevant severe somatic disease; (4) no active substance dependence. The control group ($n = 10$) was recruited from the local community. Healthy controls were screened by experienced clinicians with the Mini-International Neuropsychiatric Interview (M.I.N.I.) (Sheehan et al., 1998).

The study was approved by the local ethics' committee.

2.2 Procedure

Speech samples for automated analysis were obtained by trained clinicians with a short semi-structured interview, the Narrative of Emotions Task (NET) (Buck et al., 2014). It includes three questions about four emotions: sadness, fear, anger and happiness: (1) What does this emotion mean to you? (2) Describe a situation where you felt this emotion. (3) Why do you think you felt this emotion in this situation? The interview is designed to prompt participants to define this range of simple emotions with the intention to "assess the richness and coherence with which one explains emotional and social events" (Buck et al., 2014, p. 235). Semi-structured interviews have already been used in studies on automated speech analysis in schizophrenia (Elvevag et al., 2007; Minor et al., 2019). The structured format of the NET interview allows direct comparison between subjects and open questions generate

	Total $N = 30$	Patients $n = 20$		Controls $n = 10$	Statistics
Word count		with positive FTD $n = 10$	without positive FTD $n = 10$		
Raw data	21,668 722.27 (468.14)[†]	10,089 1,008.9 (647.62)	4,352 435.2 (172.21)	7,227 722.27 (272.48)	$F^{\,a} = 4.72^{**}$
Without stop words	20,421 680.7 (455.31)	9,605 960.5 (625.66)	3,984 398.4 (164.42)	6,832 683.2 (271.98)	$F = 4.81^{**}$

[†] Mean (SD); [a] ANOVA; $^{**}p < .05$

Table 2: Dataset.

larger samples of free speech. All NET interviews were recorded. They were transcribed by the first and third author.

The assessment also included a test of verbal intelligence, the exploration of demographic data and the M.I.N.I. (Sheehan et al., 1998) for controls. After the session, interviewers rated patients for psychopathology.

2.3 Measures

Psychopathology: Psychopathology was rated by trained clinicians with common psychiatric rating scales: the Scale for the Assessment of Negative Symptoms (SANS) (Andreasen, 1989) and the Scale for the Assessment of Positive Symptoms (SAPS) (Andreasen, 1984). Both scales have good psychometric properties and have frequently been used in schizophrenia research (Norman et al., 1996; van Erp et al., 2014). The patient sample was divided in two groups based on SAPS ratings of global positive FTD, including ratings of incoherence or tangentiality. The group with positive FTD was defined by SAPS ratings of at least mild (≥ 2) global positive FTD and at least mild incoherence or tangentiality (≥ 2).

Severity of illness: The Clinical Global Impression – Severity Scale (CGI) (Guy, 2000) allows trained clinicians to assess the severity of a patient's illness on a scale from 1 (not at all ill) to 7 (extremely severely ill).

Verbal intelligence: "Crystallized" verbal intelligence was assessed with a German vocabulary test, the Wortschatztest (WST) (Schmidt and Metzler, 1992).

3 Data Analysis

3.1 Preparation of data

The dataset consists of 241 min 51 sec of 30 recorded NET interviews. Interview length ranged between 3 to 22 min, with an average length of 8 min. The interviewer's speech has been left out of more complex analysis because the interviewer's speech can be reduced to the questions mentioned above.

However, questions have been used to categorize participants' speech as definitions of emotions (question 1), descriptions of situations (questions 2) and reasoning why a situation evoked an emotion (question 3) (Buck et al., 2014). When interviewers deviated from the NET interview, those remarks were removed to ensure comparability. After cleaning transcripts of interviewer's speech, the dataset for baseline analysis consists of 21,668 words, ranging from 137 to 2,641 words, with an average of 722.3 words per participant.

For the other coherence models, verbal fillers and sentences only containing stop words have also been excluded from analysis, because they have been shown to bias coherence measures (Iter et al., 2018). This reduced the dataset to 20,421 words, ranging from 121 to 2,551 words, with an average of 680.7 words per participant (see Table 2).

3.2 Speech analysis of transcripts

All speech analysis uses models inspired by those of Iter et al. (2018) which they base on research by Elvevag et al. (2007) and Bedi et al. (2015). Iter et al. (2018) name these approaches the Tangentiality and the Incoherence model, following the above definitions in the SAPS (Andreasen, 1984). In the Incoherence Model (Bedi et al., 2015), the cosine similarity between pairs of adjacent sentences embeddings serves as a measure of coherence. The Tangentiality model (Elvevag et al., 2007) models coherence as the slope of a linear regression line for the cosine similarities between a question and a moving

129

fixed-sized window of the response. A steeper negative slope means that the response is becoming less similar to the question over time. A steeper positive slope indicates that the response is getting more similar to the question over time, i.e. what psychiatry calls a circumstantial response (Andreasen, 1984). In either case, incoherent responses are characterized by steeper slopes. The differentiation of positive and negative slopes and the following necessity to calculate with absolute values has not been emphasized by Elvevag et al. (2007) or Iter et al. (2018).

Both the Incoherence and the Tangentiality model define coherence "as the concept overlap between two texts" (Iter et al., 2018) – either between utterances of the same speaker or between a question and the following response. These definitions reflect the intuition that, in order to be deemed coherent, a contribution to a verbal interaction is expected to adhere to the topic mutually established by the participants at any given stage of the conversation. The word distributions that form the basis for this kind of analysis are thus to be conceived of as a kind of epiphenomenon of more general principles of communication.

Baseline coherence model: The first step of speech analysis aims to test the Incoherence and Tangentiality model on the raw dataset. No filtering of stop words or fillers was performed except for the unavoidable loss of words not covered by vocabulary of the respective models. Baseline models use mean vector sentence embeddings, i.e. the mean of all word vectors per sentence or window of tokens (Iter et al., 2018). The vectors are given by a word2vec model (Mikolov et al., 2013) and a GloVe model (Pennington et al., 2014) trained on German data. The Tangentiality model at baseline uses a fixed-size window of four tokens.

In contrast to Iter et al. (2018), we refrained from using LSA in our analysis due to the lack of availability of such a model that has not already performed a TF-IDF-weighting (Lintean et al., 2010) at the stage of training. Additionally, the weighting scheme used at the training of the model at hand differs from that adopted by Iter et al. (2018). Consequently, in order to preserve a certain level of comparability, we decided not to use the available LSA model. However, the use of word2vec for our baseline is justifiable by the fact

that the main improvement from baseline to any of the other embeddings is not so much the choice of model but rather the filtering of stop words and fillers as well as the different weighting schemes.

New coherence models: Following Iter et al. (2018), we test mean of word vectors and three types of sentence embeddings on our preprocessed dataset: TF-IDF (Lintean et al., 2010), Smooth Inverse Frequency (SIF) (Arora et al., 2016) and Sent2Vec (Pagliardini et al., 2018).

For TF-IDF, we use the parameterization of Lintean et al. (2010), also used by Iter et al. (2018): multiplying each word embedding by the raw (non-logged) term frequency (#of times that word occurs in the sentence) and dividing by the (non-logged) document frequency (#of documents in which the term is used in a corpus). As a reference corpus for document frequencies we used a lemmatized dump of German Wikipedia (2011). Words not appearing in any document of the reference corpus were discarded, as closer investigation revealed them to be artifacts of the preprocessing steps rather than very uncommon and highly predictive words. Sent2Vec can be seen as an extension of Word2Vec in that its objective has been modified to encompass whole sentences rendering their embeddings predictive of the sentences surrounding them. Finally, SIF starts out by representing sentences by a weighted average of their word embeddings. In a further step, the projections of the average vectors on their first singular vector are removed, the effect of which is intended to be the removal of biases along directions reflecting idiosyncrasies of the underlying data. The principal goal of such weighting schemes lies in reducing the influence of very common words that contribute little to nothing semantically to the overall meaning of the sentence.

4 Results

4.1 Sample characteristics

Patient groups and controls did not differ significantly regarding age and verbal IQ. Patients with and without signs of positive FTD did not differ significantly regarding duration of illness. Patients with positive FTD were more often inpatients and rated to be more severely ill than those without positive FTD, as measured by CGI. As expected, patients with positive FTD had higher clinical ratings for a number of symptoms than patients without positive FTD, including SAPS global positive FTD, incoherence, and tangentiality. See Table 1 for an overview of ratings of psychopathology and significant differences.

4.2 NET interviews

Interview length and word count differed significantly between groups: Patients with positive FTD had longer interviews and used more words than controls. Patients without positive FTD had shorter interviews and used less words than controls. This difference persisted after cleaning transcripts of stop words. The amount of verbal fillers and sentences only containing stop words did not differ significantly between groups. The dataset is presented in Table 2.

4.3 Coherence models

Incoherence model: Mean values for cosine similarities were calculated per interview. Group means were compared by ANOVA after testing for normal distribution (results for all models are presented in Table 3, extended results can be found in Appendix A). Group differences were only significant for TF-IDF term weighting using GloVe word embeddings: healthy controls showed higher coherence scores than patients without ratings of positive FTD who in turn exhibited higher coherence scores than patients with ratings of positive FTD. Coherence metrics were significantly negatively correlated with SAPS ratings of various positive symptoms: clothing and appearance ($r = -.62$; $p < .05$), social and sexual behavior ($r = -.5$; $p < .05$), global severity of bizarre behavior ($r = -.48$; $p < .05$), and symptoms of positive FTD: derailment ($r = -.5$; $p < .05$), tangentiality, ($r = -.4$; $p < .1$), incoherence ($r = -.45$; $p < .05$), illogicality ($r = -.48$; $p < .05$), clanging ($r = -.41$; $p < .1$), and inappropriate affect ($r = -.5$; $p < .05$). SANS ratings of negative symptoms were not significantly correlated with coherence metrics. As Iter et al. (2018), we did not detect any significant group differences at baseline for the Incoherence model. Removing verbal fillers and sentences composed entirely of stop words did not change this result for mean vector sentence embeddings, which were also used at baseline. Sent2Vec and SIF embeddings, and TF-IDF weighting using word2vec word embeddings also did not yield significantly different coherence metrics between groups.

Incoherence model		
Sentence	**Word**	F^a
Baseline	Word2Vec	.510
Mean Vector	GloVe	.338
	Word2Vec	.109
TF-IDF	GloVe	**4.735****
	Word2Vec	.857
SIF	GloVe	2.012
	Word2Vec	2.068
Sent2Vec	Sent2Vec	.300
Tangentiality model		
Sentence	**Word**	F
Baseline	Word2Vec	2.273
Mean Vector	GloVe	.334
	Word2Vec	.547
TF-IDF	GloVe	.594
	Word2Vec	1.777
SIF	GloVe	.719
	Word2Vec	.821
Sent2Vec	Sent2Vec	1.517

[a] ANOVA; **$p < .05$

Table 3: Group differences in coherence metrics.

Tangentiality model: First, absolute values of the computed slopes in the Tangentiality model were determined. This is necessary as high (negative or positive) values for slopes indicate incoherence. Thus, calculating means without absolute values could lead to false interpretations. Second, mean slopes were calculated per individual response, i.e. per question and emotion, yielding 12 values per interview. Those were further combined to mean values per each of the three questions and per each of the four emotions as well as to one overall mean slope per interview. Group means were compared by ANOVA after testing for normal distribution. Since results did not differ for comparisons of overall means versus means per question/emotion, we only report results for overall means. Overall mean slopes did not differ significantly between groups for any of the embeddings.

4.4 Improving coherence models in schizophrenia

Following observations of abnormalities in referential meaning made by Hinzen and Rosselló (2015), Iter et al. (2018) incorporate the presence of ambiguous pronouns in the data into their means of classification. They define ambiguous pronouns as either referring cataphorically or not having a referent at all. In contrast to Iter et al. (2018), we refrained from using automated coreference resolution which appeared to be relatively error-prone. We believe the evaluation on the basis of manual annotation to be more informative. We therefore manually marked ambiguous pronouns throughout the interview transcripts which allowed for determining a total number of ambiguous pronouns per interview. The average number of ambiguous pronouns was significantly higher for patients with ratings of positive FTD than for the other two groups ($F = 4.79$; $p < .05$). There was no significant difference between controls and patients without ratings of positive FTD. However, since pressured speech and word count differed significantly between groups, we repeated the comparison controlling for word count by only analyzing a window of the first 120 words per transcript. With this adjustment, the significant difference disappeared.

More detailed analysis of results revealed significant group differences in the amount of unknown words that were discarded before coherence metrics were computed because they were not contained in the respective model: patients with ratings of positive FTD used significantly more unknown words than patients without ratings of positive FTD or controls ($F = 5.85$; $p < 0.05$). When controlling for word count, this significant difference disappeared. However, it is worth differentiating unknown words: They can either be uncommon or quite specific actual words (e.g. exacerbation) or neologisms that are more or less intelligible (e.g. Rotwut: "red-rage"; e.g. vergehlich: approx. "fleeting", no exact translation possible). While no control subject and only one patient without positive FTD used neologisms, five patients with ratings of positive FTD used neologisms ($\chi^2 = 8.75$; $p < .05$). This difference remained significant after controlling for word count ($\chi^2 = 6.67$; $p < .05$).

Closer investigation of transcripts revealed that participants with high scores for the Incoherence model (TF-IDF, GloVe) often repeated target words such as "sad" or "fear". Low scores coincided with less repetitions – in some but not all cases. This is a mere qualitative observation.

5 Discussion

This study tested different computational linguistic approaches to modeling coherence in schizophrenia. The Incoherence model, using TF-IDF sentence embeddings and GloVe word embeddings, was able to distinguish between healthy controls and patients with or without ratings of positive FTD. Results from other approaches were not significant which demands for cautious interpretation. Although the significant group difference matches clinical impression, we argue to treat the result with caution. When judging the performance of a coherence model in schizophrenia, it might be misleading to merely base it on significant group differences. This approach by Iter et al. (2018) is based on the assumption that the speech of patients with schizophrenia contains less contextual overlap than the speech of healthy controls and that a model that detects this difference is correct and "outperforming" models that lack significant results. However, this basic assumption also requires critical evaluation. Perseveration poses one potential problem: This symptom of positive FTD involves constant repetitions and thus, influences models that are based on similarity between sets of key words, without actually accounting for whether the speech is intelligible (Iter et al., 2018). Since perseveration indicates speech abnormality but does not have to impede coherence (Liddle et al., 2002), it can bias automated coherence models. Future studies should clinically assess perseveration when recruiting patients to ensure that it is equally distributed. In comparison with controls though, the problem would remain. Based on our results, it remains unclear whether coherence modelled as context overlap differs significantly between patients with schizophrenia and healthy controls.

We would also like to emphasize that, while interpretation of the Incoherence model is intuitive, results in the Tangentiality model are substantially more complex. To our knowledge, the differentiation between negative and positive slopes has not been made in previous studies (Elvevag et al., 2007; Iter et al., 2018), albeit its relevance for interpretation. It remains open whether this measure accurately models incoherent features of speech in schizophrenia.

It should be noted that another possibility for the lack of significant results might be the quality of the trained models. In contrast to other studies testing automated coherence analysis in schizophrenia (Bedi et al., 2015; Elvevag et al., 2007; Iter et al., 2018), our models were trained on the German version of Wikipedia and may be inferior to models trained in English. Under the reasonable assumption that the English models were trained on the respective Wikipedia dump, the training data nearly triples that available for a German model.[1] The resulting differences in representational quality are likely to be substantial.

Additionally, German morphology may have to be taken into account as an aggravating factor as training was performed without any preprocessing beyond conversion to lower case letters. Being considerably richer than its English counterpart, it makes the demand for greater amounts of training data even more pressing, since the model has to generalize over a wider morphological spread. This problem is illustrated in Table 4 containing a sample of the cosine similarities computed with our GloVe model. Here the word pair *anger/ happiness* achieves a higher score than the noun *anger* and its derivate *angry,* and similarities are even lower between the inflections of the adjective *angry*. Furthermore, the grammar of German famously features a productive rule of noun composition that in some cases leads to the exacerbation of the problem of *out-of-vocabulary-words*. For example, in one instance, 14 out of 31 the words not covered by our model were instances of such compound nouns.

More detailed analysis of our results inspires to improve coherence models by taking into account other ways of modelling coherence than context overlap and by controlling for possible confounding variables in the speech of patients with schizophrenia. We agree with Iter et al. (2018) that quantifying ambiguous pronoun use can be a valid approach to operationalizing a characteristic of incoherent speech in schizophrenia that has been frequently described (Ditman and Kuperberg, 2010; Docherty et al., 1998; Hinzen and Rosselló, 2015; Rochester, 2013). Moreover, unknown words that are automatically removed from analysis because they are not contained in the vocabulary of the coherence model might confound results. In our sample, patients with high

Word Pair	Cosine Similarity
Wut, Freude	0.5278492
Wut, wütend	0.48702702
wütende, wütend	0.29909012
wütenden, wütend	0.28667736

Table 4: Sample word pairs with their corresponding cosine similarities.

ratings of positive FTD did use more uncommon, specific or neologized words. While incomprehensible neologisms can be associated with "schizophasia" (Lecours and Vanier-Clément, 1976), they are not necessarily a marker for incoherence. They can even enrich (therapeutic) discourse, e.g. as descriptions of novel, otherwise inexpressible ideas (Bleuler, 1911/1975; Covington et al., 2005). Thus, discarding them without further analysis might over- or underestimate the coherence of speech of thought disordered patients. Plus, we point out the importance of controlling for word count when examining prevalence of speech abnormalities. Iter et al. (2018) missed the opportunity of this adjustment despite large differences in word count between patients and controls, thereby possibly overseeing a confounding variable. Pressured speech is a common symptom of positive FTD in schizophrenia and can be correlated with incoherence (Andreasen, 1979b, 1984) – still, mere higher production of speech is no sign of incoherence. This limited our analysis on the first 120 words of responses – future research could test whether markers of incoherence vary depending on which part of the response is examined.

In conclusion, while automated coherence models can further improve understanding of incoherent speech in schizophrenia, our results emphasize the importance of carefully analyzing the data at hand while considering potential relationships between incoherence and other relevant variables. Moreover, they underline the necessity for the establishment of some standard with regards to the vector models underlying analysis. Nevertheless, this interdisciplinary approach can enable mutual stimulation between linguistics and psychiatry.

---

[1] The English Wikipedia constitutes 11.7% of the articles of all language editions combined whereas the German version represents only 4.6%.
(see en.wikipedia.org/wiki/Wikipedia:Size_of_Wikipedia)

References

Abu-Akel, A. (1997). A study of cohesive patterns and dynamic choices utilized by two schizophrenic patients in dialog, pre- and post-medication. *Lang Speech, 40* (Pt 4), 331-351. doi:10.1177/002383099704000402

Allen, H. A., Liddle, P. F., and Frith, C. D. (1993). Negative features, retrieval processes and verbal fluency in schizophrenia. *The British Journal of Psychiatry, 163*(6), 769-775.

American Psychiatric Association. (2013). *Diagnostic and statistical manual of mental disorders: DSM-5* (5. ed.). Washington, DC u.a.: American Psychiatric Publ.

Andreasen, N. C. (1979a). Thought, language, and communication disorders. I. Clinical assessment, definition of terms, and evaluation of their reliability. *Arch Gen Psychiatry, 36*(12), 1315-1321.

Andreasen, N. C. (1979b). Thought, language, and communication disorders: II. Diagnostic significance. *Archives of general Psychiatry, 36*(12), 1325-1330.

Andreasen, N. C. (1984). *Scale for the assessment of positive symptoms (SAPS)*: University of Iowa Iowa City.

Andreasen, N. C. (1989). The Scale for the Assessment of Negative Symptoms (SANS): Conceptual and Theoretical Foundations. *British Journal of Psychiatry, 155*(S7), 49-52. doi:10.1192/S0007125000291496

Andreasen, N. C., and Grove, W. M. (1986). Thought, language, and communication in schizophrenia: diagnosis and prognosis. *Schizophrenia bulletin, 12*(3), 348-359.

Arora, S., Liang, Y., and Ma, T. (2016). A simple but tough-to-beat baseline for sentence embeddings.

Bartolucci, G., and Fine, J. (1987). The frequency of cohesion weakness in psychiatric syndromes. *Applied Psycholinguistics, 8*(1), 67-74. doi:10.1017/S0142716400000072

Bedi, G., Carrillo, F., Cecchi, G. A., Slezak, D. F., Sigman, M., Mota, N. B., . . . Corcoran, C. M. (2015). Automated analysis of free speech predicts psychosis onset in high-risk youths. *NPJ Schizophr, 1*, 15030. doi:10.1038/npjschz.2015.30

Bleuler, E. (1911/1950). Dementia praecox or the group of schizophrenias.

Bleuler, E. (1911/1975). » Lehrbuch der Psychiatrie «, 13., von Manfred Bleuler neubearb. *Aufl., Berlin/Heidelberg/New York.*

Broome, M. R., Bottlender, R., Rösler, M., and Stieglitz, R. (2017). *The AMDP System: Manual for Assesment and Documentation of Psychopathology in Psychiatry*: Hogrefe Publishing.

Buck, B., Ludwig, K., Meyer, P. S., and Penn, D. L. (2014). The use of narrative sampling in the assessment of social cognition: the Narrative of Emotions Task (NET). *Psychiatry Res, 217*(3), 233-239. doi:10.1016/j.psychres.2014.03.014

Chaika, E. (1974). A linguist looks at schizophrenic language. *Brain and Language, 1*(3), 257-276. doi:10.1016/0093-934X(74)90040-6

Chaika, E., and Lambe, R. A. (1989). Cohesion in schizophrenic narratives, revisited. *J Commun Disord, 22*(6), 407-421.

Covington, M. A., He, C., Brown, C., Naçi, L., McClain, J. T., Fjordbak, B. S., . . . Brown, J. (2005). Schizophrenia and the structure of language: the linguist's view. *Schizophrenia research, 77*(1), 85-98.

Ditman, T., and Kuperberg, G. R. (2010). Building coherence: A framework for exploring the breakdown of links across clause boundaries in schizophrenia. *J Neurolinguistics, 23*(3), 254-269. doi:10.1016/j.jneuroling.2009.03.003

Docherty, N. M., Rhinewine, J. P., Labhart, R. P., and Gordinier, S. W. (1998). Communication disturbances and family psychiatric history in parents of schizophrenic patients. *The Journal of nervous and mental disease, 186*(12), 761-768.

Elvevag, B., Foltz, P. W., Weinberger, D. R., and Goldberg, T. E. (2007). Quantifying incoherence in speech: an automated methodology and novel application to schizophrenia. *Schizophr Res, 93*(1-3), 304-316. doi:10.1016/j.schres.2007.03.001

Fang, A., Macdonald, C., Ounis, I., and Habel, P. (2016). *Using Word Embedding to Evaluate the Coherence of Topics from Twitter Data*. Paper presented at the Proceedings of the 39th International ACM SIGIR conference on Research and Development in Information Retrieval, Pisa, Italy.

Guy, W. (2000). Clinical global impressions (CGI) scale. *Handbook of Psychiatric Measures. Washington, DC: American Psychiatric Association*, 100-102.

Hinzen, W., and Rosselló, J. (2015). The linguistics of schizophrenia: thought disturbance as language pathology across positive symptoms. *Frontiers in Psychology, 6*(971). doi:10.3389/fpsyg.2015.00971

Iter, D., Yoon, J., and Jurafsky, D. (2018). *Automatic Detection of Incoherent Speech for Diagnosing Schizophrenia*.

Kuperberg, G. R. (2010). Language in schizophrenia Part 1: an Introduction. *Lang Linguist Compass, 4*(8), 576-589. doi:10.1111/j.1749-818X.2010.00216.x

Landauer, T. K., and Dumais, S. T. (1997). A solution to Plato's problem: The latent semantic analysis theory of acquisition, induction, and representation of knowledge. *Psychological Review, 104*(2), 211-240. doi:10.1037/0033-295X.104.2.211

Lecours, A., and Vanier-Clément, M. (1976). Schizophasia and jargonaphasia: A comparative description with comments on Chaika's and Fromkin's respective looks at "schizophrenic" language. *Brain and language, 3*(4), 516-565.

Levy, O., Goldberg, Y., and Dagan, I. (2015). Improving Distributional Similarity with Lessons Learned from Word Embeddings. *Transactions of the Association for Computational Linguistics, 3*, 211-225.

Liddle, P. F., Ngan, E. T., Caissie, S. L., Anderson, C. M., Bates, A. T., Quested, D. J., . . . Weg, R. (2002). Thought and Language Index: an instrument for assessing thought and language in schizophrenia. *Br J Psychiatry, 181*, 326-330.

Lintean, M., Moldovan, C., Rus, V., and McNamara, D. (2010). *The role of local and global weighting in assessing the semantic similarity of texts using latent semantic analysis.* Paper presented at the Twenty-Third International FLAIRS Conference.

Mann, W. C., and Thompson, S. A. (1987). Rhetorical Structure Theory: Description and Construction of Text Structures. In G. Kempen (Ed.), *Natural Language Generation: New Results in Artificial Intelligence, Psychology and Linguistics* (pp. 85-95). Dordrecht: Springer Netherlands.

Mercado, C. L., Johannesen, J. K., and Bell, M. D. (2011). Thought disorder severity in compromised, deteriorated, and preserved intellectual course of schizophrenia. *J Nerv Ment Dis, 199*(2), 111-116. doi:10.1097/NMD.0b013e3182083bae

Mikolov, T., Chen, K., Corrado, G., and Dean, J. (2013). Efficient estimation of word representations in vector space. *arXiv preprint arXiv:1301.3781.*

Minor, K. S., Willits, J. A., Marggraf, M. P., Jones, M. N., and Lysaker, P. H. (2019). Measuring disorganized speech in schizophrenia: automated analysis explains variance in cognitive deficits beyond clinician-rated scales. *Psychol Med, 49*(3), 440-448. doi:10.1017/S0033291718001046

Norman, R. M. G., Malla, A. K., Cortese, L., and Diaz, F. (1996). A study of the interrelationship between and comparative interrater reliability of the SAPS, SANS and PANSS. *Schizophrenia Research, 19*(1), 73-85. doi:https://doi.org/10.1016/0920-9964(95)00055-0

Pagliardini, M., Gupta, P., and Jaggi, M. (2018). *Unsupervised learning of sentence embeddings using compositional n-gram features.* Paper presented at the Proceedings of the 2018 Conference of the North American Chapter of the Association for Computational Linguistics: Human Language Technologies, Volume 1 (Long Papers).

Pennington, J., Socher, R., and Manning, C. (2014). *Glove: Global vectors for word representation.* Paper presented at the Proceedings of the 2014 conference on empirical methods in natural language processing (EMNLP).

Roche, E., Creed, L., MacMahon, D., Brennan, D., and Clarke, M. (2015). The Epidemiology and Associated Phenomenology of Formal Thought Disorder: A Systematic Review. *Schizophr Bull, 41*(4), 951-962. doi:10.1093/schbul/sbu129

Rochester, S. (2013). *Crazy talk: A study of the discourse of schizophrenic speakers*: Springer Science & Business Media.

Schmidt, K., and Metzler, P. (1992). Wortschatztest (WST). Beltz. In: Weinheim.

Sheehan, D. V., Lecrubier, Y., Sheehan, K. H., Amorim, P., Janavs, J., Weiller, E., . . . Dunbar, G. C. (1998). The Mini-International Neuropsychiatric Interview (MINI): the development and validation of a structured diagnostic psychiatric interview for DSM-IV and ICD-10. *The Journal of clinical psychiatry.*

Stede, M. (2007). *Korpusgestützte Textanalyse: Grundzüge der Ebenen-orientierten Textlinguistik.* Tübingen: Narr.

van Erp, T. G. M., Preda, A., Nguyen, D., Faziola, L., Turner, J., Bustillo, J., . . . Fbirn. (2014). Converting positive and negative symptom scores between PANSS and SAPS/SANS. *Schizophrenia research, 152*(1), 289-294. doi:10.1016/j.schres.2013.11.013

A Appendices

Incoherence model						
Sentence	**Word**	**Patients with positive FTD**	**Patients without positive FTD**	**Controls**	**F [a]**	**p**
Baseline	Word2Vec	.740 (.071) [†]	.721 (.057)	.748 (.057)	.510	.606
Mean Vector	Glove	.827 (.05)	.806 (.075)	.814 (.045)	.338	.716
	Word2Vec	.778 (.048)	.769 (.045)	.775 (.046)	.109	.897
TF-IDF	Glove	.228 (.054)	.249 (.046)	.291 (.037)	**4.735****	**.017**
	Word2Vec	.587 (.07)	.558 (.082)	.597 (.052)	.857	.435
SIF	Glove	.103 (.05)	.061 (.059)	.064 (.045)	2.012	.153
	Word2Vec	.097 (.053)	.046 (.062)	.073 (.054)	2.068	.146
Sent2Vec	Sent2Vec	.164 (.021)	.157 (.025)	.163 (.018)	.300	.743
Tangentiality model						
Sentence	**Word**	**Patients with positive FTD**	**Patients without positive FTD**	**Controls**	**F**	**p**
Baseline	Word2Vec	.263 (.217)	.444 (.338)	.221 (.156)	2.273	.122
Mean Vector	Glove	2.022 (1.481)	2.534 (1.755)	2.326 (.822)	.334	.719
	Word2Vec	1.577 (.852)	2.058 (1.285)	1.857 (.909)	.547	.585
TF-IDF	Glove	5.512 (1.874)	5.823 (2.784)	6.812 (3.465)	.594	.559
	Word2Vec	3.89 (.933)	4.965 (1.704)	5.479 (2.707)	1.777	.188
SIF	Glove	4.709 (1.293)	5.143 (1.196)	4.76 (1.75)	.275	.762
	Word2Vec	4.1 (1.044)	5.008 (2.195)	4.256 (1.642)	.821	.451
Sent2Vec	Sent2Vec	2.889 (.776)	2.381 (.873)	2.979 (.834)	1.517	.237

[†] Mean (SD); [a] ANOVA; **$p < .05$

Appendix A: Extended experimental results.

Overcoming the bottleneck in traditional assessments of verbal memory: Modeling human ratings and classifying clinical group membership

Chelsea Chandler[1], Peter W. Foltz[1,2], Jian Cheng[3], Jared C. Bernstein[3], Elizabeth P. Rosenfeld[3], Alex S. Cohen[4], Terje B. Holmlund[5], and Brita Elevåg[5,6]

[1]University of Colorado Boulder, {chelsea.chandler, peter.foltz}@colorado.edu
[2]Pearson [3]Analytic Measures Inc.
[4]Louisiana State University
[5]University of Tromsø [6]Norwegian Centre for eHealth Research

Abstract

Verbal memory is affected by numerous clinical conditions and most neuropsychological and clinical examinations evaluate it. However, a bottleneck exists in such endeavors because traditional methods require expert human review, and usually only a couple of test versions exist, thus limiting the frequency of administration and clinical applications. The present study overcomes this bottleneck by automating the administration, transcription, analysis and scoring of story recall. A large group of healthy participants (n = 120) and patients with mental illness (n = 105) interacted with a mobile application that administered a wide range of assessments, including verbal memory. The resulting speech generated by participants when retelling stories from the memory task was transcribed using automatic speech recognition tools, which was compared with human transcriptions (overall word error rate = 21%). An assortment of surface-level and semantic language-based features were extracted from the verbal recalls. A final set of three features were used to both predict expert human ratings with a ridge regression model (r = 0.88) and to differentiate patients from healthy individuals with an ensemble of logistic regression classifiers (accuracy = 76%). This is the first 'outside of the laboratory' study to showcase the viability of the complete pipeline of automated assessment of verbal memory in naturalistic settings.

1 Introduction

Assessing human memory is one of the most important ways in which neurocognitive function is established. Memory is of central interest in numerous neurodevelopmental, neurodegenerative and neuropsychiatric conditions, as well as in brain injuries that affect cortical and subcortical brain systems (Baddeley and Wilson, 2002).

Given the importance of verbal memory, it is a core component of the globally employed Wechsler Memory Scale (Wechsler, 1997). The Logical Memory subtest requires the repetition of short stories that have been spoken by the examiner, both immediately and after a delay. Administering these assessments requires participants to be physically present with the examiner, who then gives scores manually by assigning points for key words or thematic units correctly recalled. The required time-consuming human review combined with the availability of only a couple of test versions limits their use and as such contributes to the bottleneck in the assessment of verbal memory. This unfortunately translates into infrequent assessments.

Automating certain aspects of such assessments holds promise of enabling more regular assessments as well as remote ones, which may be beneficial for monitoring treatment effectiveness and may also avert tragedy. Given that the verbal memory task is spoken, it is well-suited for automatization by leveraging recent advances in speech technology and machine learning. It has become possible to assess not just the words generated, but deeper measures of semantic understanding, which can be used to develop objective and sensitive metrics from the speech of patients with dementia (Fraser and Hirst, 2016; Yancheva and Rudzicz, 2016; Zhou et al., 2016), aphasia (Fraser et al., 2013), autism (Losh and Gordon, 2014; Prud'hommeaux et al., 2017; Goodkind et al., 2018), and mental illness (Elevåg et al., 2007, 2010; Rosenstein et al., 2015; Bedi et al., 2015; Iter et al., 2018; Corcoran et al., 2018; Holmlund et al., 2019b). However, it is now time to move beyond simple proof of concept and translate such findings into viable clinical tools (Foltz et al., 2016). Indeed, since machine learning based approaches make it possible to mimic the actual assessment processes employed by expert humans,

Proceedings of the Sixth Workshop on Computational Linguistics and Clinical Psychology, pages 137–147
Minneapolis, Minnesota, June 6, 2019. ©2019 Association for Computational Linguistics

the modeling and prediction of cognitive functions can be done by a machine in much the same manner as by humans. Thus, the entire pipeline can be automated from administration and transcription, to analysis and actual scoring of memory recall.

In the present research, we applied computational approaches to the speech generated from participants retelling stories from the verbal recall task in order to characterize the quality of their recall and determine the accuracy of this characterization. The approach developed natural language processing (NLP) measures that were designed to align with features related to verbal memory and story recall in order to best assess the data. This study focused on two computational tasks: 1) automatically assigning ratings to participants' retells based on how much of the content from the original story they remembered, and 2) performing a classification task to distinguish psychiatric patients from healthy participants. The study further examined how well these measures can be incorporated into a full analysis pipeline starting from data collection on a mobile platform outside of the traditional laboratory (thus in the real-world, perhaps noisy, environment), to automated speech recognition (ASR), and then to the conversion of the language to predictions of recall quality.

2 Related Work

NLP has been used in a range of clinical applications from detecting depression in twitter feeds (Coppersmith et al., 2015) to analyzing coherence in patient-clinician interactions (Elvevåg et al., 2007). In each case, text is reduced to a set of variables to relate to clinical measures of interest.

There are several classes of variables that can encode characteristics of texts. One class of measures are considered surface features of language. This includes counts of words, phrases, and words related to cognitive and affective processes (Pennebaker et al., 2015; Prud'hommeaux and Roark, 2011). A second class of measures examines structural features of language, such as parses of the syntactic structure, the probabilities that word pairs would likely occur together (e.g., n-grams), and the cohesion and coherence of a text. Finally, semantic features assess the meaning expressed in texts, such as choice of words as they relate to a specific topic, as well as encoding the underlying meaning of words, sentences, or whole passages. Such measures are often based on corpora that en-

code general knowledge of the world or a domain to measure meaning at a conceptual level rather than through the counting of direct overlap.

Previous studies have measured story recall by computing the distance between two pieces of text (Lehr et al., 2012, 2013). For example, a participant's retell can be compared against the original story to determine the amount of information retained. One approach to measuring this distance is computing a word alignment between the texts, which relies on participants using *exact* words and phrases to achieve a high memory score (Prud'hommeaux and Roark, 2011). A more robust approach is to measure the distance in a derived embedding space between two pieces of text. Latent Semantic Analysis (LSA) (Landauer et al., 1998) applies a singular value decomposition to a matrix of word-document co-occurrences in a large corpus. It then uses the cosine distance between representations which is able to account for semantic relationships in which a participant may make small changes in concepts such as "store" and "market". Recent studies (Dunn et al., 2002; Rosenstein et al., 2014) have used LSA to successfully model recall data from the Logical Memory subtest of the Wechsler Memory Scale to quantify the degradation of performance with increasing retrieval intervals.

More recently, word embedding models have been applied to assessing clinical discourse. Iter et al. (2018) modeled the coherence of patient discourse using LSA, word2vec (Mikolov et al., 2013) and GloVe (Pennington et al., 2014). While LSA derives its semantic context from a bag-of-words across documents, the word2vec word embedding model derives its representation by considering the contexts in which each word appears by examining the window of words around each word. This window measures context, either taking into account the order of the words in that window or independent of word order. An advantage of the latter approach is that the method learns both semantic associations and syntactic word order.

3 Data

The present study was the result of data collection through a mobile application for the purpose of longitudinally tracking the mental state of psychiatric patients (Cohen et al., 2019; Holmlund et al., 2019a). The application is composed of a number of assessment tasks that engage participants in spoken and touch-based interactions in order to capture daily measures of cognition, affect and clinical state.

As part of the overall examination, participants' verbal memory was assessed. Stories were presented orally in a male voice and the participant was then immediately asked to retell the story with as many details as possible. After a delay of approximately one day, they were prompted to retell the same story. Each participant was presented with one new story per session and all stories were sampled across participants. There were a total of 24 different stories developed to be structurally similar to the Logical Memory subtest of the Wechsler Memory Scale-III (Wechsler, 1997). Multiple versions were created to enable frequent administration, as there exist only two test versions in the Wechsler Memory Scale which limits the frequency of administration and hence its clinical application. The stories were narrative in nature and ranged from 61 to 82 words in length. They each had two characters, a setting, an action that caused a problem, and a resolution. An example story is as follows:

> "On Monday morning, the woman woke up more tired than usual. When she walked downstairs to make herself a cup of coffee, she found her husband in the kitchen. She was surprised because he usually left an hour before she woke up. Her husband greeted her and reminded her that daylight savings time was over. Realizing the clocks were wrong, she happily ran upstairs and jumped back into bed."

Since this research concerned itself with evaluating the viability of leveraging speech technologies to automate a traditional verbal memory task, our focus was on usability engineering to ensure a robust design that could be implemented on a large scale, out of the controlled laboratory, and self-administered by the participant themselves. Therefore, the traditional matching of groups was not considered a priority, and nor is this feasible in machine learning studies that seek sample sizes in the thousands. Our participants comprised 105 stable patients with mental illness at a substance use treatment program and 120 undergraduate students at Louisiana State University presumed to be healthy (henceforth termed 'healthy participants'; see Holmlund et al., 2019a for details). This research program was approved by the relevant ethics committee (LSU Institutional Review Board #3618) and participants provided their informed written consent to this study. The 105 patients produced 750 retell responses, of which 575 were immediate retells and 175 were delayed retells. Each patient produced between 2 and 19 retells, with an average of 7.35 and standard deviation of 4.50. The 120 healthy participants produced 427 retell responses, of which 216 were immediate retells and 211 were delayed responses. Each produced between 2 and 15 retells with an average of 4.97 and standard deviation of 2.76. The scale of the collected data was impressive in size and quality given that an experimenter was not present during administration.

4 Human Rating of Story Recall

The audio of the memory recalls were transcribed by humans. Trained human raters read the transcriptions and assigned scores on the quality and amount of concepts and themes recalled, including characters, events, dates, descriptors, and feelings. The scores assigned were on a scale from 1 to 6, with 1 indicating no details were recalled, and 6 indicating all major and almost all minor concepts and themes were recalled. The responses were rated by three trained human raters with clinical experience. A subset (326) of the responses were rated by two independent raters in order to verify inter-rater reliability (r = 0.92). The high degree of agreement suggests that the rating rubric was reliable and thus appropriate for use in training a machine learning algorithm.

Over all the ratings, healthy participants generally received higher ratings for the amount of content recalled from the original story. For the immediate retells, they received an average rating of 4.31 (SD = 1.38) as compared to patients' average rating of 3.15 (SD = 1.44, t = 9.5, p < .001). The biggest differentiator between the two groups was in delayed retell (healthy participants average = 3.95, SD = 1.45; patients average = 2.24, SD = 1.66; t = 9.8, p < .001). Figure 1 shows that the average ratings assigned to patients on both the immediate and delayed retell were significantly lower than the average ratings assigned to healthy participants. The wide error bars indicate a large variability in the averages among both groups.

The two groups of participants differed both in

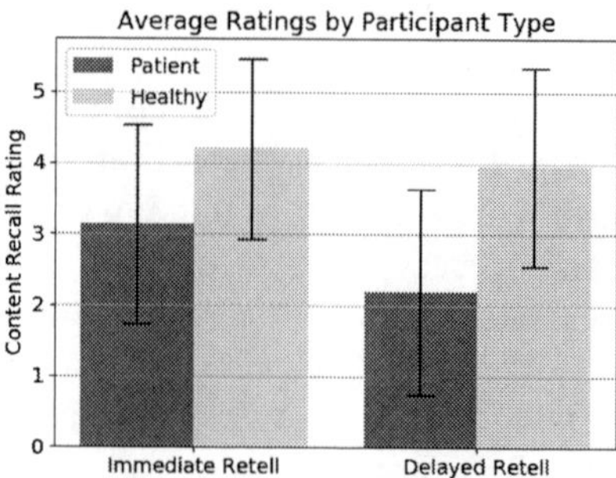

Figure 1: Average ratings by participant type. Error bars represent standard deviations of the samples.

the number of words typically spoken in a retell, and also in the relevance of their retell to the original story. While the histograms in figure 2 are somewhat biased since there was an uneven breakdown in the number of samples analyzed between the two groups, they do show that the peak of the distribution for patients is skewed more to the lower word counts than the peak for the healthy participants.

A noteworthy observation from the data is the amount of missing or silent responses. The tasks were self-administered by the participant outside of a traditional controlled setting, and there were several responses that were either silent or along the lines of "I don't remember". As expected, this type of missing data was more prevalent among patients than healthy participants, with 5% of the patient immediate retells and 19% of the patient delayed retells being less than 5 words or silent. While this is a constraint in live data collection in uncontrolled real world settings, it is a trait of realistic data that it will never be perfect and forced the creation of models capable of generating predictions on imperfect data. Instead of including silent responses (and thus allowing a classifier to learn that this is a trait common to patients), all silent responses were eliminated in order to create models that learn based on the language production, not the lack of any language.

5 Overview of Analysis Approach

There were four major components to this study. The first was feature engineering in order to determine a set of features that could be instantiated through computational NLP approaches and would assess important aspects of recall. We narrowed the large feature set down to only those most relevant to the constructs of story memory. Second, we built a regression model that could predict the ratings an expert human would assign to a story recall. Third, in order to show the predictive power of our data, we used the same features in a classification model to predict whether a participant was a patient or healthy participant. Fourth, in order to fully automate the pipeline, these analyses were completed on transcripts derived using ASR rather than the human transcriptions.

6 Feature Engineering

In designing NLP-based features to assess recall, it was critical to consider what aspects were most significant. A retell can be characterized by the amount of information recalled, the level of detail, changes in structure, as well as the quality of expressed language. Linguistic surface features provide indications of the overall amount of information recalled. Overuse of particular parts of speech, such as determiners, have been shown to provide indications of language ability, in that certain language constructions may indicate more sophisticated ability (Bedi et al., 2015). Retells, however, are affected by transformations of words within semantic memory (Kintsch, 1988). Indeed, surface features of a story (e.g., exact wording) are quickly lost in memory, but the gist is retained. Although a story may contain the word "market", a person may recall it as a "store". Thus, features that can account for semantics may be more effective at measuring the degree to which a memory has changed, with subtle effects of synonymy. Therefore, we investigated a variety of feature types ranging from linguistic surface features such as word counts to semantic features like cosine similarity between embedded representations.

The surface features included either raw or normalized counts of the number of tokens (word count), types (unique word count), n-grams (counts of word sequences of length n), or particular parts of speech. The surface features, while not the most sophisticated, nonetheless proved to be highly predictive. For instance, a simple count of the tokens informed how detailed the retell was. Whether the details aligned with the original story or not was revealed by the more advanced surface and semantic features. We further explored the use of specific parts of speech and ambiguous pronoun usage as Iter et al. (2018) concluded these are

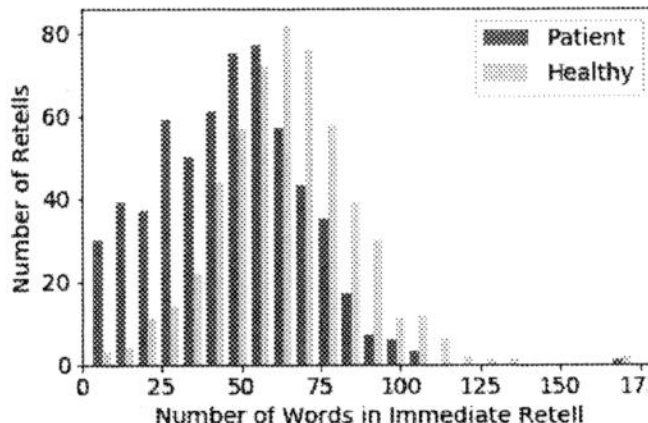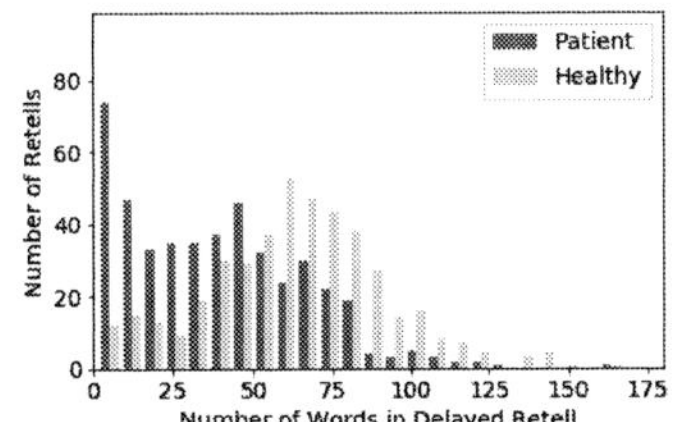

Figure 2: Immediate (left) and delayed (right) retell word count histograms by participant type.

traits of disordered speech. Since our data were composed of short responses that were fairly constrained in content, these features did not prove to be especially useful.

A step beyond raw counts is overlapping surface features, e.g. alignment, between the original story and the retell (Prud'hommeaux and Roark, 2012). The number of overlapping types between the original story and the retell measured how many concepts were remembered. For instance, if a retell stated that the event took place on "an afternoon" when it was actually "a rainy afternoon" in the original story, the type overlap can pick up on a missing detail. These counts offered a semantic relatedness indicator since recall of words from the original prompt was a good measure of memory, however, more interesting were metrics that could measure semantic similarity directly, somewhat independent of surface features.

Semantic features can be analyzed by using different types of embedded representations and metrics to score the distance between these representations. Word embeddings are widely employed to represent the semantic content as well as syntactic relationships of variable-length pieces of text. In this study, we tested pre-trained word embeddings, including word2vec and GloVe, and found that the pre-trained word2vec Google News corpus word embedding model (3 million 300-dimensional embeddings) produced results most correlated with our data.

Calculating the cosine distance between the average (both tf-idf weighted and unweighted) of the word embedding representations of two documents is a standard metric in NLP. We tested this in the current study, as well as the word mover's distance (WMD). Cosine distance was not as effective as WMD as it tends to smooth out the importance of individual words.

WMD is a good metric for analysing recall data as it captures word meaning and how semantically

distant each word in a document is to its closest aligned word in another document. Thus, for verbal memory assessment, it provides a way to characterize how much semantic change there is from the original story to the recalled story. Put simply, WMD finds a mapping from each word in a document to its closest counterpart in the other and the distance is calculated as the sum of all Euclidean distances between matched words. Figure 3 illustrates the WMD calculation on document 1 (D1) and document 2 (D2) from a single source document (D0). Ignoring stop words, the model first finds a pairing between the most semantically similar words in the two documents. The arrows drawn between words in the documents represent a matching and are labeled with their distance contribution. WMD calculates a total distance as a function of all word pairings. D1 and D2 have an equal 'bag of words' distance of 0 from D0 as there are no overlapping content tokens, but semantically, D1 is much closer than D2. WMD is a more sophisticated method than cosine distance and has been shown to outperform it in many classification tasks (Kusner et al., 2015). For example, we compared the embedded representation for each participant's retell to the embedded representation for each original story using both the cosine metric and WMD, and overall the WMD metric correlated -0.82 with the human ratings while the cosine metric correlated -0.72.

A final feature considered was retell structure. Prior work has shown that language coherence can be useful clinically and predict risk of psychosis onset. To measure coherence, word embeddings are generated of n contiguous words in the retell and the semantic similarity to the embedded representation of the next n words is computed. Then the window is moved ahead by one word to make the next comparison, and then all the semantic similarities are averaged (Elvevåg et al., 2007). This approach provides a smoothed metric of the

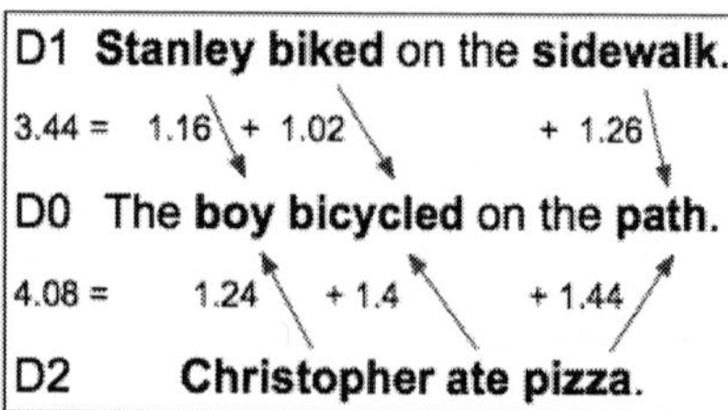

Figure 3: Example adapted from original paper by Kusner et al. (2015). The source sentence D0 and two query sentences, D1 and D2, aligned by words, and document distance computed by some function of total word pair distances.

cohesiveness of a retell, in that if the response is tangential or switches topics, it was assigned a lower overall coherence. In the present study, using a window size of four words on the retells correlated at 0.39 to the human ratings of the retells, indicating that better retells tended to be more coherent.

7   Human Rating Prediction Models

To fully automate the modeling of memory recall, a regression model was created that assigned a performance score to a story retell, treating immediate and delayed retells as the same task. Using a combination of univariate statistical tests and recursive feature elimination on the feature set, we identified the best combination of 3 features. They were not collinear and accounted for aspects of the rating task that align well with attributes that trained humans look for when rating recall. The features assessed the overall amount of content generated, the direct overlap of word types with the original story and the overall semantic change.

A ridge regression model was trained with a regularization parameter set to 0.01. We chose only three features to incorporate into the model in order to derive a system that is simple and interpretable. The three features used to generate ratings were the common types between the original story and retell (mean regression coefficient of 3.14), the word type count in the retell (mean regression coefficient of 2.47), and the word mover's distance between the original story and retell (mean regression coefficient of -2.71). 4 shows the correlations of each of the features to the rating given to the retell. The overall average correlation (Pearson r) with the human rating over 10-fold cross-validation through the data was 0.88. This average correlation of 0.88 of the model

to the average human rating was in line with the 0.9 correlation between human raters. The implication is that automated assessment performs on par with humans, and additionally is an unbiased and convenient method. Success notwithstanding, it should be noted that the model performed poorly on responses that should have received low scores because key details of the original story were not recalled, but achieved a high word count, token overlap, and a reasonable word mover's distance. For instance, when a participant was prompted to retell the "balloon story" yet could not remember much, since prompted to talk about balloons, they were nonetheless able to ramble on about balloons, in essence 'fooling' the regression model.

8   Classification of Clinical Group Membership

The ability to automatically score recall is most definitely noteworthy, but the predictive power of the features was additionally demonstrated with a classification task which successfully identified the clinical group membership of the participant. Given that participants recalled each story twice, three classes of features were derived from the data: (i) how similar the initial retell was to the original story, (ii) how similar the delayed retell was to the original story, and (iii) how similar the initial retell was to the delayed retell.

As mentioned in the data section above, a goal of the current study was for the model to perform well in participants who were unable to complete both parts of the task. Therefore, an ensemble classifier was necessary to retain data for partial task completion. Each classifier made a classification based on features derived from a single session and the resulting subject-level classification was made from a combination of the individual session's prediction probabilities. This allowed silent or missing retells to be discarded yet still make predictions based on language data.

Prior applications of computational approaches in the cognitive health field have tended to perform classifications on a session-level (Prud'hommeaux and Roark, 2011; Rosenstein et al., 2014) rather than examining recall over multiple sessions. It was a goal of this research program to build a longitudinal model of behavior of an individual participant, so while the classifiers generated probability calculations at the session-level, all of these probabilities were aggregated over time and

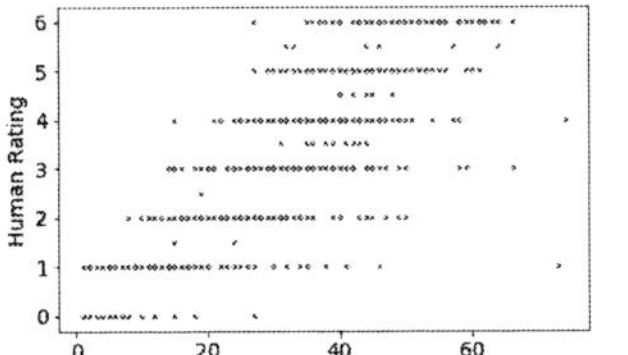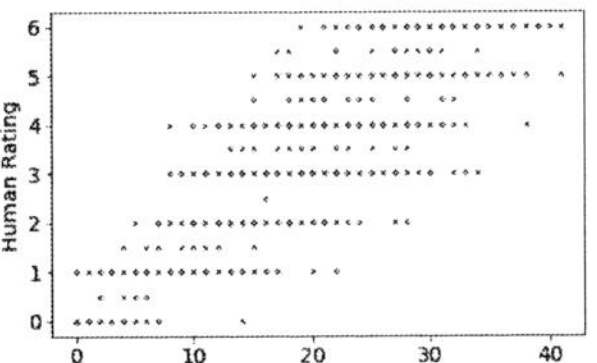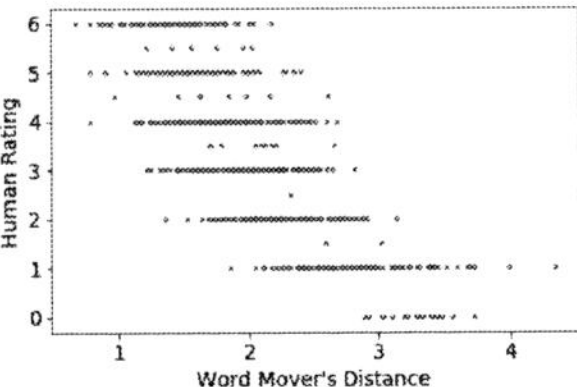

Figure 4: Scatter plots of our top features with human ratings. The number of common word types between the original story and the retell has a Pearson r correlation to average human ratings of 0.86, the number of word types in the retell has a Pearson r correlation of 0.82, and the word mover's distance between the original story and the retell has a Pearson r correlation of -0.82.

	Patient	Healthy
# immediate retells	575	216
# delayed retells	175	211
Average retells per participant	7.35, SD = 4.50	4.97, SD = 2.76
Range of retells per participant	[2,19]	[2,15]

Table 1: Breakdown of retell counts.

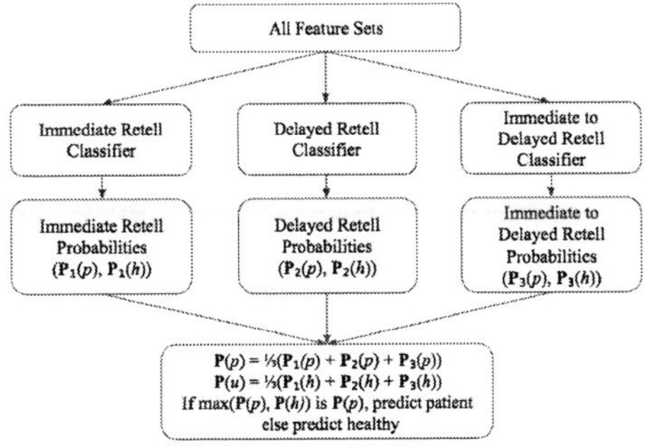

Figure 5: Ensemble Classifier Diagram.

tasks to make a final prediction at the subject-level. Leave-one-out cross-validation was performed across data from individual participants, training on all participants but one, and then subsequently testing on the one who was left out. Table 1 below contains a detailed breakdown of how many retells constituted profiles of the different groups, excluding any silent responses or responses with less than 5 words, which resulted in a disproportionate loss of delayed retells in patients.

The features used in the retell classifier were the number of unique types in the retell, the number of overlapping types between the original and retell, and the word mover's distance between the original and retell. Unsurprisingly, word mover's distance proved to be the most significant feature in the classifier. The delayed retell classifier was composed of the same features, but with calculations made on the delayed retell *in lieu* of the immediate retell. The last classifier, which focused on the change between the initial and delayed retell utilized two features: the number of common types between the immediate and delayed retell and the word mover's distance between the immediate and delayed retell.

The workflow for the ensemble classifier is shown in figure 5. The three classifiers were logistic regression classifiers optimized individually at a session-level. For instance, the retell classifier was trained on all retell features in the data, and predicted only on these inputs. Each classifier returned a tuple for each session of the probability that the session belonged to a healthy participant, $P_x(h)$, and to a patient, $P_x(p)$, where h represented the healthy class and p represented the patient class, x represented the classifier type, either *retell*, *reretell*, or *change*, and $P_x(h) + P_x(p) = 1$. The model then summed the probabilities coming from each classifier and normalized the summation to reach a final class membership probability.

$$P(p) = P_{retell}(p) + P_{reretell}(p) + P_{change}(p)$$
$$P(h) = P_{retell}(h) + P_{reretell}(h) + P_{change}(h)$$

The final prediction was a patient if $P(p) > P(h)$, otherwise it was a healthy participant. Put simply, the class with the largest overall probability was taken as the prediction.

The model correctly classified 78% of the patients and 74% of the healthy participants. Table 2 shows the confusion matrix from this classification model. The delayed retell classifier was the most accurate of the three as it was the biggest differentiator in performance between the two classes.

Patients misclassified as healthy had highly

	Predicted Healthy	Predicted Patient	Total
True Healthy	**89**	31	120
True Patient	23	**82**	105
Total	112	113	

Table 2: Confusion matrix of ensemble classification model. Model accuracy = 76%, precision = 73%, recall = 80%, F1 score = 76%.

rated retells that overlapped both semantically and at a word level with the original stories. Healthy participants misclassified as patients had multiple "I don't remember" or silent responses so their memory performance was characterised as poor.

9 Automated Speech Recognition

These results demonstrate that transcribed retells can be accurately characterized through computational methods. However, humans transcriptions in real-time are not practically viable. Therefore, to test how the methods would work in a fully automated pipeline, the same retells as generated by ASR were assessed. The audio files were run through two systems: (i) the latest Google Speech API (https://cloud.google.com/speech-to-text) which is a deep learning-based model trained on general English language, and (ii) a task-specific model that used a Deep Neural Network - Hidden Markov Model (Zhang et al., 2014) containing 5 hidden layers and 350 p-norm (p = 2) non-linearity neurons with a group size G = 10 per hidden layer, trained using Librispeech's (Panayotov et al., 2015) 960 hours of clean native (L1) reading data (Cheng, 2018). No speech data from the current dataset were used to train this acoustic model, but a 5-gram model based on the retells was used for the recognition.

Using Google's acoustic model, the average word error rate compared to human transcriptions across all patient retells was 26.51% and 16.38% across all healthy participant retells, totalling 20.90% on average. Using the task-specific model, the average patient word error rate was substantially less at 13.36% and the average healthy participant word error rate was 5.90% with an overall average of 10.79%. Some word errors were due to different word normalizations, for example "hashbrowns" versus "hash browns". Although transcriptions derived via ASR strayed somewhat from human transcriptions, the same ridge regression model described above, employing the same parameters, was then applied to the ASR-derived transcripts. As compared to the correlation r = 0.88 on a human transcription trained and tested regression model, the Google ASR trained and tested model achieved an r = 0.86, and the custom ASR trained and tested model achieved an r = 0.87. The change in performance on the classification ensemble model was similar; compared with an accuracy of 76% on the human model, both the Google ASR model and the custom ASR model achieved an accuracy of 74%. Thus, even with 10-20% word error rate, the model's predictive performance only lost a few percentage points, likely because it captures multiple aspects of the expressed language and so is highly robust to small errors if the overall sense is retained. The important implication is that audio collected from participants over mobile devices in realistic environments can be automatically transcribed with sufficient accuracy to provide useful predictions. Of note however, the nature of the current task and the fact that the retells had all been transcribed by humans who could screen for any potentially identifying information, ensured that there was zero risk of any identifying information being uploaded to the Google ASR system and thus critically maintained participant privacy. However, research that includes sensitive information (e.g., discussion of symptoms or things of a personal nature) must take additional measures to comply with relevant legislation and privacy protection rules.

10 Assessment Pipeline

This study - as illustrated in Figure 6 - demonstrates the solution to the bottleneck caused by time-consuming human review that is required in traditional settings and the resulting infrequent administration of verbal memory tests in current assessment practice. Our methodology enables the frequent and remote assessment of verbal memory and provides metrics of significant value in the new era of personalized medicine (Insel, 2017).

11 Conclusion

In conclusion, this study has overcome a classic bottleneck in traditional assessment practice and demonstrated that the promise of a truly personalized medicine approach to verbal memory assessment is realistic. The current study has validated the metrics on scores from expert human

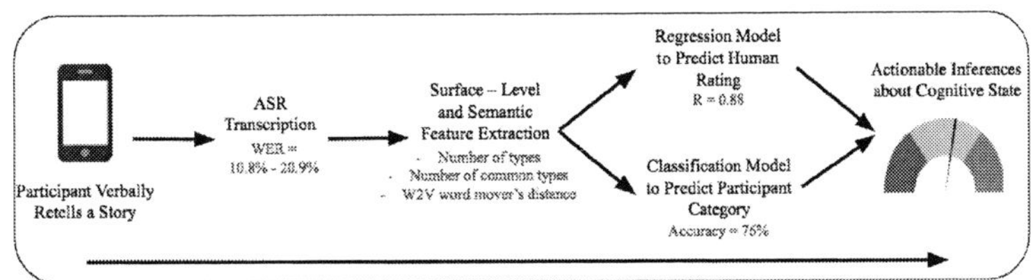

Figure 6: The complete pipeline of automated verbal memory assessment: It begins with a participant verbally retelling a story previously presented. Next, the retell is transcribed by an automatic speech recognition system. Once the speech is converted to text, various features are extracted, and sent to both a regression and classification model for ratings and categorization. Once complete, actionable inferences about cognitive state can be taken.

raters, and validated the actual assessment tool in terms of its functionality and usability. The design is demonstrably and sufficiently robust that this assessment tool is now ready to be applied within clinical settings to track patients longitudinally and inform clinicians accordingly. Future studies need to 'close the triage' by providing semi-immediate feedback from the assessment to the relevant entity. However, establishing the clinical and behavioral implications of such new metrics - such that they are calibrated correctly - remains an extremely complex empirical task which will necessitate the incorporation and modeling of multiple and dynamic data streams, as variables should not be interpreted in isolation when actionable clinical inferences are to be made.

12   Acknowledgements

This project was funded by grant 231395 from the Research Council of Norway awarded to Brita Elvevåg.

References

Alan Baddeley and Barbara A. Wilson. 2002. Prose recall and amnesia: implications for the structure of working memory. *Neuropsychologia*, (40(10)):1737–1743.

Gillinder Bedi, Facundo Carrillo, Guillermo A. Cecchi, Diego Fernández-Slezak, Mariano Sigman, Natalia B. Mota, Sidarta Ribeiro, Daniel C. Javitt, Mauro Copelli, and Cheryl M. Corcoran. 2015. Automated analysis of free speech predicts psychosis onset in high-risk youths. *npj Schizophrenia*, (1:15030).

Jian Cheng. 2018. Real-time scoring of an oral reading assessment on mobile devices. In *Proceedings Interspeech, Hyderabad, India, September 2–6*, pages 1621–1625.

Alex S. Cohen, Taylor Fedechko, Elana Schwartz, Thanh Le, Peter W. Foltz, Jared Bernstein, Jian Cheng, Elizabeth Rosenfeld, Terje B. Holmlund, and Brita Elvevåg. 2019. Ambulatory vocal acoustics, temporal dynamics, and serious mental illness. *Journal of Abnormal Psychology*, (128):97–105.

Glen Coppersmith, Mark Dredze, Craig Harman, and Kristy Hollingshead. 2015. From ADHD to SAD: Analyzing the language of mental health on Twitter through self-reported diagnoses. In *Proceedings of the 2nd ACL Workshop on Computational Linguistics and Clinical Psychology: From Linguistic Signal to Clinical Reality*.

Cheryl M. Corcoran, Facundo Carrillo, Diego Fernández-Slezak, Gillinder Bedi, Casimir Klim, Daniel C. Javitt, Carrie E. Bearden, and Guillermo A. Cecchi. 2018. Prediction of psychosis across protocols and risk cohorts using automated language analysis. *World Psychiatry*, (17(1)):67–75.

John C. Dunn, Osvaldo P. Almeida, Lee Barclay, Anna Waterreus, and Leon Flicker. 2002. Latent semantic analysis: A new method to measure prose recall. *Journal of Clinical and Experimental Neuropsychology*, (24(1)):26–35.

Brita Elvevåg, Peter W. Foltz, Mark Rosenstein, and Lynn DeLisi. 2010. An automated method to analyze language use in patients with schizophrenia and their first-degree relatives. *Journal of Neurolinguistics*, (23):270–284.

Brita Elvevåg, Peter W. Foltz, Daniel R. Weinberger, and Terry E. Goldberg. 2007. Quantifying incoherence in speech: An automated methodology and novel application to schizophrenia. *Schizophrenia Research*, (93(1-3)):304–316.

Peter W. Foltz, Mark Rosenstein, and Brita Elvevåg. 2016. Detecting clinically significant events through automated language analysis: Quo imus? *npj Schizophrenia*, (2:15054).

Kathleen Fraser and Graeme Hirst. 2016. Detecting semantic changes in alzheimer's disease with vector space models. In *LREC*.

Kathleen Fraser, Frank Rudzicz, Naida Graham, and Elizabeth Rochon. 2013. Automatic speech recognition in the diagnosis of primary progressive aphasia.

In *4th Workshop on Speech and Language Processing for Assistive Technologies*, pages 47–54.

Adam Goodkind, Michelle Lee, Gary E. Martin, Molly Losh, and Klinton Bicknell. 2018. Detecting language impairments in autism: A computational analysis of semi-structured conversations with vector semantics. In *Proceedings of the Society for Computation in Linguistics*, volume 1.

Terje B. Holmlund, Jian Cheng, Peter W. Foltz, Alex S. Cohen, and Brita Elvevåg. 2019b. Updating verbal fluency analysis for the 21st century: Applications for psychiatry. *Psychiatry Research*.

Terje B. Holmlund, Peter W. Foltz, Alex S. Cohen, H. D. Johansen, R. Sigurdsen, P. Fugelli, D. Bergsager, Jian Cheng, Jared Bernstein, Elizabeth Rosenfeld, and Brita Elvevåg. 2019a. Moving psychological assessment out of the controlled laboratory setting and into the hands of the individual: Practical challenges. *Psychological Assessment*, (31(3)):292–303.

Thomas R Insel. 2017. Digital phenotyping: Technology for a new science of behavior. In *JAMA*, 318(13), pages 1215–1216.

Dan Iter, Jong H. Yoon, and Dan Jurafsky. 2018. Automatic detection of incoherent speech for diagnosing schizophrenia. In *Proceedings of the Fifth Workshop on Computational Linguistics and Clinical Psychology*, pages 136–146.

Walter Kintsch. 1988. The role of knowledge in discourse comprehension: a construction-integration model. *Psychological Review*, (95):163–182.

Matt Kusner, Yu Sun, Nicholas I. Kolkin, and Kilian Q. Weinberger. 2015. From word embeddings to document distances. In *Proceedings of the 32nd International Conference on Machine Learning*, pages 957–966.

Thomas K. Landauer, Peter W. Foltz, and Darrell Laham. 1998. Introduction to latent semantic analysis. *Discourse Processes*, (25):259–284.

Maider Lehr, Emily Prud'hommeaux, Izhak Shafran, and Brian Roark. 2012. Fully automated neuropsychological assessment for detecting mild cognitive impairment. *Proceedings of the 13th Annual Conference of the International Speech Communication Association.*, 2.

Maider Lehr, Izhak Shafran, Emily Prud'hommeaux, and Brian Roark. 2013. Discriminative joint modeling of lexical variation and acoustic confusion for automated narrative retelling assessment. In *Proceedings of the 2013 Conference of the North American Chapter of the Association for Computational Linguistics: Human Language Technologies*, pages 211–220.

Molly Losh and Peter C. Gordon. 2014. Quantifying narrative ability in autism spectrum disorder: A computational linguistic analysis of narrative coherence. *Journal of Autism and Developmental Disorders*.

Thomas Mikolov, Kai Chen, Greg Corrado, and Jeffrey Dean. 2013. Efficient estimation of word representations in vector space. In *Proceedings of Workshop at ICLR*.

Vassil Panayotov, Guoguo Chen, Daniel Povey, and Sanjeev Khudanpur. 2015. Librispeech: an ASR corpus based on public domain audio books. In *ICASSP*, pages 5206–5210.

James W. Pennebaker, Ryan L. Boyd, Kayla Jordan, and Kate Blackburn. 2015. *The development and psychometric properties of LIWC2015*. Austin, TX: University of Texas at Austin.

Jeffrey Pennington, Richard Socher, and Christopher Manning. 2014. Glove: Global vectors for word representation. In *Proceedings of the 2014 Empirical Methods in Natural Language Processing*, pages 1532–1543.

Emily T. Prud'hommeaux and Brian Roark. 2011. Extraction of narrative recall patterns for neuropsychological assessment. In *Proceedings of the 12th Annual Conference of the International Speech Communication Association (Interspeech)*, pages 3021–3024.

Emily T. Prud'hommeaux and Brian Roark. 2012. Graph-based alignment of narratives for automated neurological assessment. In *Proceedings of the 2012 Workshop on Biomedical Natural Language Processing*.

Emily T. Prud'hommeaux, Jan van Santen, and Douglas Gliner. 2017. Vector space models for evaluating semantic fluency in autism. In *Proceedings of the 55th Annual Meeting of the Association for Computational Linguistics (Short Papers)*, pages 32–37.

Mark Rosenstein, Catherine Diaz-Asper, Peter W. Foltz, and Brita Elvevåg. 2014. A computational language approach to modeling prose recall in schizophrenia. *Cortex*, (55):148–166.

Mark Rosenstein, Peter W. Foltz, Lynn E DeLisi, and Brita Elvevåg. 2015. Language as a biomarker in those at high-risk for psychosis. *Schizophrenia Research*, (165):249–250.

David Wechsler. 1997. *Wechsler Memory Scale - Third Edition, WMS-III: Administration and scoring manual*. The Psychological Corporation.

Maria Yancheva and Frank Rudzicz. 2016. Vector-space topic models for detecting alzheimer's disease. In *Proceedings of the 54th Annual Meeting of the Association for Computational Linguistics*, pages 2337–2346.

Xiaohui Zhang, Jan Trmal, Daniel Povey, and Sanjeev Khudanpur. 2014. Improving deep neural network acoustic models using generalized maxout networks. In *ICASSP*, pages 215–219.

Luke Zhou, Kathleen Fraser, and Frank Rudzicz. 2016. Speech recognition in alzheimers disease and in its assessment. In *Interspeech 2016*, pages 1948–1952.

Analyzing the use of existing systems for the CLPsych 2019 Shared Task

Alejandro González Hevia, **Rebeca Cerezo Menéndez** and **Daniel Gayo-Avello**

University of Oviedo, Spain
{UO251513, cerezorebeca, dani}@uniovi.es

Abstract

In this paper we describe the UniOvi-WESO classification systems proposed for the 2019 Computational Linguistics and Clinical Psychology (CLPsych) Shared Task. We explore the use of two systems trained with ReachOut data from the 2016 CLPsych task, and compare them to a baseline system trained with the data provided for this task. All the classifiers were trained with features extracted just from the text of each post, without using any other metadata. We found out that the baseline system performs slightly better than the pretrained systems, mainly due to the differences in labeling between the two tasks. However, they still work reasonably well and can detect if a user is at risk of suicide or not.

1 Introduction

The objective of this shared task is to predict the degree of suicide risk of a person given the posts that they have made on Reddit. Participants can take part in three different subtasks, which simulate multiple scenarios related to this kind of problems. We will be participating in task A, where we need to assess the level of risk of users given the posts that they have made in the r/SuicideWatch subreddit. In order to participate in this task, all the ethical review criteria mentioned in the shared task paper (Zirikly et al., 2019) were met.

Our main objective is to try to reuse two systems that we have developed and trained for the CLPsych 2016 shared task (Milne et al., 2016), and to evaluate how these systems perform compared to a baseline model trained specifically for this task. We also want to evaluate the use of cross-lingual word embeddings, which could be useful in similar tasks which use posts from forums written in different languages besides English.

The remainder of the paper is organized as follows. In Section 2 we are going to present the data

used for these models. In Section 3 we will describe the systems that we have submitted for the task. In Section 4 we will present the results that we have obtained for each submitted model. Finally, we will summarize our conclusions in Section 5 .

2 Data

2.1 Baseline system

The baseline system was trained using the data provided for this shared task, which is an adaptation of the University of Maryland Reddit Suicidality Dataset (Shing et al., 2018), constructed using posts from Reddit. For task A, there are 847 labeled posts made by 496 different users on the SuicideWatch subreddit. Each user is annotated with one of the following 4 labels: *No risk, Low risk, Moderate risk* and *Severe risk*, indicating the degree of suicide risk of the user. In order to obtain the final label of the user's level of risk his posts are divided into several annotation units, and the highest risk level of the annotation units is assigned to the user. However, for this task we only rely on the final label of the user in order to train the systems.

2.2 Pretrained systems

The other two systems presented in this paper were trained using the data provided for the CLPsych 2016 shared task. This data is a collection of posts obtained from ReachOut, an Australian mental heath forum dedicated to help young people. It consists of 65,024 posts from the site structured in XML format, with 1,227 of them being labeled. Each post is annotated with one of the following 4 labels: *Green, Amber, Red* and *Crisis*, which describe how much a post requires the attention of a mental health professional.

Proceedings of the Sixth Workshop on Computational Linguistics and Clinical Psychology, pages 148–151
Minneapolis, Minnesota, June 6, 2019. ©2019 Association for Computational Linguistics

Label	Frequency	
	RO	SW
No Risk / Green	549	127
Low Risk / Amber	249	50
Moderate Risk / Red	110	113
Severe Risk / Crisis	39	206
Total	**947**	**496**

Table 1: Frequency of labels in the data.

2.3 Comparing both datasets

In order to reuse the systems trained for the CLPsych 2016 Shared Task, we can establish the following mapping between the labels provided for SW users and the ones from RO posts:

- No Risk - Green

- Low Risk - Amber

- Moderate Risk - Red

- Severe Risk - Crisis

However, while ReachOut posts were labeled taking into account the need of a mental health professional to assist the user, SuicideWatch posts were labeled based on the user's degree of suicide risk. While these labels can be similar, the annotation process and criteria was not the same in both cases, which can lead to some differences between them. Furthermore, ReachOut labels are assigned at a post level, while SuicideWatch ones are at a user level.

As we can see in table 1, 549 of the 947 posts in the ReachOut dataset belong to the *Green* class, while 206 of the 496 users in the SuicideWatch dataset belong to the *Severe Risk* class. Both datasets are imbalanced in different ways: the most frequent label in the SW dataset (*Severe Risk*) is the least frequent in the RO one, and the most present label in the RO dataset (*Green*), is not as frequent in the SW one.

3 Systems description

3.1 Text preprocessing

Some preprocessing steps were performed before extracting the features from the text in order to reduce the noise of the original data. All HTML special characters (e.g. ">") and stopwords were removed, each post was tokenized into words using spaCy (Honnibal and Montani, 2017), and all tokens were lowercased.

3.2 Features used

In order to train the models we relied just on features extracted from the body of each post, without relying on the title of the post or any other metadata. We used 4 different kind of features in our systems:

- TF-IDF: We generated TF-IDF feature vectors from the labeled dataset. We explored the use of different n-gram sizes for the TF-IDF representation, but unigrams led to better results.

- Word embeddings: One of the systems was trained using pre-trained multilingual word embeddings aligned in a common vector space (Conneau et al., 2017). A system trained with this kind of features can work reasonably well with posts written in different languages besides English (Lample et al., 2017). One of our objectives was to see if there was a significant decrease in performance between the models trained just for English data and the cross-lingual one.

- Document embeddings: We also used doc2vec (Le and Mikolov, 2014) to obtain document level embeddings for each post. We explored different kind of parameters for the vector representation, and found out that a window of 2 and a vector size of 100 gave the best results.

- VAD score of the post: Finally, we also used the NRC Valence, Arousal, and Dominance Lexicon (Mohammad, 2018) to obtain a normalized VAD score for each post. This score consists of three different values: the level of pleasure/displeasure of the post (*Valence*), the active/passive dimension (*Arousal*) and the powerful/weak dimension (*Dominance*).

3.3 Systems

Using the features described before, we have submitted the following 3 systems:

- *pretrained_svm*: This system consists of a Support Vector Machine (SVM) trained on the ReachOut data, using as features a combination of the TF-IDF representation of the post, its document embedding and its value for each dimension of the VAD score. The document embeddings were trained using the whole collection of posts provided in the CLPsych 2016 Shared Task, which consists of 65,000 unique posts. We used this classifier to annotate the degree of risk of every post of each user. After that, all the labels obtained for each user were normalized and fed as input to a logistic regression classifier that returned the final score of the user.

- *pretrained_rnn*: This system consists of a Recurrent Neural Network (RNN) trained on the ReachOut data, using as features the cross-lingual aligned word embeddings. The RNN is composed of gated recurrent units (GRU), which are shown to be better than traditional units and comparable to more complex units like LSTMs, while being faster to train (Chung et al., 2014). In order to avoid overfitting, we apply dropout and layer normalization (Ba et al., 2016) to the network. This classifier was used to annotate the posts of each user, and these annotations were normalized and fed to a logistic regression classifier, following the same process as with the *pretrained_svm* system.

- *custom_svm*: The final system that serves as a baseline is a SVM trained on the Suicide-Watch data, using as features the TF-IDF representation of the post and its VAD score. In order to train the model, we first asigned to every post of each user the same label as the final one of the user. After that, we trained the SVM on this data. The model works exactly the same as the first SVM: it annotates each post of the user, and then we aggregate these labels using a logistic regression classifier to obtain the final label of the user.

The hyper-parameters of the models were tuned using an exhaustive grid search over a subset of the possible parameters with 5-fold cross validation on the train set. Both SVMs use an rbf kernel, while the RNN is composed of one layer of 256 GRU cells.

We used available scikit-learn (Pedregosa et al., 2011) implementations of both the SVM and Logistic Regression classifiers, while the recurrent neural network was implemented specifically for this task using Tensorflow (Abadi et al., 2015).

System	Accuracy	F1
pretrained_svm	0.53	0.28
pretrained_rnn	0.51	0.27
custom_svm	0.61	0.32

Table 2: Macro-averaged results of each system using 5-fold cross validation on the train data.

4 Results

In order to obtain the results shown in this section, we performed 5-fold cross-validation on the training data. In table 2 we can see the accuracy and macro-averaged f1 score of each of the submitted systems. As we can see, the results of the models trained on ReachOut data are quite similar, with the SVM obtaining better accuracy and f1 scores than the RNN with cross-lingual embeddings. Our baseline SVM trained on the SuicideWatch data performed better than the other two systems both in terms of accuracy and f1-score.

In table 3 we can observe the results of the submitted systems for the test set. The three systems have difficulties distinguishing between the three levels of risk (*Low*, *Moderate* and *Severe*), which made them obtain a low macro-averaged f1-score and accuracy. However, the systems performed significantly better in terms of flagged (no risk vs risk) and urgent (moderate and severe risk vs low and no risk) f1-scores, with the best systems obtaining a score of 0.89 and 0.88 respectively.

5 Conclusions

In this paper we evaluated the use of systems trained on ReachOut data from previous CLPsych shared tasks for the current 2019 task. We observed a small decrease in performance with respect to a baseline system trained on this task's data, mostly related to the different annotation instructions and criteria used in both tasks. However, there are still some similarities in the tasks that make the pretrained systems perform reasonably well for this task.

We also explored the performance of cross-lingual word embeddings for this kind of problems. Using this type of embeddings we observed that the performance is pretty similar to other systems trained on different features. It could be interesting to explore these systems, which could work on data from many other forums written in different languages.

System	Accuracy	F1	Urgent f1	Flagged f1
pretrained_svm	0.49	0.27	0.87	0.79
pretrained_rnn	**0.52**	0.30	**0.88**	0.84
custom_svm	0.51	**0.31**	0.82	**0.89**

Table 3: Results of the systems for the test set.

Acknowledgments

We would like to thank the organizers for their work and effort dedicated to this shared task. This work is partially funded by the Spanish Ministry of Economy and Competitiveness (Society challenges: TIN2017-88877-R).

References

Martín Abadi, Ashish Agarwal, Paul Barham, Eugene Brevdo, Zhifeng Chen, Craig Citro, Greg S. Corrado, Andy Davis, Jeffrey Dean, Matthieu Devin, Sanjay Ghemawat, Ian Goodfellow, Andrew Harp, Geoffrey Irving, Michael Isard, Yangqing Jia, Rafal Jozefowicz, Lukasz Kaiser, Manjunath Kudlur, Josh Levenberg, Dan Mané, Rajat Monga, Sherry Moore, Derek Murray, Chris Olah, Mike Schuster, Jonathon Shlens, Benoit Steiner, Ilya Sutskever, Kunal Talwar, Paul Tucker, Vincent Vanhoucke, Vijay Vasudevan, Fernanda Viégas, Oriol Vinyals, Pete Warden, Martin Wattenberg, Martin Wicke, Yuan Yu, and Xiaoqiang Zheng. 2015. Tensorflow: Large-scale machine learning on heterogeneous systems. Software available from tensorflow.org.

Jimmy Lei Ba, Jamie Ryan Kiros, and Geoffrey E Hinton. 2016. Layer normalization. *arXiv preprint arXiv:1607.06450*.

Junyoung Chung, Caglar Gulcehre, KyungHyun Cho, and Yoshua Bengio. 2014. Empirical evaluation of gated recurrent neural networks on sequence modeling. *CoRR*, abs/1412.3555.

Alexis Conneau, Guillaume Lample, Marc'Aurelio Ranzato, Ludovic Denoyer, and Hervé Jégou. 2017. Word translation without parallel data. *arXiv preprint arXiv:1710.04087*.

Matthew Honnibal and Ines Montani. 2017. spacy 2: Natural language understanding with bloom embeddings, convolutional neural networks and incremental parsing. *To appear*.

Guillaume Lample, Alexis Conneau, Ludovic Denoyer, and Marc'Aurelio Ranzato. 2017. Unsupervised machine translation using monolingual corpora only. *arXiv preprint arXiv:1711.00043*.

Quoc Le and Tomas Mikolov. 2014. Distributed representations of sentences and documents. In *International conference on machine learning*, pages 1188–1196.

David N Milne, Glen Pink, Ben Hachey, and Rafael A Calvo. 2016. Clpsych 2016 shared task: Triaging content in online peer-support forums. In *Proceedings of the Third Workshop on Computational Lingusitics and Clinical Psychology*, pages 118–127.

Saif Mohammad. 2018. Obtaining reliable human ratings of valence, arousal, and dominance for 20,000 english words. In *Proceedings of the 56th Annual Meeting of the Association for Computational Linguistics (Volume 1: Long Papers)*, pages 174–184. Association for Computational Linguistics.

F. Pedregosa, G. Varoquaux, A. Gramfort, V. Michel, B. Thirion, O. Grisel, M. Blondel, P. Prettenhofer, R. Weiss, V. Dubourg, J. Vanderplas, A. Passos, D. Cournapeau, M. Brucher, M. Perrot, and E. Duchesnay. 2011. Scikit-learn: Machine learning in python. *Journal of Machine Learning Research*, 12:2825–2830.

Han-Chin Shing, Suraj Nair, Ayah Zirikly, Meir Friedenberg, Hal Daumé III, and Philip Resnik. 2018. Expert, crowdsourced, and machine assessment of suicide risk via online postings. In *Proceedings of the Fifth Workshop on Computational Linguistics and Clinical Psychology: From Keyboard to Clinic*, pages 25–36.

Ayah Zirikly, Philip Resnik, Özlem Uzuner, and Kristy Hollingshead. 2019. CLPsych 2019 shared task: Predicting the degree of suicide risk in Reddit posts. In *Proceedings of the Sixth Workshop on Computational Linguistics and Clinical Psychology: From Keyboard to Clinic*.

Similar Minds Post Alike: Assessment of Suicide Risk by Hybrid Language and Behavioral Model

Lushi Chen[*] **Abeer Aldayel**[*] **Nikolay Bogoychev** [*] **Tao Gong**[+]
[*] School of Informatics, University of Edinburgh, Edinburgh, United Kingdom
[+] Educational Testing Service, Princeton, New Jersey, USA
{Lushi.Chen, a.aldayel, n.bogoych}@ed.ac.uk, gtojty@gmail.com

Abstract

This paper describes our system submission for the CLPsych 2019 shared task B on suicide risk assessment. We approached the problem with three separate models: a behaviour model; a language model and a hybrid model. For the behavioral model approach, we model each user's behaviour and thoughts with four groups of features: posting behaviour, sentiment, motivation, and content of the user's posting. We use these features as an input in a support vector machine (SVM). For the language model approach, we trained a language model for each risk level using all the posts from the users as the training corpora. Then, we computed the perplexity of each user's posts to determine how likely his/her posts were to belong to each risk level. Finally, we built a hybrid model that combines both the language model and the behavioral model, which demonstrates the best performance in detecting the suicide risk level.

1 Introduction

Every year, there are over 800,000 people who die of suicide (WHO, 2019). Although health care systems play a major role in assessment of suicide risk, given limited time, clinicians are unable to assess thoroughly all the risk factors. One of the most important warning signs for suicide is the expressions of suicidal thoughts. The standard practice of clinicians asking people about suicidal thoughts cannot effectively predict and prevent suicide, because most patients who died of suicide did not report any suicidal thoughts when asked by a doctor (McHugh et al., 2019; Chan et al., 2016), therefore, many of them were assessed to have a low or moderate risk before their suicide attempts (Powell et al., 2000).

The CLpsych 2019 shared task B (Zirikly et al., 2019) attempts to address the challenge of automatic suicide risks asssessment using people's forum postings. The aim of the task is to distinguish the levels of suicide risks among users who posted any contents in the suicide watch (SW) subreddit. The dataset includes all the posts (N = 31,553) in any subreddit from 621 users who had posted on SW. One of the four risk levels ranging from "No Risk" to "Severe Risk" was assigned to each user according to their SW posts. The annotation process is described in Shing et al. (2018).

We treat the task as a multi-classification problem. We approach it with three models: a behavioural model (BM), a suicide language model (SLM) and a hybrid model (HM_{BM_SLM}) that combines the (BM) and (SLM) models. The SLM offers good classification accuracy, but it does not provide any human interpretable reason for its classification decisions. Hence, we define a collection of features to better capture users' posting behaviours and thoughts, then we use these features in the BM. The overall results show that the hybrid model (HM_{BM_SLM}) performs the best in identifying the risk level with a f1 score 38% for the CLpsych task B.

2 Related work

Suicide is a complex behaviour involving biological, psychological and social factors. For psychological factors, a large amount of literature suggests that a history of psychiatric disorders, especially affective disorders, is a strong predictor of suicide (Angst et al., 2002; Brent et al., 1993; Bostwick and Pankratz, 2002). Another important precursor of suicide is self-harm or previous attempt. Biological and social factors that contribute to suicide include: substance abuse (Vijayakumar et al., 2011; Hawton et al.; Bergen et al., 2012; Chan et al., 2016; Joiner, 2007), gender (males have a higher suicide risk) and living alone(Joiner, 2007).

The suicidal behaviour model by Wilson et al. (2005); Cukrowicz et al. (2011) proposed that the unmet need of belonging (e.g. relationship

Proceedings of the Sixth Workshop on Computational Linguistics and Clinical Psychology, pages 152–157
Minneapolis, Minnesota, June 6, 2019. ©2019 Association for Computational Linguistics

breakup) and the self perceived burden were the major motivations for suicidal behaviors (Trout, 1980). Other motivations include: having a negative self-image, hopelessness (Kovacs and Garrison, 1985), and having a plan of the suicidal attempt. The duration, intensity, and frequency of the suicidal desires also indicate the pertinacity to the attempt.

The majority of the prior work on the suicide risk detection focuses on manually generated (BoW) features centering only around the textual cues of the user's post (Varathan and Talib, 2014; O'Dea et al., 2015), such as the LIWC pre-trained word embeddings (Husseini Orabi et al.) or supervised learning topics (e.g., latent Dirichlet allocation) (Ji et al., 2018). Unlike these studies, we design a model that leverages user's behavioural data in combination with a suicide language model to detect the suicide risk level. Our features intend to capture the language and behavioral characteristics proposed by clinical literature as suicide risk factors. For example, we develop a feature vector that represent suicide motivations. Examining the validity of these features in our experimental model provides us a way to understand the prevalence of these characteristics in people with different suicide risk levels.

3 Suicide risk identification models

In this study, we propose three models to measure suicide risk levels. BM uses user's posting behaviours and manual selected language characteristics to predict suicidal risk level. SLM learns the language characteristics of each risk level. The hybrid model (HM_{BM_SLM}) combines the advantages of the BM and SLM models.

3.1 Behavioral model

Most of the existing studies focus on the language used in expressing suicide thoughts, and only a small number of them examine the behavioral and thought patterns on social media. For instance, Colombo et al. (2016) use twitter followers, friends, and number of retweets to represent the connectivity between users having suicide ideas. Based on the clinical literature, we engineer four sets of features that capture user behaviors and thoughts for the Behavioural model (BM), including: posting behaviour, sentiment, content, and motivation for suicide. Posting behaviours consist of users' posting patterns in SW, mental

health related subreddits and all the other subreddits. Sentiment features consist of a sentiment profile for each user, user's sentiment towards selected topics (e.g., friends and family). Content features consist of Linguistic Inquiry and Word Count (LIWC) (Pennebaker et al., 2001), EMPATH (Fast et al., 2016) and count vectors normalized by TF-IDF (Salton and McGill, 1986). For the motivation features, we use a word count approach to define whether the user have suggested any motivations.

Some of these features were constructed using Suicide Watch (SW) posts only, while others were constructed using all the reddit posts from the users. Although many of these posts might not be directly related to suicide thoughts, we hypothesized that using irrelevant posts to define a user's interaction behaviour and emotional magnitude would help to identify the virtual community of the users with suicide risk.

3.1.1 Sentiment

Sentiment profile. The sentiment of each user's previous posting was used to identify the similarity between users' postings. This set of features are represented as a vector of sentiment value corresponding to a user's previous posting. Then, we use the Levenshtein Distance to compute the similarity between two such vectors (Yujian and Bo, 2007).

Topic Sentiment. We inspect the sentiment of specific topics in the SW posts. We extract the sentences containing keywords related to family members (e.g. mom, dad), partners (e.g. boyfriend), and self (e.g. myself). We then use sentiStrength (Thelwall et al., 2010) to detect the sentiment values of these sentences and aggregate the topic sentiment at a user level.

3.1.2 Posting behaviours

Frequency of posting We use the number of posts, word count in each post, whether and when a user posts more frequently as features. To check whether a user has recently started posting more frequently, we define a posting frequency vector by computing the average posting time interval between any two posts from a user. We use a sliding window from the head to the tail of the frequency vector to identify which time interval(s) are at least one standard deviation below the mean of all intervals. Users are highly likely to post more frequently if the last window is one standard deviation below the mean. Frequency of posting is inspected in the SW posts, all user posts, and posts involving mental illnesses and drugs use. To extract the posts involving mental illnesses and drugs use, we compile a dictionary of mental illnesses names and symptoms. Posts that contain words from this dictionary are selected. Meanwhile, posts from subreddits that are associated with mental illnesses self help groups (e.g., self-harm, TwoXADHD) are also extracted.

3.1.3 Motivation factors

Financial problems, drug use, mental illness history, relationship break up, hopelessness, suicide tools and self-harm have been found to be predictive to suicidal behaviors (Kessler et al., 1999). In our study, we compile dictionaries for each of the motivation factors. Terms in drug use, mental illness and suicide tools dictionaries are extracted from websites using the webscraping techniques.

3.1.4 Content feature

We use both the open and closed BoW approaches to generate the content feature. For the open vocabulary approach, we counted the term frequency and normalized it with tf-idf. For the closed vocabulary approach, we used LIWC and Empath. Both tools are used to count words from predefined psychologically meaningful categories.

3.1.5 Clustering

We use model-based clustering (Banfield and Raftery, 1993) to group sentiment, posting behaviour and motivation factors. Model-based clustering assumes that the data are formed by multiple Gaussians. The clustering algorithm tries to recover the models that generate the data. The best model is selected according to the Bayesian information criterion (BIC). We adopt five clusters as our solution.

3.2 Suicide language model

The behavioural model (BM) enables us to observe the behavioral and thought differences among individuals with various suicide risk levels. However, one disadvantage of the BM approach is that we might miss some relevant cases that do not contain words in the manually selected dictionary, or include irrelevant cases but contain the dictionary words.

With this challenge in mind, we also tackle the suicide risk classification problem with sui-cide language modeling (SLM). Language modeling is used in domains such as machine translation, speech recognition and text classification (McCallum et al., 1998; Brants et al., 2007; Coppersmith et al., 2014). The principle of language modeling is to compute a probability distribution over words in order to determine how likely a specific language model is to generate a given document. In our case, we train one model for each risk level. Then, we calculate a document's likelihood (perplexity) for all the models, and select the model with the best score.

4 Dataset and experiment setup

The dataset used for training the models is provided by the CLPsych shared task B (Zirikly et al., 2019). It contains 621 reddit users who had posted on SW with an overall of 31,553 posts. The users are labeled as "no risk" (class A), "low risk" (class B), "moderate risk" (class C), and "severe risk" (class D). Dataset statistics is presented in table 1. From the training set, it is shown that nearly half of the posts were labeled as "severe risk", class B only accounts for less than 10% of the posts. Nearly half of the posts in both the training and testing sets did not have any contents in the post body.

Table 1: Basic Statistics for train and test set

Train	postNum/%	WC	U	P/U	SW/U	emP
A	10662 (34%)	52	127	84	1.28	6070
B	2715 (9%)	101	50	54	1.18	984
C	5726 (18%)	79	113	51	1.36	2556
D	12450 (39)	72	206	60	2.64	5344
Test	9610	63	125	77	1.49	4704

Note: A:no risk, B:mild risk, C: moderate risk, D: severe risk. postNum: number of posts. WC: average word count in posts. U: users. P/U: post per user. SW/U: suicideWatch post per user. emP: posts without content in the post body.

4.1 Suicide language model setup

We train the (SLM) language model with the minimal processed data (raw text), and tokenized and truecased data. For the raw text model, the data are preprocessed as follows: Sentences are split by the NLTK sentence splitter and then spaces are inserted around each full stop to make sure mis-spelled cases are parsed correctly. For example, "tomorrow.And today" is processed as "tomorrow . And today". For the tokenized and truecased model, we apply the tokenizer and true-caser from the Moses machine translation toolkit

(Koehn et al., 2007).

The language model is trained with KenLM's default settings (modified Kneser-Nay smoothing) (Heafield et al., 2013). In each model, all the posts from a redditor and annotated with a specific risk level are used as the training corpora. All the posts from a redditor are treated as a single document. To assign a risk level to the document, we calculate its perplexity for each language model, and assign the document's class based on the language model that produces the lowest perplexity score. We experiment with the context windows of 3 to 6-gram, and find that 4-gram works the best.

5   Experiments

In the SLM, for each document, the model with the lowest perplexity is assigned to the document. Perplexity is the inverse probability of a test set, normalized by the number of words, a low perplexity indicates that the probability distribution is good at predicting the sentence (Sennrich, 2012). Given a sample test, we calculate its likelihood for all the models, and select the model with the best score.

In the BM, we use random forest to select the top 300 features to use in the final prediction. We validate our BM features on the multi-classification problem using support vector machines (SVM) in scikitlearn [1]. We use the 5-fold cross validation on training data and grid-search parameters to explore both the kernels and margin of the hyperplane (C parameter).

Furthermore, we construct a hybrid model based on our observations on the prediction results from the SLM and the BM. In the training process, we observe the BM is weak in distinguishing classes B and C, but the SLM is better in identifying class B. Therefore, we adopt the class B results from the SLM. We also find that some posts in class A are suicide experiences from someone associated with the authors, but not the authors themselves. The BM is better than the language model in identifying these cases, so we use the BM for class A. However, if the confidence score is lower than 0.4, the SLM becomes better at identifying class A. Therefore, we replace the results with confidence score lower than 0.4 with those from the SLM model.

[1] https://scikit-learn.org/stable/

6   Results

Table 2 shows the test set results of the three models. Table 3 shows f1 for flagged vs. non-flagged and urgent vs. non-urgent. Flagged vs. non-flagged distinguished class A from the rest of the classes. Urgent vs. non-urgent distinguished classes A, B with classes C, D. The hybrid model had the best average f1 macro in the risk assessment task.

Table 2: Results for risk assessment task

Model	Risk level	P	R	F
BM	A	53	78	63
	B	22	15	18
	C	14	14	14
	D	55	42	48
	Fl_{AVG}			36
SLM	A	73	25	37
	B	27	23	25
	C	12	7	9
	D	49	83	62
	Fl_{AVG}			33
HM_{BM_SLM}	A	56	72	63
	B	25	39	30
	C	12	11	11
	D	55	42	48
	Fl_{AVG}			**38**

P: precision (%), R: recall (%), F: f1 macro average (%).
Fl_{AVG}: f1 (%) macro average of four classes.

Table 3: Results for flagged and urgent cases

	Flagged			Urgent		
	P	R	F	P	R	F
BM	91	76	83	80	69	74
SLM	79	97	87	69	89	78
HM_{BM_SLM}	89	81	85	81	65	72

P: precision (%), R: recall (%), F: f1 macro average (%).

In our test set result, we find that SLM is overfitting. SLM classifies most of the posts to class D in the testing set. Whereas, the BM has consistent good performances on classes A and D, but poor performances on classes B and C.

7   Conclusion

Our results demonstrate that suicide risk can be gauged by user's posting behaviors. Suicide risk factors identified by clinical literature are useful in automatic detection of suicide risks. Suicide language can be modeled by statistical language model, especially for risk level B and D, in which cases it surpasses the behavioral model. Hence, a combination of the two models results in a more accurate user classification. As a future work, a further analysis of each feature would gauge its contribution towards identifying suicide risk levels.

References

F Angst, H. H Stassen, P. J Clayton, and J Angst. 2002. Mortality of patients with mood disorders: follow-up over 3438 years. 68(2):167–181.

Jeffrey D Banfield and Adrian E Raftery. 1993. Model-based gaussian and non-gaussian clustering. *Biometrics*, pages 803–821.

Helen Bergen, Keith Hawton, Keith Waters, Jennifer Ness, Jayne Cooper, Sarah Steeg, and Navneet Kapur. 2012. Premature death after self-harm: a multi-centre cohort study. 380(9853):1568–1574.

John Michael Bostwick and V. Shane Pankratz. 2002. Affective disorders and suicide risk: A reexamination. 157(12):1925–1932.

Thorsten Brants, Ashok C Popat, Peng Xu, Franz J Och, and Jeffrey Dean. 2007. Large language models in machine translation. In *Proceedings of the 2007 Joint Conference on Empirical Methods in Natural Language Processing and Computational Natural Language Learning (EMNLP-CoNLL)*.

DAVID A. Brent, JOSHUA A. Perper, GRACE Moritz, CHRIS Allman, AMY Friend, CLAUDIA Roth, JOY Schweers, LISA Balach, and MARIANNE Baugher. 1993. Psychiatric risk factors for adolescent suicide: A case-control study. 32(3):521–529.

Melissa KY Chan, Henna Bhatti, Nick Meader, Sarah Stockton, Jonathan Evans, Rory C O'Connor, Nav Kapur, and Tim Kendall. 2016. Predicting suicide following self-harm: systematic review of risk factors and risk scales. *The British Journal of Psychiatry*, 209(4):277–283.

Gualtiero B Colombo, Pete Burnap, Andrei Hodorog, and Jonathan Scourfield. 2016. Analysing the connectivity and communication of suicidal users on twitter. *Computer communications*, 73:291–300.

Glen Coppersmith, Mark Dredze, and Craig Harman. 2014. Quantifying mental health signals in twitter. In *Proceedings of the workshop on computational linguistics and clinical psychology: From linguistic signal to clinical reality*, pages 51–60.

Kelly C Cukrowicz, Jennifer S Cheavens, Kimberly A Van Orden, R Michael Ragain, and Ronald L Cook. 2011. Perceived burdensomeness and suicide ideation in older adults. *Psychology and aging*, 26(2):331.

Ethan Fast, Binbin Chen, and Michael S Bernstein. 2016. Empath: Understanding topic signals in large-scale text. In *Proceedings of the 2016 CHI Conference on Human Factors in Computing Systems*, pages 4647–4657. ACM.

Keith Hawton, Kate EA Saunders, and Rory C O'Connor. Self-harm and suicide in adolescents. 379(9834):2373–2382.

Kenneth Heafield, Ivan Pouzyrevsky, Jonathan H Clark, and Philipp Koehn. 2013. Scalable modified kneser-ney language model estimation. In *Proceedings of the 51st Annual Meeting of the Association for Computational Linguistics (Volume 2: Short Papers)*, volume 2, pages 690–696.

Ahmed Husseini Orabi, Prasadith Buddhitha, Mahmoud Husseini Orabi, and Diana Inkpen. Deep learning for depression detection of twitter users. In *Proceedings of the Fifth Workshop on Computational Linguistics and Clinical Psychology: From Keyboard to Clinic*, pages 88–97. Association for Computational Linguistics.

Shaoxiong Ji, Celina Ping Yu, Sai-fu Fung, Shirui Pan, and Guodong Long. 2018. Supervised learning for suicidal ideation detection in online user content. *Complexity*, 2018.

Thomas Joiner. 2007. *Why people die by suicide*. Harvard University Press.

Ronald C Kessler, Guilherme Borges, and Ellen E Walters. 1999. Prevalence of and risk factors for lifetime suicide attempts in the national comorbidity survey. *Archives of general psychiatry*, 56(7):617–626.

Philipp Koehn, Hieu Hoang, Alexandra Birch, Chris Callison-Burch, Marcello Federico, Nicola Bertoldi, Brooke Cowan, Wade Shen, Christine Moran, Richard Zens, et al. 2007. Moses: Open source toolkit for statistical machine translation. In *Proceedings of the 45th annual meeting of the association for computational linguistics companion volume proceedings of the demo and poster sessions*, pages 177–180.

Maria Kovacs and Betsy Garrison. 1985. Hopelessness and eventual suicide: a 10-year prospective study of patients hospitalized with suicidal ideation. *American journal of Psychiatry*, 1(42):559–563.

Andrew McCallum, Kamal Nigam, et al. 1998. A comparison of event models for naive bayes text classification. In *AAAI-98 workshop on learning for text categorization*, volume 752, pages 41–48. Citeseer.

Catherine M McHugh, Amy Corderoy, Christopher James Ryan, Ian B Hickie, and Matthew Michael Large. 2019. Association between suicidal ideation and suicide: meta-analyses of odds ratios, sensitivity, specificity and positive predictive value. *BJPsych open*, 5(2).

Bridianne O'Dea, Stephen Wan, Philip J. Batterham, Alison L. Calear, Cecile Paris, and Helen Christensen. 2015. Detecting suicidality on twitter. 2(2):183–188.

James W Pennebaker, Martha E Francis, and Roger J Booth. 2001. Linguistic inquiry and word count: Liwc 2001. *Mahway: Lawrence Erlbaum Associates*, 71(2001):2001.

John Powell, John Geddes, Jonathan Deeks, Michael Goldacre, and Keith Hawton. 2000. Suicide in psychiatric hospital in-patients: risk factors and their predictive power. *The British Journal of Psychiatry*, 176(3):266–272.

Gerard Salton and Michael J McGill. 1986. Introduction to modern information retrieval.

Rico Sennrich. 2012. Perplexity minimization for translation model domain adaptation in statistical machine translation. In *Proceedings of the 13th Conference of the European Chapter of the Association for Computational Linguistics*, pages 539–549. Association for Computational Linguistics.

Han-Chin Shing, Suraj Nair, Ayah Zirikly, Meir Friedenberg, Hal Daumé III, and Philip Resnik. 2018. Expert, crowdsourced, and machine assessment of suicide risk via online postings. In *Proceedings of the Fifth Workshop on Computational Linguistics and Clinical Psychology: From Keyboard to Clinic*, pages 25–36.

Mike Thelwall, Kevan Buckley, Georgios Paltoglou, Di Cai, and Arvid Kappas. 2010. Sentiment strength detection in short informal text. *Journal of the American Society for Information Science and Technology*, 61(12):2544–2558.

Deborah L Trout. 1980. The role of social isolation in suicide. *Suicide and Life-Threatening Behavior*, 10(1):10–23.

K. D. Varathan and N. Talib. 2014. Suicide detection system based on twitter. In *2014 Science and Information Conference*, pages 785–788.

Lakshmi Vijayakumar, M Suresh Kumar, and Vinayak Vijayakumar. 2011. Substance use and suicide:. 24(3):197–202.

WHO. 2019. Who.int.

Keith G. Wilson, Dorothyann Curran, and Christine J. McPherson. 2005. A burden to others: A common source of distress for the terminally ill. 34(2):115–123.

Li Yujian and Liu Bo. 2007. A normalized levenshtein distance metric. *IEEE transactions on pattern analysis and machine intelligence*, 29(6):1091–1095.

Ayah Zirikly, Philip Resnik, Özlem Uzuner, and Kristy Hollingshead. 2019. CLPsych 2019 shared task: Predicting the degree of suicide risk in Reddit posts. In *Proceedings of the Sixth Workshop on Computational Linguistics and Clinical Psychology: From Keyboard to Clinic*.

Predicting Suicide Risk from Online Postings in Reddit
The UGent-IDLab submission to the CLPysch 2019 Shared Task A

Semere Kiros Bitew, Giannis Bekoulis, Johannes Deleu, Lucas Sterckx, Klim Zaporojets
Thomas Demeester and Chris Develder

IDLab, Ghent University - imec
{semerekiros.bitew, firstname.lastname}@ugent.be

Abstract

This paper describes IDLab's text classification systems submitted to Task A as part of the CLPsych 2019 shared task. The aim of this shared task was to develop automated systems that predict the degree of suicide risk of people based on their posts on Reddit.[1] Bag-of-words features, emotion features and post-level predictions are used to derive user-level predictions. Linear models and ensembles of these models are used to predict final scores. We find that predicting fine-grained risk levels is much more difficult than flagging potentially at-risk users. Furthermore, we do not find clear added value from building richer ensembles compared to simple baselines, given the available training data and the nature of the prediction task.

1 Introduction

The goal of the CLPysch 2019 shared task is to predict the degree of suicide risk based on online postings of users. This shared task is motivated by the long-term lack of progress in predicting suicide risk. McHugh et al. (2019), after reviewing more than 70 studies, argues that suicidality cannot be predicted effectively using traditional standard procedures, e.g., questions of clinicians about suicidal thoughts: the authors claim that a large fraction of patients (i.e., 80%) who committed suicide, did not admit contemplating suicide when asked by a general practitioner. Another study by Franklin et al. (2017) also concludes that prediction of suicide risks has not improved over the last 50 years and suggests that machine learning learning methods can contribute towards solving that challenge.

Typically, there are long periods of time between clinical encounters of patients. During these periods, some patients are engaged in frequent use of social media. Coppersmith et al. (2017) states

that such usage of social media can be exploited to build binary risk classifiers. However, when such systems are deployed, the number of people flagged as "at risk" will exceed clinical capacity for intervention. This in turn motivates the design of more fine-grained prediction models, predicting various risk levels, as proposed for the current shared task.

Our system uses a combination of (i) bag-of–word features, (ii) emotion labels, and (iii) information derived from post-level risk features (see Section 3.1 for more details). Using these features, we apply linear models to predict the scores. We explore different combinations to evaluate the performance of the different models.

The remainder of the paper is organized as follows: Section 2 describes the data and the shared task. Section 3 presents the details of the implemented system and the features. Section 4 shows the experimental results obtained from the test data. To compare our results to other participants in the shared task, we refer the reader to Zirikly et al. (2019). To conclude, we summarize our findings and present future directions in Section 5.

2 Data and Task A

The dataset used in the shared task is sampled from the University of Maryland Reddit Suicidality Dataset (Shing et al., 2018). It is constructed using data from Reddit, an online site for anonymous discussion on a wide variety of topics. Specifically, the UMD dataset was extracted from the 2015 Full Reddit Submission Corpus[2],using postings in the r/SuicideWatch subreddit (henceforth simply SuicideWatch or SW) to identify anonymous users who might represent positive instances of suicidality and including a comparable number of non-SuicideWatch controls. The dataset is annotated at user level, using a four-

---

[1] www.reddit.com

[2] https://www.reddit.com/r/datasets/comments/3mg812/full_reddit_submission_corpus_now_available_2006/

Proceedings of the Sixth Workshop on Computational Linguistics and Clinical Psychology, pages 158–161
Minneapolis, Minnesota, June 6, 2019. ©2019 Association for Computational Linguistics

point scale indicating the likelihood of a user to commit suicide: (a) no risk, (b) low risk, (c) moderate risk, and (d) severe risk. The corpus includes posts from 21,518 users and is subdivided into 993 labelled users and 20,525 unlabelled users. Out of the 993 labeled users, 496 have at least posted once on the SuicideWatch subreddit. The remaining 497 users are control users (i.e., they have not posted in SuicideWatch or any mental health related subreddits). The data is provided in a comma-separated values file that includes the post titles, content, timestamps, and anonymized unique user ids. The goal of shared Task A is to predict users' suicide risk into one of the four classes (i.e., (a)-(d)) given the fact that he/she has posted on SuicideWatch.

3 Systems Description

This section provides an overview of features extracted from posts, followed by a short system description of our submitted runs.

3.1 Features

TF-IDF features: We used the TF-IDF weighting scheme as text representation. The TF-IDF feature vectors of n-grams were generated for our dataset. We experimented with n-grams for n ranging from 1 to 5. In our preliminary investigations, we explored various kinds of features, such as character level n-grams, or textual statistical features (such as the total number of posts), but these did not lead to increased performance metrics.

Emotion features: We hypothesize that individuals contemplating suicide will tend to express emotions with negative sentiment, more than individuals without suicidal thoughts. Therefore, we use a pre-trained model called *DeepMoji*[3] that predicts emotions from text (Felbo et al., 2017). For an individual post of a user, a 64-dimensional emotion feature vector is generated by the model, with each dimension corresponding to the probability for one out of 64 different emojis. We take the element-wise maximum, average and standard deviation of this vector as features to represent a user's emotions.

Suicide risk features: We reason that post-level binary risk estimates can help in making the user-level risk level prediction. To achieve this, we semi-manually annotated 605 posts from the unlabelled dataset as follows. First, we trained a TF-

---

IDF based logistic regression classifier to predict the four class labels (a)–(d), using labelled data for 496 users. We adopt that classifier to assign four probabilities, one for each class (a)–(d), to each post in the unlabelled dataset. We take a random sub-sample of the automatically labelled posts, order it in terms of no-risk probability, and manually label posts taken in turn from the top and bottom of the ordered list. We thus obtain a balanced set of 605 annotated posts (302 'risk', 303 'no-risk'), spending a total annotation time of 5 hours. Subsequently, a TF-IDF based logistic regression binary classifier was trained on these manually annotated posts. Finally, the post-level binary predictions were then aggregated into user-level suicide risk features by taking the maximum, mean, and standard deviation of the predicted post-level scores. The motivation behind this annotation experiment was to investigate the effectiveness of a cheap additional annotation effort in boosting the final model's prediction accuracy. By 'cheap' annotation effort, we refer to annotations on the *post-level* as opposed to user-level, *binary* as opposed to 4-label, and *directly balanced* as opposed to a larger random sample to obtain the same amount of at-risk posts.

3.2 Models

Three different systems were explored for our submission to the shared task. A logistic regression classifier and two ensemble-based classifiers.

1. **Baseline classifier**: a logistic regression classifier (Pedregosa et al., 2011) is trained based on TF-IDF weighted bag-of-word features.

2. **Ensemble without Risk classifier**: this ensemble combines the scores from the baseline logistic regression classifier, a linear SVM classifier and the emotion classifier. The linear SVM, included in scikit-learn (Pedregosa et al., 2011) is trained on the TF-IDF representations. This ensemble uses an additional logistic regression classifier (at the next level) to predict the final classes.

3. **Ensemble (all)**: this model combines the scores from all classifiers as illustrated in Fig. 1. This ensemble uses a second level Logistic Regression classifier similar to the previous ensemble.

With this system choice, we are able to measure the impact of combining linear classifiers

---

[3]https://github.com/bfelbo/DeepMoji

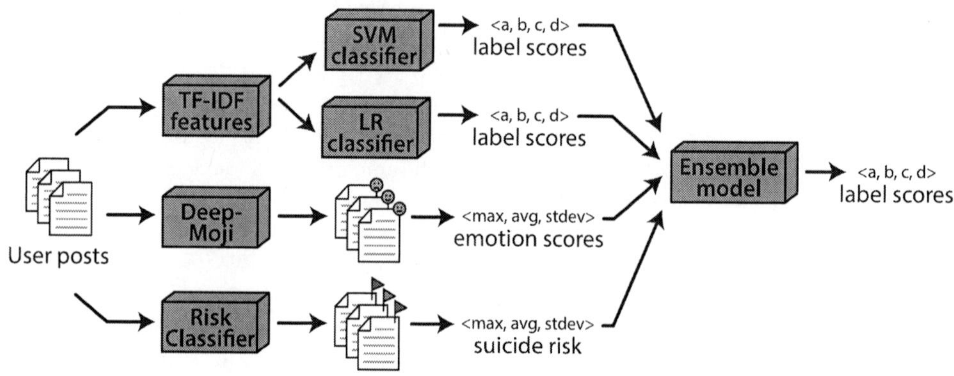

Figure 1: Main elements of the presented system setup.

with emotion features compared to a simple linear model (second vs. first run), and to measure the added value from the additional post-level annotations (third vs. second run).

4 Experimental Results

In this section, we present the final test results of the three submitted systems on the official test set. The test set consists of a total of 189 posts from 125 different users. The official evaluation metric used in the shared task is the macro F_1 score on all four classes. Table 1 depicts the official models' performance on the test data. Our baseline classifier outperforms the ensemble models. This can be explained by (i) bias in the training/test split during development, (ii) the small number of annotated training instances, or (iii) the partly subjective nature of the task, and in particular the distinction between fine-grained levels such as 'low risk' and 'moderate risk'. Note that, however, our most advanced model did perform best for the simpler task of detecting potentially at-risk ('flagged') users. Further research is required to investigate these potential issues.

Models	Precision	Recall	F_1
Baseline	0.444	0.457	**0.445**
Ensemble w/o Risk	0.428	0.402	0.407
Ensemble (all)	0.445	0.419	0.426

Table 1: Official results

In addition, two more metrics were used. The first metric is the F_1 score for *flagged versus non-flagged* users. The flagged vs. non-flagged F_1 is relevant for a use case in which the goal is to distinguish users that can be safely ignored (category (a), no risk) from those that require attention (i.e., categories (b), (c), (d)), such as when human moderators need to investigate the risk further. Table 2 shows the performance of the models in binary classification of flagged and non-flagged users, whereby the ensemble with sentiment features ('Ensemble w/o Risk) outperforms the linear baseline, but the overall ensemble with binary post-level risk predictions performs slightly better still. Given the much higher scores, the task of flagging potentially at-risk users appears much simpler than making fine-grained risk-level predictions.

Models	Precision	Recall	F_1
Baseline	0.904	0.806	0.852
Ensemble w/o Risk	0.848	0.903	0.875
Ensemble (all)	0.850	0.914	**0.881**

Table 2: Flagged vs Non-flagged

The second metric is the *urgent versus non-urgent* F_1 score that measures distinction between users who are at a severe risk of suicide (category (c) and (d)) and other users. Table 3 shows the models' performance for classifying users into urgent and non-urgent classes. The overall higher scores in Table 3 indicate that the binary classification of urgent from non urgent users is fairly simpler task when compared to the fine-grained risk level classification.

Models	Precision	Recall	F_1
Baseline	0.833	0.750	**0.789**
Ensemble w/o Risk	0.795	0.725	0.758
Ensemble (all)	0.792	0.762	0.777

Table 3: Urgent vs Non-urgent

5   Conclusion and Future work

In this paper, we described the Ghent University-IDLab submission to the CLPysch 2019 shared Task A. We found that the baseline classifier based on logistic regression outperformed the ensemble of classifiers. Specifically, our baseline model obtained a macro F_1-score of 0.445 on the shared task. Our system also achieves a macro F_1-score of 0.881 and 0.789 on flagging non-risk users and distinguishing urgent from non-urgent users, respectively. The more advanced models (i.e., ensembles) did not bring any added value in the fine-grained user level risk prediction. This can be due to the limited number of training examples in the provided dataset, bias in train/test splits during development and the subjective nature of the task.

As next steps, we plan on investigating alternative ways of splitting train from test data such as stratified cross-validation (i.e., to avoid different distributions of the target variable in the train/test splits). We also want to explore more sophisticated ways of ensembling and stacking techniques while also taking into account the time stamp meta-data of posts.

Acknowledgments

We would like to thank the CLPsych 2019 shared task organizers for organizing the competition and providing us with the online postings of users data from Reddit.

Ethical Review

To meet the ethical review criteria as discussed in the Zirikly et al. (2019) overview paper, this study was evaluated by the Ethics Committee of the faculty of Psychology and Educational Sciences of Ghent University. The committee concluded that ethical approval was not needed for conducting the research.

References

Glen Coppersmith, Casey Hilland, Ophir Frieder, and Ryan Leary. 2017. Scalable mental health analysis in the clinical whitespace via natural language processing. In *2017 IEEE EMBS International Conference on Biomedical & Health Informatics (BHI)*, pages 393–396. IEEE.

Bjarke Felbo, Alan Mislove, Anders Søgaard, Iyad Rahwan, and Sune Lehmann. 2017. Using millions of emoji occurrences to learn any-domain representations for detecting sentiment, emotion and sarcasm. In *Conference on Empirical Methods in Natural Language Processing (EMNLP)*.

Joseph C Franklin, Jessica D Ribeiro, Kathryn R Fox, Kate H Bentley, Evan M Kleiman, Xieyining Huang, Katherine M Musacchio, Adam C Jaroszewski, Bernard P Chang, and Matthew K Nock. 2017. Risk factors for suicidal thoughts and behaviors: a meta-analysis of 50 years of research. *Psychological Bulletin*, 143(2):187.

Catherine M McHugh, Amy Corderoy, Christopher James Ryan, Ian B Hickie, and Matthew Michael Large. 2019. Association between suicidal ideation and suicide: meta-analyses of odds ratios, sensitivity, specificity and positive predictive value. *BJPsych open*, 5(2).

F. Pedregosa, G. Varoquaux, A. Gramfort, V. Michel, B. Thirion, O. Grisel, M. Blondel, P. Prettenhofer, R. Weiss, V. Dubourg, J. Vanderplas, A. Passos, D. Cournapeau, M. Brucher, M. Perrot, and E. Duchesnay. 2011. Scikit-learn: Machine learning in Python. *Journal of Machine Learning Research*, 12:2825–2830.

Han-Chin Shing, Suraj Nair, Ayah Zirikly, Meir Friedenberg, Hal Daumé III, and Philip Resnik. 2018. Expert, crowdsourced, and machine assessment of suicide risk via online postings. In *Proceedings of the Fifth Workshop on Computational Linguistics and Clinical Psychology: From Keyboard to Clinic*, pages 25–36.

Ayah Zirikly, Philip Resnik, Özlem Uzuner, and Kristy Hollingshead. 2019. CLPsych 2019 shared task: Predicting the degree of suicide risk in Reddit posts. In *Proceedings of the Sixth Workshop on Computational Linguistics and Clinical Psychology: From Keyboard to Clinic*.

CLPsych2019 Shared Task: Predicting Users' Suicide Risk Levels from Their Reddit Posts on Multiple Forums

Victor Ruiz[a*], Lingyun Shi[a*], Jorge Guerra[a,b], Wei Quan[a,c], Neal Ryan[d], Candice Biernesser[d], David Brent[d], and Fuchiang Tsui[a-c]

[a]*Tsui Laboratory, Children's Hospital of Philadelphia, Philadelphia, PA, USA,*
[b]*Institute for Biomedical Informatics, University of Pennsylvania, Philadelphia, PA, USA,*
[c]*Drexel University, Philadelphia, PA, USA,*
[d]*Department of Psychiatry, University of Pittsburgh, Pittsburgh, PA, USA*
*: *Authors contributed equally*

Abstract

We aimed to predict an individual suicide risk level from longitudinal posts on Reddit discussion forums. Through participating in a shared task competition hosted by CLPsych2019, we received two annotated datasets: a training dataset with 496 users (31,553 posts) and a test dataset with 125 users (9610 posts). We submitted results from our three best-performing machine-learning models: SVM, Naïve Bayes, and an ensemble model. Each model provided a user's suicide risk level in four categories, i.e., no risk, low risk, moderate risk, and severe risk. Among the three models, the ensemble model had the best macro-averaged F1 score 0.379 when tested on the holdout test dataset. The NB model had the best performance in two additional binary-classification tasks, i.e., no risk vs. flagged risk (any risk level other than no risk) with F1 score 0.836 and no or low risk vs. urgent risk (moderate or severe risk) with F1 score 0.736. We conclude that the NB model may serve as a tool for identifying users with flagged or urgent suicide risk based on longitudinal posts on Reddit discussion forums.

Keywords: suicide, Reddit, machine learning, predictive modeling

I. Introduction

Suicide poses a challenge to our society. It is the 10th leading cause of death in the United States for all ages, and most importantly it is the second leading cause of death for 64 millions of youths between the ages of 10 and 24.(NIMH, 2018)(Howden and Meyer, 2011) Meanwhile, the use of social media among the young population is getting more poupular.

Social media websites such as Reddit discussion forums serve as a common platform for people to express their thoughts, and many people feel more comfortable discussing or sharing their mental state including suicidal thoughts on social media than they are in person. Moreover, people who can get access to the internet may not have adequate resources for mental health care. In contrast to the electronic health records that recorded the interactions between patients and clinical care providers, on-line social media posts illustrate conversations between a user and an online audience mostly comprised of non-clinicians. In March 2019, Reddit was estimated to have 542 million monthly visitors and 234 million unique users, 53.9% of which with bases in the United States.(Wikipedia,) There is a need to study potential suicide risks based on social media posts as a part of public health surveillance.(De Choudhury et al., 2017)

Current state-of-art approaches for mental health condition prediction leveraged machine learning (ML) and natural language processing (NLP). Common ML algorithms include support vector machines, Naïve Bayes, etc. NLP techniques include part of speech, bag-of-words modeling, word embeddings, etc. The performance of those models measured by micro-averaged F1 score ranged between 0.4 and 0.76,(Calvo et al., 2017) and by macro-averaged F1 score ranged between 0.5 and 0.84.(Shing et al., 2018) A macro-averaged score computes the metric independently for each risk level (class) and then takes the average across all levels regardless of the number of samples in each risk-level group, whereas micro-average treats each post equally regardless of class. Thus, a macro-averaged score carries more per-post weight for those risk levels (categories) with fewer posts.

In this study, we hypothesized that we can develop advanced data-driven predictive models that can predict individual suicide risk level from longitudinal posts on Reddit discussion forums.

Our study has three key contributions. First, we developed 10 feature domains based on NLP and feature engineering, described in Section II.2, including clinical findings and semantic role labeling (those were not commonly included in previous shared tasks competition for social media data(Shing et al., 2018)) for the prediction of suicide risk from Reddit posts. Second, we developed several state-of-the-art machine learning models including deep neural network models for the prediction task. Third, we developed a modeling strategy for improving prediction accuracy.

II. Methods

This section describes study datasets, text preprocessing, feature engineering, predictive modeling, and evaluation metrics.

II.1 Datasets

We received two datasets from the CLPsych2019(Zirikly et al., 2019): 1) a training dataset and 2) a test dataset. Both datasets comprised annotated posts on the Reddit discussion form and its sub-discussion forms, also known

Proceedings of the Sixth Workshop on Computational Linguistics and Clinical Psychology, pages 162–166
Minneapolis, Minnesota, June 6, 2019. ©2019 Association for Computational Linguistics

as *subreddits*. The training dataset study period is between 2005 and 2015, comprising 31,553 posts from 496 Reddit users with the cohort definition: a user had at least one post on the *SuicideWatch* subreddit; users who posted on the SuicideWatch may not be of risk to suicide. The data elements in the training dataset included a user id, a subreddit name, a post title and body from the user's posts in any subreddit, and post timestamp in a unified time zone. The CLPsych2019 organization provided the gold standard for the training dataset.(Shing et al., 2018; Zirikly et al., 2019) Following the same cohort definition, the test dataset comprised 9,610 posts from 125 Reddit users. We received the training and test datasets one month and five days before the competition deadline, respectively.

The study is approved under the Children's Hospital of Philadelphia IRB.

II.2 Natural Language processing and Feature Engineering

II.2.1. Text preprocessing

We performed a series of preprocessing pipeline including sentence splitting, tokenization, removal of stop words, part of speech tagging, and lemmatization.(Posada et al., 2017)

II.2.2 Feature domains from users' posts

Similar to the work by Shing et. al.(Shing et al., 2018), we developed the following feature domains:

Clinical findings: A social media post may contain clinical findings such as depression, schizophrenia, cancer, etc. We utilized the clinical Text Analysis and Knowledge Extraction System (cTakes)(Savova et al., 2010) developed by the Mayo clinic, to extract clinical findings from each post. cTAKES extracts each finding with a Concept Unique Identifier (CUI) represented in the standard Unified Medical Language System (UMLS) developed by the National Library of Medicine (NLM). We also flagged suicide attempt related CUIs (SA CUIs) using a pre-defined CUI list from our previous suicide attempt study with electronic health records (EHR).(Tsui et al., 2019)

Social determinants of health (SDOH): We classified each sentence into one or more of the 11 social categories that we previously developed.(Quan et al., 2019; Liu et al., 2019) The 11 categories included: 1) social environment, 2) education, 3) occupation, 4) housing, 5) economic, 6) health care, 7) interaction with legal system, 8) social support circumstances and social network, 9) transportation, 10) spirituality and 11) other (e.g., exposure to disaster, war, other hostilities, and access to weapons, etc.).

Emotion and health-disorder association: We identified posts' lemmas that matched terms in the Word-Emotion Association Lexicon developed by Mohammad et. al.(Mohammad and Turney, 2013), as well as a lexicon compiled from terms available in the list of psychological disorders(,). We identified words in a post associated with emotion categories, e.g., joy, sadness, fear, etc.

Readability score: Readability score provides a gauge for the level of understanding of a document. We used spaCy library to calculate 7 readability scores for each post: (1) automated readability index, (2) Coleman-Liau index, (3) Dale-Chall index, (4) Flesch-Kincaid grade level, (5) Flesch-Kincaid reading ease index, (6) forecast index and (7) smog index.

Semantic role labeling (SRL): SRL is a linguistic process that identifies semantic roles, e.g., subject, object and verb, of a sentence. We used two latest state-of-the-art statistical SRL models: Bidirectional Long Short-Term Memory (BiLSTM) model(He et al., 2017) and the Embeddings from Language Models (ELMo)(Peters et al., 2018), which provides deep contextualized words representations, to identify the semantic role labels and predicate-argument structure from each sentence in a user's post. The identified predicate-argument information indicates detailed semantic structure and roles, i.e., "who" did "what" to "whom" at "where" and "when". Table 1 shows an example. SRL plays a critical role for revealing self-referential thinking.

Table 1. Semantics analysis of a sample sentence from a Reddit forum. The right column in the table demonstrates the identified argument labels (subject and object labels), predicate and negation labels from the sentence on the left column after applying SRL process; the *arg0* tag, the *arg1* tag, and *argm-negation* tag represent the subject "I", the object "the loneliness and pain", and the sentence negation, respectively.

Sentence in a post	Predicate-argument structure
"I can't handle the loneliness and pain anymore."	"arg0": " I", "argm-mod": " ca", "argm-negation": " n't", "predicate": " handle", "arg1": "the loneliness and pain"

Sentiment levels: A sentiment level provides a gauge for the level of sentiment of a sentence. We used Stanford CoreNLP(Manning et al., 2014) to identify 5 sentiments: *"Very Negative"*, *"Negative"*, *"Neutral"*, *"Positive"*, *"Very Positive"* for each post. To create the features, per user, we calculated the following averages: 1) micro average: the sum of all sentiments across all the post of a user divided it by the total number of sentences across those post per that user; 2) macro average: the sum of each post level sentiment vector of a user divided by the total number of post by that use; 3) post-level vector: the sum of all sentiment vectors in a post divided by the total number of the sentence in that post.

Topic modeling: Topic modeling provides an unsupervised-based learning to map each post into a predefined number of topics. We used the unsupervised learning Latent Dirichlet Allocation (LDA) to identify 10, 20 and 30 topics from all the posts.

Empathy topics: We used Empathy text analysis tool to identify 196 pre-defined topics(Fast et al., 2016) from each of the posts, e.g., death, negative emotion, sadness, etc. Each post has an empathy vector, $E_i^{196 \times 1}$, where i represents a post, and each topic, $e_{i,j} \in Z, [0,100]$.

Doc2Vec model: We built a Doc2Vec model via distributed bag of words (DBOW) based on the training Reddit posts, and represented each Reddit post as a 300x1vector.

Aggregate Statistics (AS): We created summary statistics features that characterize users' posting habits. Table 2 summarizes the list.

Table 2. Aggregate statistics based on feature domains

Feature Domain	Statistics at the post and user levels
Clinical Finding	• Individual CUI counts from all posts • Average count of each CUI per post • Average count of each CUI per CUI-post (CUI-post refers to the post with at least one identified CUI) • Total count of distinct CUIs from all posts • Total count of SA CUIs per user • Total count of SA CUI-posts per user (SA CUI-post refers to the post with at least one identified SA CUI) • Total count of distinct SA CUIs per user
Semantic Role Labeling (SRL)	• Average count of each *arg0* and *arg1* per post • Minimum/Maximum counts of each *arg0* and *arg1* in one post • Average count of "negative"-*arg0* per post (An "negative"-*arg0* refers to the arg0 with an *argm-negation* modifier for the predicate as shown in Table 1) • Minimum/Maximum count of each "negative"-Arg0 in one post • Count of distinct *arg0* and distinct *arg1* values per user • Minimum/Maximum count of distinct *arg0* and *arg1* values in one post • Average number of part-of-speech tags (nouns, verbs, adjectives, adverbs, etc.) in the last two years
SDOH	• Total number and percentage of sentences in each social determinants of health category
Forum Posting Behavior (FPB)	• Number of total posts for the user in all subreddits • Number of total posts for the user in in the last two years • Number of weeks with posts to the *SuicideWatch* subreddit • Number of active days between the first and last posts • Average post time difference between 2am (EST) and the post time in the last two years • Average length (characters) of posts in the last two years • Days since last post to the *SuicideWatch* subreddit • proportion of the user's posts containing the word 'edit' in the last two years • Proportion of posts made between 2a and 6m EST • proportion of posts made during weekends (Saturday and Sunday) in the last two years • Maximum number of consecutive weeks in which users' made posts to *SuicideWatch* in the last two years. • All subreddits that the user posted to in the last two years • Number of posts to *SuicideWatch* by week in the last two years (1x104 vector) • Number of posts made by users to *SuicideWatch* in the last two years.
Sentiment	• Proportions of sentiment score at post and sentence levels
Readability	• Averages of 7 readability at post and sentence levels
Emotion	• Average count of each emotion-related term across all posts
Topic modeling	• Average count of each topic across all posts

II.3 Predictive modeling and evaluation

We developed seven machine learning models: Naïve Bayes (NB), gradient boosting (GB), random forest (RF), support vector machine (SVM), and deep neural networks including augmented convolution neural networks (CNN) and long short-term memory neural networks (LSTM). Unlike conventional deep neural networks, we developed augmented deep neural networks included input not only from freetext posts (Doc2Vec) but also the user-level aggregate statistics defined in Section II.2.

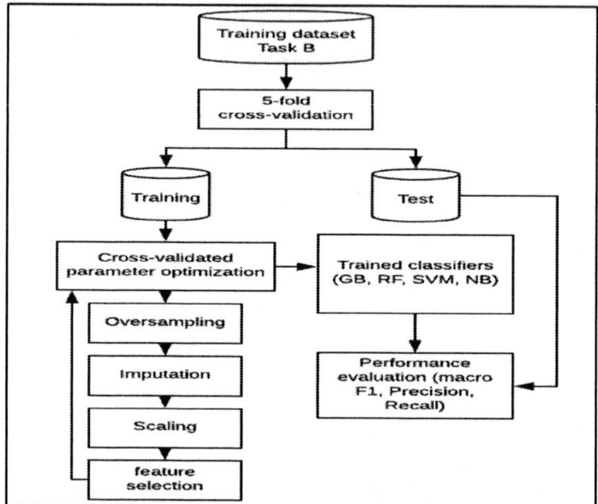

Figure 1. Experimental design and modeling process.

Figure 1 outlines our modeling and evaluation process. We trained and optimized models in a nested 5-fold cross-validation approach, and each model was optimized based on macro-averaged F1 score, which is the main measure for ranking models developed by the shared-task participants. First, we oversampled the sparsely-represented classes to alleviate the existing class imbalance.(Chawla et al., 2002) Then, we conducted imputed missing values with variable means, and either did not scale variables, or scaled values $\in R$ to [0, 1]. Then we performed a two-phase feature selection process. First, we applied a correlation-based feature-selection filter(Hall, 1999), and then conducted a forward greedy search over an increasing number of features selected based on information gain feature ranking.(Tsui et al., 2017)

For the competition, each team was limited to submit up to three models' results, we chose top two models and added an ensemble model based on our three best-performing models. We used the 5-fold average of macro-averaged F1 scores to evaluate each model. The models used to submit results to the competition were re-trained with the full training dataset following the same approach used during cross-validation.

We created an ensemble classifier from our best 3 performing models. Predictions from these models were used to tally votes and generate the final predictions of the ensemble classifier. Since there were more risk categories (4) than the number of classifiers (3) in the ensemble, it is possible that all models produce different predictions. In this scenario, we created a rule by favoring the classes that were likely to be misclassified.

Besides macro-averaged F1, our evaluation metrics include macro-averaged accuracy, precision and recall. We further compared the performance based on binary classifications, i.e., flagged risk (low, moderate, and severe risks) vs. no risk, and urgent risk (moderate, and severe risks) vs. others.

III. Results

Table 3. Risk level distributions in two datasets.

	Training Dataset	Test Dataset
No risk	127 (25.6%)	32 (25.6%)
low risk	50 (10.08%)	13 (10.4%)
moderate risk	113 (22.78%)	28 (22.4%)
Severe risk	206 (41.53%)	52 (41.6%)
Number Subreddits covered	3662	1593

Table 3 shows the distributions of users in 4 different risk categories in the training and test datasets. Both datasets have low counts in the low risk level and share almost the same distribution.

Table 4. Average 5-fold predictive model performance from the <u>training</u> dataset, measured by the macro-averaged F1 score followed by the number of variables (features) used by a model in parentheses.

	NB	GB	RF	SVM	CNN	LSTM
Marco-F1 score	0.422	0.412	0.395	**0.432**	0.367	0.147
# of variables	75	100	100	100	796	796

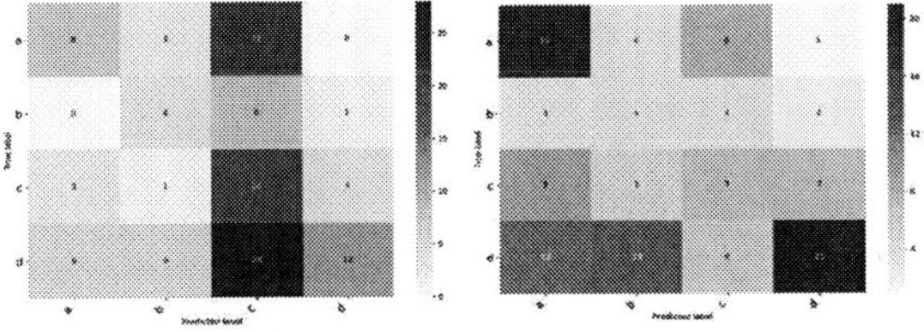

Figure 2. Confusion matrix of the NB model (left) and the SVM model (right).

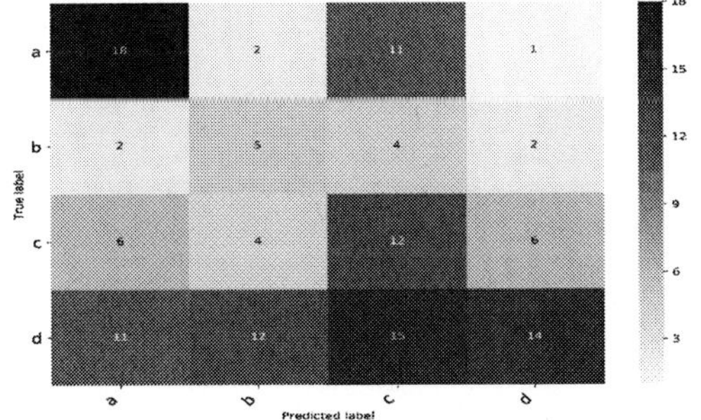

Figure 3. Confusion matrix of the ensemble model.

Our macro-averaged F1 scores from the training dataset ranged from 0.147 to 0.43. Table 4 summarizes the performance of the 6 models. SVM and NB had best F1 scores. Based on these results, we applied three models to the test dataset: SVM, NB, and an ensemble model built from the top three models: NB, SVM, and GB. The rule for breaking the tie in the ensemble model was to set the order of the preference: B (highest), C, A, then D (lowest).

Figures 2-3 show the confusion matrices of the 3 models, and Table 5 summarizes the performance of the three models submitted to the competition. These models' macro-averaged F1 scores on the holdout test dataset ranged from 0.338 to 0.379. The ensemle model had the best macro-F1 score 0.379, which was ranked 3rd among the particpating teams for this shared task competition.(Zirikly et al., 2019)

Table 5. Model performance from the test dataset. Level A-D represent no risk, low risk, moderate risk, and severe risk, respectively.

	NB	SVM	Ensemble
Marco-F1 score (4 risk levels)	0.338	0.370	**0.379**
Accuracy	0.352	**0.408**	0.392
F1 score (Flagged vs. no risk)	**0.836**	0.789	0.818
F1 score (Urgent vs. non-Urgent)	**0.736**	0.603	0.648
Level-A Precision/Recall/F1	0.471/0.250/0.327	0.442/0.594/0.507	0.486/0.562/**0.522**
Level-B Precision/Recall/F1	0.286/0.308/**0.296**	0.154/0.308/0.205	0.217/0.385/0.278
Level-C Precision/Recall/F1	0.260/0.714/**0.381**	0.280/0.250/0.264	0.286/0.429/0.343
Level-D Precision/Recall/F1	0.706/0.231/0.348	0.677/0.404/**0.506**	0.609/0.269/0.373

Table 6. Top 10 features from the feature space

Rank	Domain	Feature Description
1	SRL	Max. count of arg1 with value '*I*' in one post
2	SRL	Max. count of arg1 with value '*me*' in one post
3	FPB	Number of posts to SuicideWatch in the last two years
4	FPB	Number of weeks with any SuicideWatch posts in the last two years
5	SRL	Max. count of arg1 with value '*myself*' in one post
6	Empathy	Max. value of negative emotion in a post
7	SRL	Average count of arg1 with value '*I*'
8	Emotion	Average count of '*disgust*'-related terms across all posts
9	Empathy	Max. value of '*death*' topic across all posts
10	FPB	Max. number of SuicideWatch posts in any week in the last two years

The NB model had the best performance in two additional binary-classification tasks, i.e., no risk vs. flagged risk (any risk level other than no risk) with F1 score 0.836 and no or low risk vs. urgent risk (moderate or severe risk) with F1 score 0.736.

We started modeling from a total of 7,603 features from 10 feature domains in Section II.2, and Table 6 lists top 10 features from the whole training dataset ranked in the order of information gain. Among the top 100 features, there were 35 clinical finding features, 25 Empathy features, 17 SRL features, 14 user post-pattern features from forum posting behavior (FPB), 6 Readability features, and

3 Emotion features. Among 17 SRL features, 6 of them were related to self-referencing.

IV. Discussion and Limitations

In this study, we developed a wealth of structured features from longitudinal freetext posts, built 6 state-of-the-art machine learning models, and tested 3 models in a test dataset from the CLPsych2019 organizers. We demonstrated that data-driven machine learning models identified users with risk of suicide based on their Reddit posts. The SVM model had best macro-averaged F1 score for classifying 4 categories of suicide risk, which could be attributed by its hyperspace parameters and nonlinearity; the NB model had accurate macro-averaged F1 scores for classifying binary groups: flagged vs. no risk, and urgent risk vs. non-urgent risk groups. The NB performance may be attributed by its simple assumption and a relatively smaller number (75) of variables compared with others.

Based on the top 100 features used by the SVM model, we found that SRL, Empathy, Readability, Clinical findings, and user post patterns identified in FPB were important for classification. Most importantly, our top findings revealed that frequent self-referencing like 'I', 'me', and myself' (ranked 1, 2, 5, 7, 19) and negated self-referencing (ranked 35) posed an elevated risk as illustrated in literatures.(Burke et al., 2017; Quevedo et al., 2016)

On the other hand, LDA topic modeling, sentiment analysis, and social determinants of health did not play critical roles for classification in our experiments. We attributed its low impact due to the variety of subreddits in the cohort, which possibly makes it challenge to effectively group certain topics for classifying suicide risk levels. Our sentiment tool was based on the context of movie reviews, which may not be applicable to the suicide prediction task from Reddit posts. For social determinants of health, we built the model based on clinical data, which may be limited for social media data.

The oversampling strategy for model training improved predictive performance. Our conjecture is that oversampling enables a classifier to better tune its parameters for those rare occurrences.

The deep neural networks (CNN and LSTM) did not perform well. Both DNNs employed all the features identified in the feature engineering section. The potential explanation is that there were limited number of users in low and moderate risk levels and there were many input variables. Another factor we may consider in the future is the development of more complicated DNN structure and/or the use of multiple DNNs to catch the temporal, wide variety of feature space, and system non-linearity.

V. Conclusions

In this study, the ensemble model had best macro-averaged F1 score, and Naïve Bayes performed best for identifying users with flagged or urgent suicide risk based on longitudinal posts on Reddit discussion forums in conjunction with features from clinical findings, empathy categories, semantic role labeling, user post-patterns, readability, and emotion.

Correspondence: Fuchiang (Rich) Tsui: *tsuif@chop.edu*

References

Taylor A Burke, Samantha L Connolly, Jessica L Hamilton, Jonathan P Stange, Y Lyn, and Lauren B Alloy. 2017. Two Year Longitudinal Study in Adolescence. , 44(6):1145–1160.

Rafael A. Calvo, David N. Milne, M. Sazzad Hussain, and Helen Christensen. 2017. Natural language processing in mental health applications using non-clinical texts. *Natural Language Engineering*.

Nitesh V Chawla, Kevin W. Bowyer, Lawrence O. Hall, and W. Philip Kegelmeyer. 2002. SMOTE: synthetic minority over-sampling technique. *Journal of Artificial Intelligence Research*, 16:321–357.

Munmun De Choudhury, Emre Kiciman, Mark Dredze, Glen Coppersmith, and Mrinal Kumar. 2017. Discovering shifts to suicide ideation from mental health content in social media. In *Proc SIGCHI Conf Hum Factor Comput Syst.*, pages 2098–2110.

Ethan Fast, Binbin Chen, and Michael Bernstein. 2016. Empath: Understanding Topic Signals in Large-Scale Text. Technical report.

Ma Hall. 1999. Correlation-based feature selection for machine learning. *Diss. The University of Waikato*.

Luheng He, Kenton Lee, Mike Lewis, and Luke Zettlemoyer. 2017. Deep Semantic Role Labeling: What Works and What's Next. In

Lindsay M Howden and Julie A Meyer. 2011. 2010 Census Briefs; Age and sex composition: 2010. Technical Report May.

Haixia Liu, Lingyun Shi, Neal Ryan, David Brent, and Fuchiang Rich Tsui. 2019. Developing an annotation guideline for the classification of social determinants of health from electronic healthrRecords. Technical report.

Christopher Manning, Mihai Surdeanu, John Bauer, Jenny Finkel, Steven Bethard, and David McClosky. 2014. The Stanford CoreNLP Natural Language Processing Toolkit. In *Proceedings of 52nd Annual Meeting of the Association for Computational Linguistics: System Demonstrations*, pages 55–60.

Saif M. Mohammad and Peter D. Turney. 2013. Crowdsourcing a word-emotion association lexicon. In *Computational Intelligence*.

NIMH. 2018. Suicide.

Matthew E. Peters, Mark Neumann, Mohit Iyyer, and Matt Gardner. 2018. Deep contextualized word representations. In *Proceedings of NAACL-HLT 2018*, pages 2227–2237.

Jose D. Posada, Amie J. Barda, Lingyun Shi, Diyang Xue, Victor Ruiz, Pei-Han Kuan, Neal D. Ryan, and Fuchiang (Rich) Tsui. 2017. Predictive Modeling for Classification of Positive Valence System Symptom Severity from Initial Psychiatric Evaluation Records. *Journal of Biomedical Informatics*.

Wei Quan, Haixia Liu, Lingyun Shi, Neal Ryan, David Brent, and Fuchiang Tsui. 2019. Classifying social determinants of health from electronic health records using deep neural networks. Technical report.

Karina Quevedo, Rowena Ng, Hannah Scott, Jodi Martin, Garry Smyda, Matt Keener, and Caroline W. Oppenheimer. 2016. The neurobiology of self-face recognition in depressed adolescents with low or high suicidality. *J Abnorm Psychol.*, 125(8):1185–1200.

Guergana K Savova, James J Masanz, Philip V Ogren, Jiaping Zheng, Sunghwan Sohn, Karin C Kipper-Schuler, and Christopher G Chute. 2010. Mayo clinical Text Analysis and Knowledge Extraction System (cTAKES): architecture, component evaluation and applications. *Journal of the American Medical Informatics Association*, 17(5):507–513, September.

Han-Chin Shing, Suraj Nair, Ayah Zirikly, Meir Friedenberg, Hal Daumé III, and Philip Resnik. 2018. Expert, Crowdsourced, and Machine Assessment of Suicide Risk via Online Postings. In *Proceedings of the Fifth Workshop on Computational Linguistics and Clinical Psychology: From Keyboard to Clinic*, pages 25–36.

Fuchiang Rich Tsui, Victor Ruiz, Amie Barda, Ye Ye, Srinivasan Suresh, and Andrew Urbach. 2017. Retrospective and Prospective Evaluations of the System for Hospital Adaptive Readmission Prediction and Management (SHARP) for All-Cause 30-Day Pediatric Readmission Prediction Children ' s Hospital of Pittsburgh of UPMC , Pittsburgh , PA. In *AMIA 2017*.

Fuchiang Rich Tsui, Lingyun Shi, Victor Ruiz, Neal Ryan, Candice Biernesser, and David Brent. 2019. A large-data-driven approach for predicting suicide attempts and suicide deaths. In *The 17th World Congress of Medical and Health Informatics*, Lyon, France.

Wikipedia. Reddit wikipedia.

Ayah Zirikly, Philip Resnik, Ozlem Uzuner, and Kristy Hollingshead. 2019. {CLPsych} 2019 Shared Task: Predicting the Degree of Suicide Risk in {Reddit} Posts. In *Proceedings of the Sixth Workshop on Computational Linguistics and Clinical Psychology: From Keyboard to Clinic*.

Psychological disorders.

Suicide Risk Assessment on Social Media:
USI-UPF at the CLPsych 2019 Shared Task

Esteban A. Ríssola[1*], **Diana Ramírez-Cifuentes**[2*], **Ana Freire**[2], and **Fabio Crestani**[1]

[1]Faculty of Informatics, Università della Svizzera italiana, Switzerland
[2]Web Science and Social Computing Research Group, Universitat Pompeu Fabra, Spain
{esteban.andres.rissola, fabio.crestani}@usi.ch
{diana.ramirez, ana.freire}@upf.edu

Abstract

This paper describes the participation of the USI-UPF team at the shared task of the 2019 Computational Linguistics and Clinical Psychology Workshop (CLPsych2019). The goal is to assess the degree of suicide risk of social media users given a labelled dataset with their posts. An appropriate suicide risk assessment, with the usage of automated methods, can assist experts on the detection of people at risk and eventually contribute to prevent suicide. We propose a set of machine learning models with features based on lexicons, word embeddings, word level n-grams, and statistics extracted from users' posts. The results show that the most effective models for the tasks are obtained integrating lexicon-based features, a selected set of n-grams, and statistical measures.

1 Introduction

According to the Center for disease Control and prevention (CDC) there is one death by suicide in the United States every twelve minutes (Stone et al., 2018). Worldwide, suicide is one of the main causes of death for those with ages between 15 and 29 years old, and Europe is the continent with the highest suicide mortality rate according to the World Health Organisation (WHO) (WHO, 2016). People requiring hospital admission for treatment of mental disorders are particularly at high risk (Mortensen et al., 2000). According to the WHO, the role of major depression in suicide is strong, having been present in 65-90% of the cases with psychiatric pathologies (WHO, 2016).

Despite having brought many advantages to society, the Web has also contributed negatively to some aspects, such as easing the access to information on how to commit suicide or stigmatising people suffering from mental disorders (Biddle et al., 2008). An evident case of these are the sites created to promote suicide or eating disorders, such as anorexia and bulimia nervosa. In fact, the link between mental health issues and social media usage has lead researchers to work on the development of automated methods to detect different mental disorders, like depression (Guntuku et al., 2017). Furthermore, several works have studied and characterised the behaviour of individuals affected by mental disorders based on the analysis of the data they generate online (De Choudhury et al., 2013; De Choudhury, 2015; Prieto et al., 2014).

This paper describes a set of models to address the shared task tracks defined at the CLPsych2019. Our approach is built upon a set of features based on psychological processes, word embeddings, and statistical and linguistic information extracted from the users' posts. Different machine learning algorithms are tested to generate models suitable for the risk assessment and screening of suicidal ideation. Our team participated in the three tasks proposed by the CLPsych2019 organisers.

The remainder of this paper is organised as follows: Section 2 describes the tasks and the dataset distributed for the shared task. Section 3 outlines the features engineering process undertook. Experimental setup is reported in Section 4, followed by the results and findings in Section 5. Finally, conclusions are summarised in Section 6.

2 Tasks and Data

The CLPsych2019 shared task goal is to study different variations on the assessment of suicide risk from online postings (Zirikly et al., 2019). To this end, the organisers propose three tasks, in which participants are asked to determine a user's degree of suicide risk based on the textual content of the posts they have produced. The

---

* These two authors equally contributed to this work

Proceedings of the Sixth Workshop on Computational Linguistics and Clinical Psychology, pages 167–171
Minneapolis, Minnesota, June 6, 2019. ©2019 Association for Computational Linguistics

main difference between the tasks concerns to the information available from each user, *i.e.*, partial or complete access to a user's posting history.

The data used in the shared task comprises of a collection of posts retrieved from Reddit[1], an online site for anonymous discussion on a wide variety of topics. Positive instances of suicidality, that are users at risk of suicide, were collected based on their participation in a discussion forum called *SuicideWatch* (SW). This corpus, known as the University of Maryland Reddit Suicidality Dataset (Shing et al., 2018), includes posts from more than $11,000$ users who posted at least once on SW and a comparable number of control users who did not.

A subset of the users who posted in SW were labelled by human annotators using a four point scale, including no risk, low risk, moderate risk, and severe risk, summarised as follows: (a) **No Risk (or "None")**: I do not see evidence that this person is at risk for suicide. (b) **Low Risk**: There may be some factors here that could suggest risk, but I do not really think this person is at much of a risk of suicide. (c) **Moderate Risk**: I see indications that there could be a genuine risk of this person making a suicide attempt. (d) **Severe Risk**: I believe this person is at high risk of attempting suicide in the near future.

A total of 993 users comprises the training set and 248 the test set. A summary of the shared task training dataset is shown in Table 1. It should be noted that ethical review criteria discussed in (Zirikly et al., 2019) had to be met in order to gain access to the dataset.

	Labels				
	a	**b**	**c**	**d**	**control**
# of Users	127	50	113	206	497
# of Posts	10,662	2,715	5,726	12,450	25,462
Avg. # of Posts/User	83.95	54.30	50.67	60.43	51.23
Avg. # of Words/Post	63.20	111.25	89.69	82.29	37.30
Avg. # of Subredd./User	27.96	22.18	20.89	20.99	13.35

Table 1: Summary of CLPsych 2019 training dataset.

3 Feature Engineering

Our approach relies on features based on psychological processes, depression related vocabulary, word embeddings and linguistic information extracted from the users' posts.

The main objective of our models is to predict the suicide risk of users based on their posts. To build our predictive models we use a set of features extracted from the concatenated posts of the users. Later, we test different combinations of these features along with some statistical machine learning methods such as Logistic Regression, Support Vector Machines and Decision Trees. In addition, we use chi-square test (Forman, 2003) as a feature selection method, which allows us to identify the most predictive n-grams for each risk level. The same features were extracted for the models of tasks A, B and C. They are described in the next sections.

3.1 Bag of words and N-grams

These type of features have been previously used for detecting depression (Tsugawa et al., 2015; Schwartz et al., 2014) and eating disorders (Ramírez-Cifuentes et al., 2018). We apply a $tf.idf$ vectorisation of (1-5)grams at a word level with the training set posts. To do so, we use the *TfIdfVectorizer* from the *scikit-learn* Python library[2]. We choose not to remove stop-words given that self-references have been proved to be predictive for depression screening (Guntuku et al., 2017). However, we remove the n-grams that appeared in less than five documents to reduce the feature space. We consider a document as the concatenation of the text in all the posts of a user. Therefore, each user is represented by a single document.

3.2 Word embeddings

We use GloVe (Pennington et al., 2014) pre-trained word embeddings. The embedding representation of the words found in each document are averaged column-wise to obtain a k-dimensional representation. In particular, we select the embeddings with 200 dimensions.

3.3 Lexicon-based features

Lexicon-based features are selected according to the frequency of words belonging to all the categories of the LIWC2007 dictionary (Pennebaker et al., 2008). We consider the frequency of terms for each category, and also test a model normalising these frequencies by the total number of words in the posts of a user. As in (Pennebaker et al., 2008), a list of antidepressants (TJ and

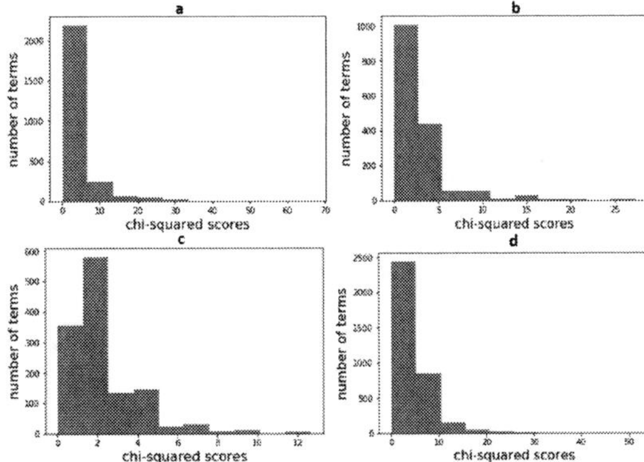

Figure 1: Number of terms per X^2 score bin for task A. The same approach was considered to select features for tasks B and C.

DR, 2017) and absolutist words categories are added. We based our work on (Al-Mosaiwi and Johnstone, 2018), who stated that the elevated use of absolutist words is a marker specific to anxiety, depression, and suicidal ideation.

3.4 Statistical features

We use as predictive features the following: total number of posts per user, size of all the users' posts given by the average post size, total number of subreddits in which each user posted and number of posts of a user per subreddit (available only for Tasks B and C).

3.5 Features Selection

Since using (1-5)grams generates a large feature space, we conduct a chi-square test (X^2) (B. S Harish, 2017) to discard those n-grams which are most likely to be irrelevant for the classification. To this end, we first rank the (1-5)grams according to how predictive they are with respect to each class. Subsequently, we analyse the distribution of the number of n-grams per score for each class, and define a threshold on the number of features to select based on this. Figure 1 depicts the number of n-grams per X^2 score bin for task A. As we observe, most of the n-grams in each category have low scores. Since the number of n-grams have been distributed in ten score bins, we choose a set of bins with the highest scores per class. The same approach is followed for tasks B and C. For task A we choose 807 n-grams, for task B 871 n-grams, and for task C 1, 596.

4 Experimental Setup

4.1 Pre-processing

We perform several text pre-processing steps prior to feature extraction in order to reduce the noise in the original posts. To this end, we use a Python library called *ekphrasis* (Baziotis et al., 2017). This library is tailored towards text from social media sites. The tool performs tokenisation, word normalisation, word segmentation (for splitting hashtags) and spell correction, using word statistics. Furthermore, it applies different regular expressions, in addition to the ones already normalised by the task organisers to extract particular units, such as percent, money, phone, number, etc., and separates them from the rest of the tokens.

We decide to keep the stop-words since words such as pronouns, articles and prepositions reveal part of people's emotional state, personality, thinking style and connection with others. As a matter of fact, such words that are called *function words*, account for less than one-tenth of one percent of an individual's vocabulary but constitute almost 60 percent of the words a person employs (Chung and Pennebaker, 2007).

4.1.1 Classifiers

We train different models combining the features proposed in Section 3 in various ways. Since the three shared task tracks are multi-class classification problems we decide to follow a One-vs-All (OvA) strategy. This approach, provides a way to leverage binary classification.

In particular, we have four possible classes, one for each suicide risk level. The OvA strategy consists in fitting four separate binary classifiers, where each class is fitted against the remaining ones. One of the main advantages of this method is its interpretability. Given that each class is represented by a single classifier, it is possible to inspect each corresponding classifier and gain knowledge about each class in particular.

We chose two different classification algorithms, Logistic Regression (LR) and Linear Support Vector Machine (SVM). To this end, we use the scikit-learn library implementation of both methods and set the corresponding parameter to perform OvA training. L2 regularisation is employed to avoid overfitting. In addition to the LR and SVM classifiers, we evaluated a Random Forest classifier. However, the performance was not competitive compared with the other methods

Task/Model	a			b			c			d		
	Precision	Recall	F_1	Precision	Recall	F_1	Precision	Recall	F_1	Precision	Recall	F_1
A: LR_Reduced_LIWC_Stats	28.10	28.10	28.10	00.00	00.00	00.00	47.60	35.70	40.80	42.40	53.80	47.50
B: LR_Reduced	28.60	31.20	29.90	45.50	38.50	41.70	11.80	07.41	08.90	40.30	48.10	43.90
C: LR_Reduced_LIWC_Stats	-	-	-	06.20	07.70	06.90	11.50	10.70	11.10	28.60	19.20	23.00

Table 2: Precision, Recall and F_1 per class for the models with the best performance on the test set, according to the macro-average F_1. "Reduced" denotes the n-grams selected following the method described in Section 3.5, "LIWC" corresponds to the lexicon-based features (Section 3.3), "Stats" represents the statistical features (Section 3.4) and, finally, LR stands for Logistic Regression.

and, therefore, we chose not to include in the final submission.

In order to select the best models for each track, we perform 5-fold stratified cross-validation on the training set (993 labelled users). In particular, we use macro-average precision, recall, and F1 to assess each classifier performance, as these are the official CLPsych2019 shared tasks evaluation metrics.

5 Results

Nine different models were selected for our submissions to the shared task. The results obtained for each task on the test set are presented in Table 2. Due to space constraints, we only show the three models that achieved the highest effectiveness for each task. In addition, Table 3 describes the macro average F_1 achieved by each of the models presented in Table 2. In this table, Training refers to the performance on the training set, Test corresponds to the performance on the test set, flagged is a F_1 measure relevant to distinguish users that can be safely ignored (class a) from those that might require attention (classes b, c and d). Urgent is a F_1 measure that identifies users that are at severe risk (classes c and d) from the others.

Task	Training	Test	Flagged	Urgent	Rank
A	47.26	29.10	75.30	70.70	11th
B	52.69	31.10	74.30	66.70	6th
C	37.00	13.67	29.40	27.00	6th

Table 3: Macro Average F_1 achieved by the selected models for each task (Table 2). The results for the training and test sets are presented. Rankings are out of 12 systems submitted for task A, 11 for task B and 8 for task C.

We observe that for Tasks A and C, class b is the hardest to predict. This could be caused by the low number of training samples in comparison with the rest of the classes and also by the fact that, as the level of suicide risk is the lowest one, the vocabulary of these users is not so different from those in class A. The inclusion of additional users' posts from other subreddits (Task B), allowed to increase the performance on class b. Although, it introduced some noise for classes c and d, as the effectiveness decreased while predicting these classes.

Users in class D make use of a vocabulary quite distinctive from the rest of the users. In fact, such vocabulary contributes to the improvement of the performance when SW posts are included. The overall effectiveness decreases by about a 50% when such content is not used to train and test the models (Task C).

Finally, regarding the n-grams selected using X^2, we notice that for task A, the X^2 scores for the predictive n-grams of classes b and c are relatively low compared with the scores obtained for those of class a and d. For task B the lowest scores are obtained by the n-grams corresponding to class c. Finally, for task C we find that "depression" is a unigram which characterise control cases.

6 Conclusions

We presented different machine learning based models for suicide risk assessment on social media. Such models were trained using several features extracted from the text and metadata of the posts generated by Reddit users. We also considered the usage of X^2 as a feature selection method. The results obtained on the test set showed that the most suitable models for the tasks were given by the combination of lexicon-based features, a selected set of n-grams, and statistical measures.

Acknowledgments

This work was partially supported by the Spanish Ministry of Economy and Competitiveness under the Maria de Maeztu Units of Excellence Programme (MDM-2015-0502).

References

Mohammed Al-Mosaiwi and Tom Johnstone. 2018. In an absolute state: Elevated use of absolutist words is a marker specific to anxiety, depression, and suicidal ideation. *Clinical Psychological Science*, 6(4):529–542.

M. B. Revanasiddappa B. S Harish. 2017. A comprehensive survey on various feature selection methods to categorize text documents. *International Journal of Computer Applications*, 164:1–7.

Christos Baziotis, Nikos Pelekis, and Christos Doulkeridis. 2017. Datastories at semeval-2017 task 4: Deep lstm with attention for message-level and topic-based sentiment analysis. In *Proceedings of the 11th International Workshop on Semantic Evaluation (SemEval-2017)*, pages 747–754, Vancouver, Canada.

Lucy Biddle, Jenny Donovan, Keith Hawton, Navneet Kapur, and David Gunnell. 2008. Suicide and the internet. *BMJ (Clinical research ed.)*, 336:800–2.

Cindy Chung and James Pennebaker. 2007. The psychological functions of function words. *Frontiers of social psychology. Social communication*.

Munmun De Choudhury. 2015. Anorexia on tumblr: A characterization study. In *Proceedings of the 5th International Conference on Digital Health, DH '15*, pages 43–50, Florence, Italy.

Munmun De Choudhury, Michael Gamon, Scott Counts, and Eric Horvitz. 2013. Predicting depression via social media. In *Proceedings of the Seventh International Conference on Weblogs and Social Media, ICWSM 2013*, Cambridge, USA.

George Forman. 2003. An extensive empirical study of feature selection metrics for text classification. *J. Mach. Learn. Res.*, 3:1289–1305.

Sharath C Guntuku, David B Yaden, Margaret L Kern, Lyle H Ungar, and Johannes C Eichstaedt. 2017. Detecting depression and mental illness on social media: an integrative review. *Current Opinion in Behavioral Sciences*, 18:43 – 49.

PB Mortensen, E Agerbo, T Erikson, P Qin, and N Westergaard-Nielsen. 2000. Psychiatric illness and risk factors for suicide in denmark. *The Lancet*, 355(9197):9 – 12.

James W. Pennebaker, Cindy K. Chung, Molly Ireland, Amy Gonzales, and Roger J. Booth. 2008. The development and psychometric properties of LIWC 2007. Technical report. UT Faculty/Researcher Works.

Jeffrey Pennington, Richard Socher, and Christopher Manning. 2014. Glove: Global vectors for word representation. In *Proceedings of the 2014 Conference on Empirical Methods in Natural Language Processing (EMNLP)*, pages 1532–1543, Doha, Qatar.

Victor M. Prieto, Sergio Matos, Manuel Alvarez, Fidel Cacheda, and Jose Luis Oliveira. 2014. Twitter: A good place to detect health conditions. *PLOS ONE*, 9(1):1–11.

Diana Ramírez-Cifuentes, Marc Mayans, and Ana Freire. 2018. Early risk detection of Anorexia on social media. In *Internet Science*, pages 3–14, Cham.

H. Andrew Schwartz, Johannes C. Eichstaedt, Margaret L. Kern, Gregory Park, Maarten Sap, David Stillwell, Michal Kosinski, and Lyle H. Ungar. 2014. Towards assessing changes in degree of depression through facebook. In *Workshop on Computational Linguistics and Clinical Psychology: From Linguistic Signal to Clinical Reality*, pages 118–125, Baltimore, Maryland USA.

Han-Chin Shing, Suraj Nair, Ayah Zirikly, Meir Friedenberg, Hal Daumé III, and Philip Resnik. 2018. Expert, crowdsourced, and machine assessment of suicide risk via online postings. In *Proceedings of the Fifth Workshop on Computational Linguistics and Clinical Psychology: From Keyboard to Clinic*, pages 25–36.

Deborah M Stone, Thomas R Simon, Katherine A Fowler, Scott R Kegler, Keming Yuan, Kristin M Holland, Asha Z Ivey-Stephenson, and Alex E Crosby. 2018. Morbidity and Mortality Weekly Report Vital Signs: Trends in State Suicide Rates - United States, 1999-2016 and Circumstances Contributing to Suicide - 27 States, 2015. *MMWR Morb Mortal Wkly Rep*, 67:617–624.

Moore TJ and Mattison DR. 2017. Adult utilization of psychiatric drugs and differences by sex, age, and race. *JAMA Internal Medicine*, 177(2):274–275.

Sho Tsugawa, Yusuke Kikuchi, Fumio Kishino, Kosuke Nakajima, Yuichi Itoh, and Hiroyuki Ohsaki. 2015. Recognizing depression from twitter activity. In *Proceedings of the 33rd Annual ACM Conference on Human Factors in Computing Systems*, CHI '15, pages 3187–3196, Seoul, Republic of Korea.

World Health Organization WHO. 2016. Suicide data.

Ayah Zirikly, Philip Resnik, Özlem Uzuner, and Kristy Hollingshead. 2019. CLPsych 2019 shared task: Predicting the degree of suicide risk in Reddit posts. In *Proceedings of the Sixth Workshop on Computational Linguistics and Clinical Psychology: From Keyboard to Clinic*.

Using Contextual Representations for Suicide Risk Assessment from Internet Forums

Ashwin Karthik Ambalavanan, Pranjali Dileep Jagtap,
Soumya Adhya, Murthy Devarakonda[*]

Arizona State University, Tempe, AZ

Abstract

Social media posts may yield clues to the subject's (usually, the writer's) suicide risk and intent, which can be used for timely intervention. This research, motivated by the CLPsych 2019 shared task, developed neural network-based methods for analyzing posts in one or more Reddit forums to assess the subject's suicide risk. One of the technical challenges this task poses is the large amount of text from multiple posts of a single user. Our neural network models use the advanced multi-headed Attention-based autoencoder architecture, called Bidirectional Encoder Representations from Transformers (BERT). Our system achieved the 2nd best performance of 0.477 macro averaged F measure on Task A of the challenge. Among the three different alternatives we developed for the challenge, the single BERT model that processed all of a user's posts performed the best on all three Tasks.

1 Introduction

Social media has become an important part of everyone's life, and in particular online discussion forums related to mental health provide opportunities for individuals to share their feelings and their state of mind. These self-documented posts are valuable in assessing suicidality and potentially offering interventions. Since the volume of posts and the time sensitivity of potential interventions, automation is critical for monitoring the forums.

The CLPysch 2019 shared task[1,2] used the University of Maryland Reddit Suicidality Dataset, which was developed using data from Reddit, a well-known online site for anonymous discussion forums on a wide variety of topics. As described in Shing et al[3] the data was extracted from the 2015 Full Reddit Submission Corpus, including postings from one specific *r/SuicideWatch* subreddit forum (denoted as SW here), to identify suicidality risk of the post subject. The data contained post id, anonymous user id, timestamp, subreddit name, post title, and post body. The training data contained four labels, denoted by a to d, with increasing risk of suicidality of low risk to the highest risk. The organizers also provided a separate control group of users and their subreddit posts with no known suicidality risk. The challenge involved three different subtasks, and they were:

- Task A: Assess the subject's suicide risk (a, b, c, or d) from a user's SW postings only;
- Task B: Similar to the Task A, but the user's posts from other subreddits are also used;
- Task C: Assess the subject's suicide risk from a user's Reddit posts other than from SW;

Note that the prediction is not for a post but for a user (actually for the subject of a user's posts). The training dataset contained 496 users with as an average of 1.85 posts per user. The test dataset contained 125 users. The gold standard was based on a consensus of human annotators.[3]

The challenge used macro average of the F measures of the four labels, a, b, c, and d as the system performance indicator. The evaluation also provided *accuracy* (number of correct predictions divided by the number of all predictions), and F measures of *flagged* and *urgent* predictions. The flagged predictions measure performance of identifying b, c, d, out of the four labels, and the urgent predictions measure the performance of identifying

---

[*]Contact author: mvd@acm.org

Proceedings of the Sixth Workshop on Computational Linguistics and Clinical Psychology, pages 172–176
Minneapolis, Minnesota, June 6, 2019. ©2019 Association for Computational Linguistics

c and d out of the four labels. We proposed three different methods based on BERT (Bidirectional Encoder Representations from Transformers).[4] An important hypothesis we considered here was if a model that is built for general domain NLP and only fine-tuned with the suicide-related training data performs on suicide prediction. As such, we did not make use of the suicide literature or the theories of suicide in our methods.

2 Methods

2.1 Pre-Processing and Data Preparation

We pre-processed the text from the posts and presented the resulting text (a sequence of words) as the input to the model. This processing was common to all three methods and Tasks, although some steps are only relevant to certain Tasks, as described below.

- Removed stop words (am, the, for, etc.) and punctuations
- Expanded contractions like *couldn't* to *could not* for easier interpretation and to avoid awkward splitting of words.
- Concatenated words from all posts in a sequence of decreasing order of timestamp so that the most recent post is considered first based on the intuition that the latest psychological state of a user is based on his or her most recent post. The first word in most recent post occupies the first word location in the sequence.
- For Task B and C the subreddit name was prepended to the corresponding posts. The intuition was that the subreddit name might provide a clue to the model.

In addition, we also made the following adjustments to the data:

- Class "b" was over sampled to class "a" instance count by random oversampling class b instances. This was because, given its low frequency in the given training data, our preliminary models couldn't identify the class "b" very well.
- For Task B, posts of the control users were ignored from the training set for the simplicity and also because it was not necessary for the model to predict them in the test set.
- For Task C, a None label was used for the control users and the model was trained using 5 labels instead of 4. Then post processing converted all None labels to class 'a' label as per the challenge requirements.

2.2 Bidirectional Encoder Representations from Transformers (BERT)

BERT[4] is a new exciting development in neural network models research, demonstrating significantly improved state-of-the-art performance on various general domain NLP tasks, including text classification. BERT pre-trained model produces sequence (i.e. sentence) level and word level representations, which can be fine-tuned for task-specific outcomes. BERT includes an advanced auto-encoder architecture to generate the representations. The pre-trained BERT model, after fine-tuning for a task, has been shown to perform well on multiple general domain NLP tasks, using only an additional, simple feed-forward network with a softmax layer.[4]

The BERT approach is distinctly different from the most biomedical NLP architectures where word2vec and similar representations of words are used as input that is processed by complex task-specific, heavily-engineered architectures containing Bidirectional LSTMs, RNNs, and CNNs. BERT provides a broadly applicable pre-trained model which only need to be fine-tuned using task-specific training data. BERT uses layers of neural network components known as Transformer encoders to generate representations of input words and sequences. See Figure 1. The Transformer encoders contain layers of bidirectional multi-headed self-attention[5] encoders with residual connection around each layer. Intuitively, an Attention layer produces output (say, a sentence representation) that is based on any arbitrary word positions of the input sequence. See Figure 2. Multi-headed attention can simultaneously optimize for various input combinations.

2.3 BERT-Fine-Tuning

The pre-processed (combined) posts of each user is given as input to the BERT Model with a Linear classifier (SoftMax) to predict the output labels (Fig 1). Fine tuning helps to tune the initial embeddings of BERT to the CLPsych downstream task with the help of error backpropagation. The BERT Implementation in PyTorch[6] was used to implement this architecture. The configuration that gave the best results on a validation subset of the training data was a maximum sentence length of

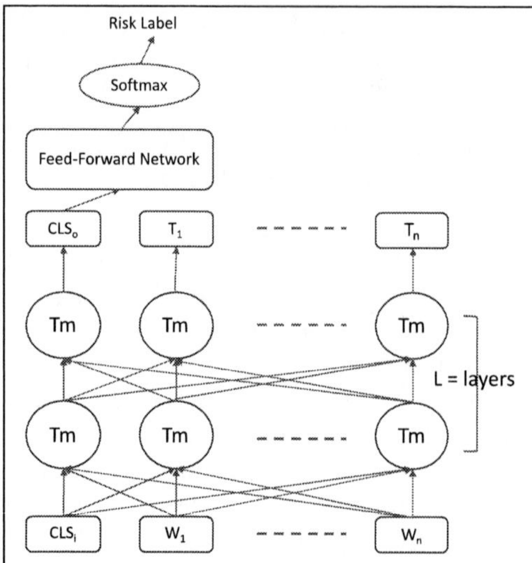

Figure 1. BERT architecture along with the feed-forward network and the softmax layer

384 tokens with a batch size of 16 and 75 epochs. Note that no theories of suicide or lexicons specific to suicide were used in the model. Instead we let BERT learn task-specific lexical clues through fine tuning on all of a user's text and make predictions on unseen posts.

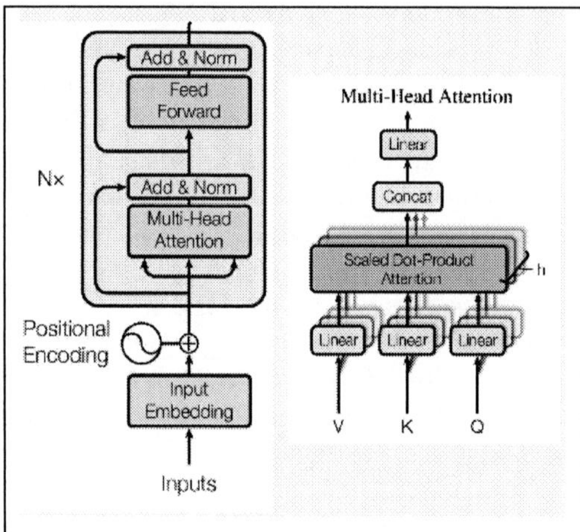

Figure 2. The transformer architecture used in BERT, with acknowledgements to Vaswani et al[5]

2.4 BERT-Sentence-Embedding + BiLSTM + ATTN

The sentences of the combined posts were extracted using NLTK. The representation for each of the sentences was obtained from the pre-trained BERT model. The [CLS$_o$] BERT output, as shown in Figure 1, provides this representation. All sentence representations of a user were concatenated

as a sequence and fed into a BiLSTM+Attention[7] model with a linear projection layer and a softmax layer. The maximum number of sentences per user was set as 50 and the model was run for 200 epochs with a batch size of 10. The intuition behind this model is to see if sentence level aggregation at the input results in better learning and prediction from a user's posts.

2.5 BERT-Multiple-Instance-Learning

Pre-Processing of a user's posts was the same as before except that the posts, ordered in decreasing order of timestamps, were kept separate. Each post was separately processed by the fine-tuned BERT model (from Task A) and a post-level representation was produced at the [CLS$_o$] output. Multiple independent BiLSTM+Attention[7] models analyzed the post-level representations. The output vectors from the BiLSTM+Attention models were concatenated, fed to a Linear projection layer with ReLU activation and dropout of 0.4, and then finally to a linear projection layer with a softmax for classification.

We configured this multi-instance learning model with five BiLSTM+Attention models because an average user had at most 5 posts. If a user had more than 5 posts, all older posts (after 5 posts) were ignored and if user had less than 5 posts, nulls were fed as input to the corresponding models.

The model takes time to fit to the data and gives poor results after the first 5 epochs but at around 10 epochs it tends to learn better and then overfits rather quickly. Thus, the use of the dropout layer was very important in this implementation to prevent overfitting. Our best configuration for this model ran 20 epochs with a batch size of 16. The intuition behind this model is to retain the word level input but to aggregate decisions from separate models each of which analyzes a single post.

3   Results

The results, determined by the organizing team from our system output, for our models and the Tasks are shown in Table 1. Generally, across the broad, the BERT-Fine-Tuning model achieved the best results in our experiments. BERT-Multi-Instance-Learning model performed close to the fine-tuning model, and in fact outperformed it on the flagged metric. We used the BERT-Sentence-Embedding model only for Task A, and its performance was the lowest of our models.

Table 1. Evaluation results for the methods proposed in this this study on the test dataset as reported by the Challenge organizing team.

Tasks	Measures	BERT-Fine-Tuning	BERT-Sentence-Embedding + BiLSTM+Attn	BERT-Multi-Instance-Learning
Task A	Macro F meas.	**0.477**	0.418	0.421
	Accuracy	**0.544**	0.512	0.504
	Flagged F meas.	0.882	0.875	**0.891**
	Urgent F meas.	**0.826**	0.795	0.812
Task B	Macro F meas.	**0.261**	---	0.169
	Accuracy	**0.368**	---	0.200
	Flagged F meas.	**0.765**	---	0.577
	Urgent F meas.	**0.691**	---	0.286
Task C	Macro F meas.	**0.159**	---	0.143
	Accuracy	**0.597**	---	0.153
	Flagged F meas.	**0.630**	---	0.455
	Urgent F meas.	**0.575**	---	0.342

Table 2. **Task A** results for all participants. The highest scores in each metric are underlined, and our system performance was shown in bold-italic.

Team	Macro F meas.	Accuracy	Flagged F meas.	Urgent F meas.
CLaC	<u>0.481</u>	0.504	<u>0.922</u>	0.776
ASU	*0.477*	*0.544*	*0.882*	*0.826*
HLAB	0.459	0.560	0.842	0.839
Text2Knowledge	0.445	0.544	0.852	0.789
CAMH	0.435	0.528	0.897	0.783
ttu	0.402	0.504	0.902	0.844
Affective_Computing	0.378	<u>0.592</u>	0.920	<u>0.862</u>
cmu	0.373	0.472	0.876	0.773
jxufe	0.364	0.464	0.882	0.779
UniOvi-WESO	0.312	0.512	0.897	0.821
usiupf	0.291	0.376	0.753	0.707
ibm_data_science	0.178	0.432	0.861	0.788

In the challenge our BERT-Fine-Tuning model achieved the 2nd place on Task A (See Table 2). However, our models did not fare well on Task B and C, in part because we could not invest enough time and resources to analyze and/or enhance our models for these tasks in the short time that was available. Across all participating teams, the best performance on Task B was 0.451 macro F measure and Task C it was 0.268. So, the performance of all participating systems was low on Task C.

4 Discussion and Conclusion

The methods we proposed here did not make use of any suicide-specific domain knowledge and yet our system finished close second on Task A. This is significant because it indicates that the BERT model that uses transfer learning from general domain NLP can perform well on a domain-specific dataset[8] after only fine-tuning for the specific domain. It suggests that the new generation of auto-encoder architectures, with pre-trained models, can potentially reduce the need for domain-specific features, lexicons, and pre-training. However, it would be interesting to explore the possibility of further customization to the domain and improvement such models can achieve from it.

In developing the methods for this task, which we mapped the task to text classification, one of the challenges was how to deal with large input text. For example, for the Task A, 27% of users had more than one post, and 4% of users had more 5 posts. Maximum length of a post (in the Task A training set) was 8457 words, and 124 posts had more than 512 words. Task B input is even larger since all posts are included not just SW. So, since our primary method (BERT-Fine-Tuning) used stop-word eliminated, truncated sequences up to 384 words as the input, we attempted to include larger parts of a user's posts with the other two methods. However, the results did not show improvement over the primary method, indicating that either the most recent words are the most important or our secondary methods do not represent larger text well.

In conclusion, this study applied the state-of-the-art, general domain, pre-trained neural network model, BERT, and achieved good performance on a domain-specific task. Future research includes error analysis, improvement of our methods with and without the use of domain-specific knowledge.

References

1. CLPsych 2019. http://clpsych.org/shared-task-2019-2/. Published 2019.

2. Zirikly A, Resnik P, Uzuner Ö, Hollingshead K. CLPsych 2019 Shared Task: Predicting the Degree of Suicide Risk in Reddit Posts. In: *Proceedings of the Sixth Workshop on Computational Linguistics and Clinical Psychology: From Keyboard to Clinic.* ; 2019.

3. Shing H-C, Nair S, Zirikly A, Friedenberg M, Daumé III H, Resnik P. Expert, crowdsourced, and machine assessment of suicide risk via online

postings. In: *Proceedings of the Fifth Workshop on Computational Linguistics and Clinical Psychology: From Keyboard to Clinic.* ; 2018:25-36.

4. Devlin J, Chang. M-W, Lee K, Toutanova K. BERT: Pre-training of Deep Bidirectional Transformers for Language Understanding. *arXiv Prepr arXiv181004805 (accepted to NAACL 2019).* 2018.

5. Vaswani A, Shazeer N, Parmar N, et al. Attention Is All You Need. In: *31st Conference on Neural Information Processing Systems (NIPS 2017).* Long Beach, CA; 2017. doi:10.1017/S0952523813000308

6. PyTorch Pre-Trained BERT. https://github.com/huggingface/pytorch-pretrained-BERT. Published 2019.

7. Dandala B, Joopudi V, Devarakonda M. Adverse Drug Events Detection in Clinical Notes by Jointly Modeling Entities and Relations using Neural Networks. *Drug Saf.*:0-2. doi:10.1007/s40264-018-0764-x

8. CLPsych 2017: Triaging content in online peer-support forum. http://clpsych.org/shared-task-2017.

An Investigation of Deep Learning Systems for Suicide Risk Assessment

Michelle Morales[1], Danny Belitz[2], Natalia Chernova[1], Prajjalita Dey[1], Thomas Theisen[3]
[1]IBM Chief Analytics Office
[2]IBM GBS Client Innovation Center Benelux
[3]IBM Rochester Data Science Team
{michelle.morales, prajjalita.dey, thomas.theisen}@ibm.com
danny.belitz-cic.netherlands@ibm.com
ncherno@us.ibm.com

Abstract

This work presents the systems explored as part of the CLPsych 2019 Shared Task. More specifically, this work explores the promise of deep learning systems for suicide risk assessment.

1 Introduction

In the United States alone, on average, approximately 1 person every 11 minutes kills themselves (Drapeau and McIntosh, 2017). In addition, the situation is worsened by the fact that 124 million Americans live in areas where there is a shortage of mental health providers (Bureau of Health Workforce, 2017). Meta-studies have shown that the ability to predict suicide attempts has been near chance for decades, and researchers have argued for the necessity to dedicate research efforts to approaches based on machine learning (Walsh et al., 2017). Machine learning systems which predict suicide risk have the potential to improve identification of people with heightened suicide risk.

This work is part of the 2019 CLPsych Shared Task[1](Zirikly et al., 2019), which focuses on predicting someone's degree of suicide risk using posts they have made on the public forum Reddit. In this paper, we present our team's results from the Shared Task. Specifically, in this work, we focused on two main objectives. The first objective is the exploration of deep learning systems for this particular task. Deep learning systems have demonstrated high performance in various NLP tasks, including text classification, however as is highlighted in past work (Shing et al., 2018), have yet to outperform more shallow machine learning models, such as Support Vector

Machines (SVM). In this work, we focus on exploring various deep learning architectures, including convolutional neural networks, long short-term memory networks, and neural network synthesis. We find that deep learning models can outperform more traditional machine learning systems for suicide risk assessment. In addition to exploring the promise of deep learning for risk assessment, we also present results for novelly tested features for this particular task.

2 Dataset

This work leverages the data provided by the 2019 CLPsych Workshop organizers (Zirikly et al., 2019). Our team's use of this data and participation in these tasks met the ethical review criteria discussed in Zirikly et al. (2019). The dataset includes a series of Reddit users who have posted on the r/SuicideWatch subreddit, with annotations from one of the following four categories: (a) No Risk, (b) Low Risk, (c) Moderate Risk, and (d) Severe Risk. For any models performing within the scope of Task A, the dataset solely includes r/SuicideWatch posts. The Task B dataset includes all of the r/SuicideWatch posts as well as each of the users' posts on any other subreddit. The Task C dataset only looks at the non-SuicideWatch posts for these same users. The dataset includes a post identifier, a user identifier, timestamp, subreddit name, title of the post, and body of the post.

3 Feature Engineering

3.1 Preprocessing

Preprocessing steps were dependent on task and model necessity. However, an overview of general preprocessing steps adopted across many of the systems included the following: joining of text title and body, lowercasing text, removal of excess punctuation/URLs/additional symbols, stop word removal, and lemmatization.

---

[1]http://clpsych.org/
shared-task-2019-2/

Proceedings of the Sixth Workshop on Computational Linguistics and Clinical Psychology, pages 177–181
Minneapolis, Minnesota, June 6, 2019. ©2019 Association for Computational Linguistics

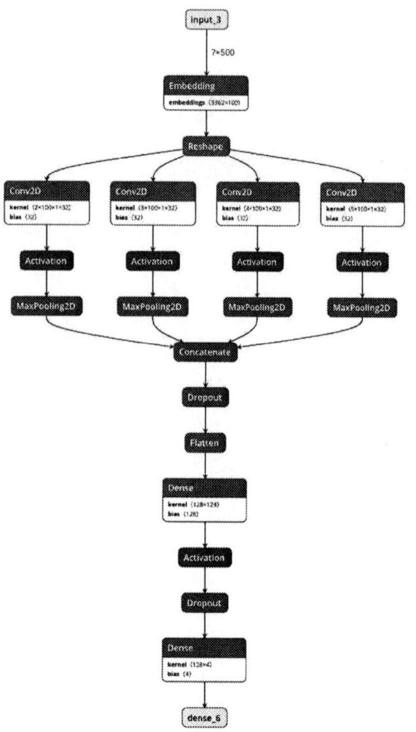

Figure 1: NeuNetS synthesized CNN architecture for Tasks A and B. The only architectural difference between both models is the input dimension.

3.2 Bag of Words

We first apply the above preprocessing steps, and then represent the concatenated post and title as a bag of words vector, including unigrams and bigrams with tf-idf weighting.

3.3 Topics

We use Gensim's LDA library to create topic models for each of the documents, one document being one post. This gave each document a topic distribution, and those distributions were used as features for the final model. We tested a range of number of topics (specifically 10, 20, 30, 40, 50, 100, 150, 200, and 250 topics) and found the macro-average precision, recall, and f1 score to remain the same, so the LDA model is ultimately trained on 50 topics.

3.4 Syntax & Named Entities

We leverage SpaCy's syntactic parser [2] to generate part-of-speech tags (POS) and named entities (NER). POS tags include both coarse-grained POS tags (Google's Universal POS tagset) and fine-grained POS tags (Penn Treebank POS tagset). Counts of each type of tag (for both sets) are taken across each post, and normalized by the word count. For NER tags, counts are taken and normalized by the number of named entities in the document.

3.5 Word Embeddings

Various word embedding architectures are explored. For each type, the same data is used for training specifically the entire task dataset (annotated and unannotated).

Skip-gram: We compute 100-dimensional embeddings for the entire Reddit corpus using a Skip-gram model (Mikolov et al., 2013), window size 5, and ignoring occurrences of words fewer than 5 times.

Retrofitted Skip-gram: For this representation, the trained Skip-gram word embeddings are optimized using the WordNet lexicon. This retrofitting approach is taken from Faruqui et al. (2014), where it was found to help improve performance on text classification tasks.

FastText: We also compute FastText embeddings (Joulin et al., 2016) for the entire Reddit corpus. FastText is an extension to the Word2Vec Skip-gram model. However, instead of training on individual words, FastText breaks words into several n-grams (sub-words). This helps capture morphological patterns and overcomes the limitation of Skip-gram when facing out-of-vocabulary words.

3.6 Novel Features

To the best of the our knowledge, the following set of features have yet to be explored for suicide risk assessment and/or screening.

Personality features: We leverage the IBM Watson Personality Insights API [3] to extract raw scores and percentiles for a variety of personality characteristics, including the Big Five (agreeableness, conscientiousness, extraversion, emotional range, and openness), as well as Needs (e.g. excitement, harmony, etc.) and Values (e.g. conservation, hedonism, etc). Important to note, that the API requires a sufficient amount of data to be provided about a user to extract personality features, namely at least 100 words per user to receive any results, at least 300 words to receive statistically significant results, but preferably even more

---

[2] https://spacy.io/

[3] www.ibm.com/watson/services/personality-insights/

System	P	R	F1
Task A			
SVM (Skip-gram)	.41	.38	.36
CNN (Skip-gram)	.38	.35	.34
NeuNetS	**.51**	**.64**	**.57**
Task B			
kNN (Personality)	.33	.33	.32
LSTM (Tone)	.42	.40	.41
NeuNetS	**.49**	**.47**	**.48**
Task C			
RF (Big 5 only)	.38	.34	.31
kNN (Big 5 only)	.33	.33	.32
kNN (Big 5 + Values)	.33	.33	.32

Table 1: Evaluation phase results. Results are reported on a 20% held out portion of the training dataset. Macro precision (P), recall (R), and F1-score reported. Only top 3 systems are reported.

- 600 or 1200 words per user. Given this limitation, these features are only explored for Task C, the screening task, where the most data about a user is given.

Tone features: We leverage the IBM Watson Tone Analyzer [4] to extract tone measures with corresponding weights (13 measures in total). The tone measures fall into 3 categories: emotion (anger, disgust, fear, joy, sadness), language (analytical, confident, tentative), and social (openness, conscientiousness, extroversion, agreeableness, emotional range). The tone measures include both the document and sentence level. The document level measures are an aggregation of the individual sentence level tone measures. Analysis on the sentence level provides insight into the range in each tone weight across the whole text body.

4 Systems

Systems are trained for three specific tasks. Two of the tasks (Task A and Task B) focus on risk assessment. The third task (Task C) focuses on screening. In addition, all tasks focus on predicting risk at the user level.

4.1 Linguistic & Personality Classification Models

Four sets of features are included in the linguistic-based system: topic distributions, syntax features, NER features, and tf-idf vectors. The various feature sets are concatenated together to train mod-

els at the post level. Majority voting is then used to aggregate the post predictions to the user level. Various machine learning algorithms are explored including: Random Forest (RF), Naive Bayes, k-Nearest Neighbors (kNN), and linear SVM. Given the imbalanced distribution across class labels, oversampling of the minority classes are performed using the SMOTE technique (Lemaître et al., 2017). During the evaluation phase, the RF model performs marginally better than the rest of the models and is therefore used as the model in the final linguistic-based system. These models are explored for Task A only. For the Personality-based models similar algorithms are explored with different subsets of the personality features tested.

4.2 Deep Learning Classification Models

4.2.1 Convolutional Neural Network

The goal of this system was to explore the potential of a Convolutional Neural Network (CNN) for risk assessment. As is highlighted in the task dataset paper (Shing et al., 2018), CNNs have been shown to be effective in many NLP tasks, especially in text classification problems. However, in past work, CNNs have not outperformed more shallow systems for suicide risk assessment. We evaluate the potential of CNN models for this task and explore the impact of various different word embedding inputs. The systems we built using CNNs focus solely on Task A, as this task presents the most challenging problem for a deep learning model, i.e. the smallest data size per user, on average ~1.8 posts per user. CNNs are built using Keras [5] and parameters are optimized using Hyperas [6]. All CNN models are trained on the post-level; user level predictions are made by averaging across the classes' probability distributions, choosing the risk label with the highest probability.

4.2.2 Long Short-Term Memory Network

The goal of this system is to transform a Reddit user's history of posts into a sequence of tone weights over time. This system was used solely for Task B. Tone data was extracted at the document level. The date/time range in post activity for each user varied widely. Some users appeared to be new to the website, while other users had been active on Reddit for years. To partially correct for

---

[4] www.ibm.com/watson/services/
tone-analyzer

[5] https://keras.io/
[6] https://github.com/maxpumperla/
hyperas

System	Accuracy	Macro F1	Flagged F1	Urgent F1
Task A				
CNN (Skip-gram)	.52	.31	.89	.83
NeuNetS	.43	.18	.86	.79
RF (Linguistic)	.40	.15	.83	.76
Task B				
LSTM (Tone)	.42	.30	.79	.75
NeuNetS	.42	.21	.82	.74
kNN (Personality)	.34	.28	.75	.67
Task C				
kNN (Big 5 + Values)	.44	.17	.55	.46
kNN (Big 5)	.42	.18	.49	.41
RF (Big 5)	.44	.12	.51	.47

Table 2: Results on CLPsych 2019 test set.

this issue only the 10 most recent posts were considered for each user. Another issue arises that the length of time between a users' most recent post and their 10th most recent post is not uniform. Thus, any relationship between a tone feature and time is not easily explained. Ultimately, for each user their features are the set of tone weights extracted on their set of maximum 10 posts. Many users had fewer than 10 posts, thus their input data was padded with zeros to maintain a constant input shape. Sequence classification modeling was performed by way of Long-Short Term Memory (LSTM) neural network. The model was utilized to predict user risk of suicide based on each users series of tone data and the corresponding risk level label for the user.

4.2.3 Neural Network Synthesis

In addition to exploring CNNs and LSTMs, we also explore Neural Network Synthesis (NeuNetS). The main objective of NeuNetS (Sood et al., 2019) is to speed up the design of a deep neural network architecture for text or image classification by synthesizing the best deep learning model for a particular dataset. NeuNetS has two main stages: Coarse-grained synthesis and fine-grained synthesis. Based on the data provided, coarse-grained synthesis automatically optimizes and determines the overall architecture of the network - how many layers there should be, how are they connected and so on. The novel step of fine-grained synthesis enables NeuNetS to take a deeper dive into each layer optimizing the individual neurons and connections, e.g. what kind of convolution filter should be applied, and which neurons and edges should be optimized. NeuNetS

is explored for both Tasks A and B. Specifically, the goal of these systems were to explore the potential for leveraging a model like NeuNetS to build a strong system for these particular tasks. As model input, the NeuNetS models take the full text (title and body) of users and generates its own word embeddings. The system is trained on the post-level; therefore, predictions for all posts of one user are aggregated into one final label to assess risk for a specific user by majority voting and choosing the higher risk label in case of a tie. The final model architecture can be seen in Figure 1.

5 Results

Results from the evaluation phase can be seen in Table 1. Although various combinations were explored, only the Top 3 systems are reported. In the evaluation phase, we explored various feature sets as well as standard and deep learning type classification models. We also explored post level vs. user level training. For both Tasks A and B, we found the NeuNetS systems to perform the highest, reporting a macro F1-score of .57 and .48 respectively. In addition, we found systems trained at the post level to outperform user-based systems.

To further test the robustness of our systems, the Top 3 performing systems are evaluated on the test set. Results from the test phase can be seen in Table 2. These results are reported for predictions made on an unseen test set which were evaluated by the Shared Task organizers. We find the CNN and the LSTM models to perform best across Tasks A and B. Unexpectedly, NeuNetS reports a low F1-score. Although NeuNetS has many procedures in place to prevent overfitting, such as dropout and regularization, it seems that it still faces the same challenges as more manually designed deep learning architectures. We believe, by design, NeuNetS is more suitable for classification tasks trained on large and balanced data sets (e.g. for text classification the training file size limit is 5GB). For Task A the training data for each label was below the minimum required to train a robust model using NeuNetS. Furthermore, the training data provided for Tasks A and B was imbalanced, providing almost 5 times more labelled posts for label *d* than for label *b*. During training this might cause the model to steer in the wrong direction. This, plus the fact that NeuNetS trains word embeddings on the input alone might be a reason that the resulting model overfits to the training data. Even though various techniques are included in NeuNetS to reduce overfitting, the training data might just be too imbalanced and too small to be a suitable use case for NeuNetS. Also interestingly, for the NeuNetS system, majority voting did not allow for any predictions of labels *b* or *c* although they appeared as intermediate results for some posts. Hence the macro-average F1 score for tasks A and B are rather low. Alternative ways to aggregate might improve these results, e.g. by averaging the confidence scores that are returned for each label. Although we see unexpected results for NeuNetS, we find other deep learning designs to perform well in the tasks, such as the results for the CNN and LSTM systems. These results suggest there is still promise in pursuing deep learning systems for tasks that face data size challenges, such as suicide risk assessment.

References

Bureau of Health Workforce. 2017. Designated health professional shortage areas: Statistics, first quarter of fiscal year 2018, designated hpsa quarterly summary. *Health Resources and Services Administration (HRSA) U.S. Department of Health and Human Services.*

Christopher W Drapeau and John L McIntosh. 2017. Usa suicide 2016: Official final data. *American Association of Suicidology.*

Manaal Faruqui, Jesse Dodge, Sujay K Jauhar, Chris Dyer, Eduard Hovy, and Noah A Smith. 2014. Retrofitting word vectors to semantic lexicons. *arXiv preprint arXiv:1411.4166.*

Armand Joulin, Edouard Grave, Piotr Bojanowski, Matthijs Douze, Hérve Jégou, and Tomas Mikolov. 2016. Fasttext.zip: Compressing text classification models. *arXiv preprint arXiv:1612.03651.*

Guillaume Lemaître, Fernando Nogueira, and Christos K Aridas. 2017. Imbalanced-learn: A python toolbox to tackle the curse of imbalanced datasets in machine learning. *The Journal of Machine Learning Research*, 18(1):559–563.

Tomas Mikolov, Ilya Sutskever, Kai Chen, Gregory S. Corrado, and Jeffrey Dean. 2013. Distributed representations of words and phrases and their compositionality. In *NIPS*.

Han-Chin Shing, Suraj Nair, Ayah Zirikly, Meir Friedenberg, Hal Daumé III, and Philip Resnik. 2018. Expert, crowdsourced, and machine assessment of suicide risk via online postings. In *Proceedings of the Fifth Workshop on Computational Linguistics and Clinical Psychology: From Keyboard to Clinic*, pages 25–36.

Atin Sood, Benjamin Elder, Benjamin Herta, Chao Xue, Costas Bekas, A. Cristiano I. Malossi, Debashish Saha, Florian Scheidegger, Ganesh Venkataraman, Gegi Thomas, Giovanni Mariani, Hendrik Strobelt, Horst Samulowitz, Martin Wistuba, Matteo Manica, Mihir R. Choudhury, Rong Yan, Roxana Istrate, Ruchir Puri, and Tejaswini Pedapati. 2019. Neunets: An automated synthesis engine for neural network design. *CoRR*, abs/1901.06261.

Colin G Walsh, Jessica D Ribeiro, and Joseph C Franklin. 2017. Predicting risk of suicide attempts over time through machine learning. *Clinical Psychological Science*, 5(3):457–469.

Ayah Zirikly, Philip Resnik, Özlem Uzuner, and Kristy Hollingshead. 2019. CLPsych 2019 shared task: Predicting the degree of suicide risk in Reddit posts. In *Proceedings of the Sixth Workshop on Computational Linguistics and Clinical Psychology: From Keyboard to Clinic.*

ConvSent at CLPsych 2019 Task A: Using Post-level Sentiment Features for Suicide Risk Prediction on Reddit

Kristen Allen *
Dept of Engineering and Public Policy
Carnegie Mellon University
Pittsburgh, PA 15213
kcallen@cmu.edu

Shrey Bagroy *
Computer Science Dept
Carnegie Mellon University
Pittsburgh, PA 15213
sbagroy@cs.cmu.edu

Alex Davis
Dept of Engineering and Public Policy
Carnegie Mellon University
Pittsburgh, PA 15213
ald1@andrew.cmu.edu

Tamar Krishnamurti
Division of General Internal Medicine
University of Pittsburgh
Pittsburgh, PA 15213
tamark@pitt.edu

Abstract

This work aims to infer mental health status from public text for early detection of suicide risk. It contributes to Shared Task A in the 2019 CLPsych workshop by predicting users' suicide risk given posts in the Reddit subforum r/SuicideWatch. We use a convolutional neural network architecture to incorporate LIWC information at the Reddit post level about topics discussed, first-person focus, emotional experience, grammatical choices, and thematic style. In sorting users into one of four risk categories, our best system's macro-averaged F1 score was 0.50 on the withheld test set. The work demonstrates the predictive power of the Linguistic Inquiry and Word Count dictionary, in conjunction with a convolutional network and holistic consideration of each post and user.

1 Introduction

Psychological distress in the form of depression, anxiety, and other mental health issues can have serious consequences for individuals and society (WHO, 2017). Unfortunately, stigma surrounding poor mental health may prevent disclosure of suicidal ideation. For example, Oexle et al. (2017) found that perceived stigma and the associated secrecy around mental illness were positively linked with feelings of hopelessness and suicidal ideation. McHugh et al. (2019) found that the standard practice of clinicians asking people about suicidal thoughts fails in many cases, as 80% of patients who ultimately died of suicide reported no suicidal thoughts when prompted by their general practitioner.

There is a need to supplement traditional methods for evaluating suicidality that minimize the need for direct disclosure from the individual. Some of those suffering from mental health challenges have adopted social media outlets, such as Reddit's r/SuicideWatch, as a means to cope (Park et al., 2012; Robinson et al., 2016). Recent research finds promising links between an individual's mental well-being and the linguistic content they share on social media (Coppersmith et al., 2014; De Choudhury et al., 2016; Vioulès et al., 2018; Shing et al., 2018).

The Sixth Annual Workshop on Computational Linguistics and Clinical Psychology (CLPsych 2019) includes a shared task on predicting a Reddit user's degree of suicide risk based on their posts in the r/SuicideWatch forum (Zirikly et al., 2019). The task involves assigning a degree of risk (no, low, moderate, or severe) to a user on Reddit based on content they have posted on Reddit. For this task, researchers were given access to the University of Maryland Reddit Suicidality Dataset (Shing et al., 2018), made available with assistance by the American Association of Suicidology. This dataset consists of ~1000 users annotated with the four-level scale, and a larger set of 20,000 unannotated users.

2 Prior work

The baseline deep learning model for classifying suicide risk on Reddit, by Shing et al. (2018), builds on the convolutional neural network (CNN) for language processing as laid out by Kim (2014). Shing et al.'s CNN makes use of unigram word embeddings, concatenated by post and then by user, then constructs an overall user score using

* These authors contributed equally

Proceedings of the Sixth Workshop on Computational Linguistics and Clinical Psychology, pages 182–187
Minneapolis, Minnesota, June 6, 2019. ©2019 Association for Computational Linguistics

Model	Precision	Recall	F1
CNN + GloVe vectors	0.55	0.43	0.42
Affect-only CNN + LIWC	0.53	0.47	0.49
Primary: CNN + all LIWC	0.65	0.55	0.56

Table 1: Average performance of our models in 10-fold cross-validation on the training set

Model	Full F1	Flagged F1	Urgent F1
Primary	0.37	0.88	0.77
Leave none out	**0.50**	**0.90**	**0.82**
Balanced classes	0.41	**0.90**	0.80

Table 2: Performance of our models by macro-averaged F1 on the test set. 'Full F1' indicates score across four classes, while 'flagged' and 'urgent' F1 reflect binary splits between no/some risk and non-severe/severe risk, respectively. All three submitted models use a convolutional network plus all LIWC features.

sliding windows over that sequence. In a separate approach, Shing et al. use an SVM to consider post-level features but make an overall risk assessment based on the most concerning individual post. Neither method incorporates distinct insights from individual posts—where, for instance, a long series of moderately concerning posts might indicate more serious risk. Our model incorporates information from multiple posts within the CNN framework.

We additionally leverage prior social media work (Braithwaite et al., 2016; Coppersmith et al., 2015) that finds suicidality can be predicted from a particular feature set, the Linguistic Inquiry and Word Count (LIWC) dictionary, as distributed by Tausczik and Pennebaker (2010).

3 Methods

All modeling methods were applied to the de-identified Reddit data as part of Shared Task A. Approval from CMU IRB was obtained on March 11 2019, and we adhered to the ethical review criteria laid out by Zirikly et al. (2019).

3.1 Modeling with word embeddings

Convolutional neural networks form the basic architecture for our models. Following Shing et al. (2018) and Kim (2014), we concatenate word embeddings for each word in a post, then concatenate these embedding sequences for all posts in order of occurrence. Our implementation uses pretrained GloVe word embeddings by Pennington

et al. (2014) and code snippets from Neubig et al. (2019).

In both of these experiments, we transform all posts by a user into a two-dimensional array of dimension $num\_total\_words \times embedding\_size$. For the CNN, filter parameters that must be trained are then $window\_size \times embedding\_size \times num\_filters$. Given the small size of the expert-annotated dataset, we next explore ways to reduce the number of features a network needs to train.

3.2 Modeling with post-level features

We next consider post-level features. In this dataset the post body field is often empty, presumably when the post comprises only an image or other embedded media, so features must be robust to this variation. In all subsequent models, each post component (title or body) is represented as a one-dimensional vector of size $num\_post\_features$. Calling each such 1-D vector $\mathbf{x}_{ij}$, we chronologically concatenate these vectors for each post title and non-empty body for user i into a longer 1-D vector:

$$\mathbf{x}_i = \mathbf{x}_{i1} \oplus \mathbf{x}_{i2} \oplus ... \oplus \mathbf{x}_{in}.$$

Thus we represent each user with the concatenated vector of all post features from posts $1 : n$, where n is their total number of post titles and non-empty post bodies. The resulting vector for user i has shape $1 \times (n * num\_post\_features)$. Users are then batched for quicker training. Each user vector is padded to the length of the longest one, resulting in a batch of k user vectors having shape $k \times (n_{max} * num\_post\_features)$. Masking prevents back-propagation of weights to padding vectors.

Others' prior work successfully incorporated LIWC features into suicidality detection (e.g. Lightman et al. (2007)). Thus, we experiment with sets of LIWC features as the summary of each post by a user, then concatenate these features from all of a user's posts. In order to maintain cross-post context while reducing the number of features, the first model considers only features from the 'affect' category. Using just these sentiments appeared likely to predict self-destructive mental state (Kumar et al., 2015). Subsequent models use all 45 features provided in the LIWC dictionary.

We next apply a convolutional neural network to this 1-D sequence of LIWC features. Our network uses the `keras` implementation of a one-dimensional CNN (Chollet et al., 2015), setting both stride length and window size equal to $num\_post\_features$ and using $num\_filters = 10$ filters. This structure means that each window looks at LIWC features from a single post title or body, and extracts relationships between these features into 10 filter representations. The model forgoes pooling (following Springenberg et al. (2014)) in favor of maintaining independent information about each post. Thus, after convolution, the batch of k users with max number of posts n_{max} has shape $k \times (n_{max} * num\_filters)$.

Convolution is followed by a dropout layer setting 30% of input units to 0 at any given timestep, intended to reduce overfitting. The next two layers are fully connected, with 250 and 100 nodes, respectively, and rectified linear activation functions; thus, after passing through the second linear layer, the data has shape $k \times 100$. Finally, labels are generated by a softmax output layer. Training seeks to minimize cross entropy, and uses 10-fold cross-validation (CV) on the training set.

'Affect-only' model

This model uses the four affect categories relating to negative sentiment: 'negative affect,' 'anger,' 'anxiety,' and 'sadness'. We selected this subset as a reasonable approximation of negative valence, and to test its predictive performance without broader information.

'Primary' model

The best-performing model on a set-aside development set serves as our primary model. This model differs from the affect-only model in incorporating all 45 LIWC categories as post features.

'Balanced classes' model

Next, we provide our model with custom weights corresponding to the penalty incurred while misclassifying each class. We provide larger weights for the underrepresented 'low risk' and 'moderate risk' classes to force the model to pay more attention to these categories while training.

'Leave none out' model

This final model used all available data for training. In the primary and balanced models, it was clear that while training set performance continues to improve, development set performance levels off somewhere around 150 epochs. That is, cross-validation results were optimized at epoch 235 for

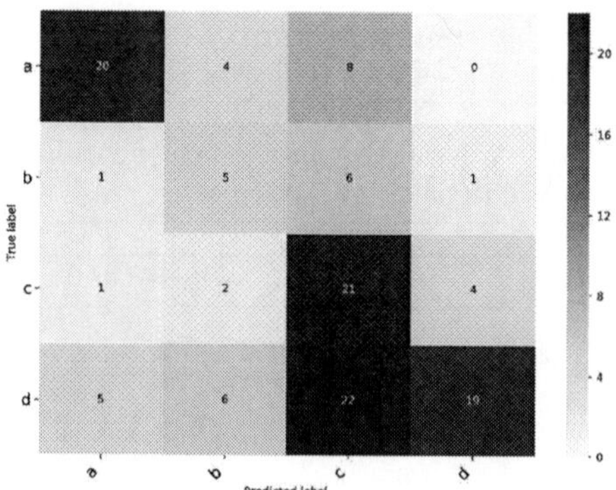

Figure 1: Confusion matrix on the test set from the best-performing model

the primary model, and 67 for the balanced classes model. Taking the average, this system uses the model state after epoch 150 to predict test set results.

Our primary evaluation metric is the resulting macro-averaged F1 score of our models; we report averages on a set-aside development set (see Table 1). For three approaches, we also present macro-averaged F1 scores on an unseen test set.

4 Results

With our initial convolutional network model, using GloVe word embeddings in a convolutional neural net in the style of Kim (2014), we confirm similar performance to Shing et al. (2018) with a macro-averaged F1 score of 0.42. We also find that this model strongly overfits the data; it performs exceptionally well on the training data (F1=0.95) but fails to generalize well on development data (F1=0.42). This overfitting is expected, since the size of our dataset is not sufficient to successfully train large models.

The high overfitting and our model's inability to further learn from the dataset encourage us to focus on simpler models, and to thoughtfully select our features.

The best-performing models all use LIWC features at the post level, concatenated by user, and run through a one-dimensional CNN with stride length and window size equal to the number of features.

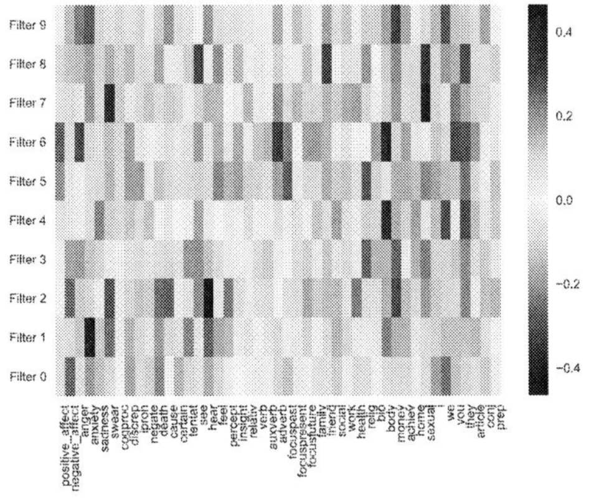
(a) Filter visualizations for each of the 10 filters

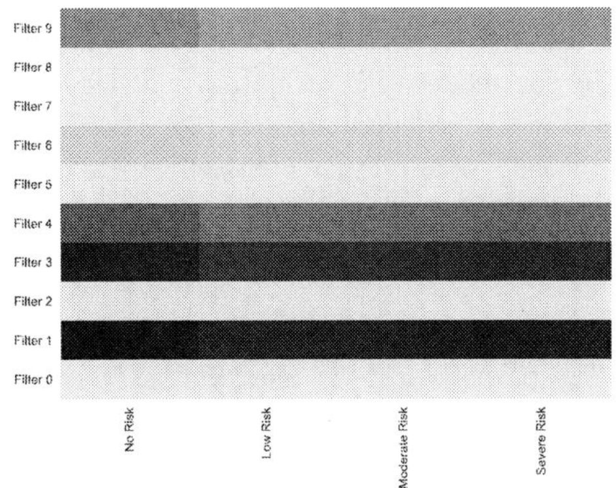
(b) Strength of average alignment between filters and the four classes.

Figure 2: Filters for the best-performing model indicate

4.1 'Affect-only' model

When representing each post as a vector of LIWC affect features, we find that the base model achieves an F1-score of 0.47 in cross-validation. We still find a significant discrepancy between our model's performance on seen/unseen data, indicating that it still suffers from overfitting. We experiment with hyperparameters like dropout and number of filters, finding that a model with 10 filters and 0.3 dropout probability outperforms all our previous models with a macro-averaged CV F1-score of 0.49.

On studying the performance of our model, we find that its behaviour is not uniform across all classes: it does well in labeling 'no risk' and 'severe risk,' but performs poorly in trying to label the intermediate risk categories.

4.2 'Primary' model

We next use variations to improve features provided while still minimizing parameters trained. For our 'primary' model, we provide all 45 LIWC category features to a CNN of the same structure.

In macro-averaging pairwise AUC scores on the development set, this model scores 0.76. On the test set, the model's macro-averaged F1 is 0.37. A random guessing strategy weighted by label frequency would yield F1=0.25.

4.3 'Balanced classes' model

We find that this change boosts the model's CV performance on our development set to an F1 score of 0.57, with a macro-averaged AUC score on the development set of 0.78. We also find that

this model performs more uniformly across the four classes than we see in the previous model, resulting in a slightly better score on the unseen test set, F1=0.40.

4.4 'Leave none out' model

With this final model and feature architecture, we train our model on the entire training dataset available for Task A, stopping after 150 epochs. This model achieves our highest score on the test set, a macro-averaged F1-score of 0.50 on this task–comparing favorably with the best-scoring system, whose F1-score is 0.53. We also note that our model achieves high F1-scores (0.90 and 0.82 respectively) for the 'flagged' and 'urgent' tasks.

This model's final confusion matrix is shown in Figure 1. We find that our model is best at identifying the 'no risk' and 'moderate risk' users, while it miscategorizes 42% of 'severe risk' cases as 'moderate risk' as well. There are fewer 'low risk' users, and about half of these are miscategorized as 'moderate risk' as well.

5 Discussion

5.1 'Affect-only' model

We can attribute this model's difficulty with intermediate labels to our usage of only the negative 'affect' category from LIWC. This category extracts counts for words associated with 'negative_affect,' 'anger,' 'anxiety,' and 'sadness', i.e, words one would typically associate with severe suicidality conditions; presence of (a large number of) these words may be common in Severe risk users, whereas their absence might be a strong indicator of No risk users. Poorer performance in the intermediate categories may indicate inconsistent use of emotion terms by those users, or may suggest a smaller range of variation between those categories as opposed to variation within the extremes.

5.2 'Primary' and 'balanced classes' models

The 'primary' and 'balanced classes' models perform similarly, with a difference in F1 scores of about 0.03. We believe that the latter model is slightly more effective because its higher weights for the intermediate categories counteracted those labels' lower representation in the training set. This is borne out in the model's slightly better performance on those classes: it categorizes $\frac{1}{13}$ of 'low risk' and $\frac{10}{28}$ 'moderate risk' users correctly, whereas the 'primary' model is right about $\frac{0}{13}$ and $\frac{8}{28}$ of such users, respectively. Macro-averaged F1 as the primary metric means that even this slight improvement is significant when comparing the two models.

It seems plausible that, because it was trained for longer, the 'primary' model was more overfitted to the training data. Because we use 10-fold cross-validation to train these models, we also note that both these models are trained using 90% of the training data; we hypothesize this missing 10% of data to be the primary reason that our leave-none-out model outperforms both of these models. A larger training dataset allows the model to "observe" more data, which helps both with getting more training data for under-represented classes (e.g. low and moderate risk) and with generalizing better on all unseen data.

5.3 'Leave none out' model

Difficulty identifying 'low risk' users may be partially explained by the fact that fewer users from the training set were in that class than any other—just 10% of examples were labeled low risk, so there was less opportunity to learn these features.

In Figure 2a, we plot the learned convolutional layer weights from our final model with respect to the input LIWC feature categories, finding that each filter is activated (or deactivated) by a subset of LIWC features. We hypothesize that each filter focuses on learning presence or absence of a particular character trait (or 'sentiment') from each post. For instance, filter 9 is inversely associated with money, anxiety, and 'we,' indicating that someone describing their stress around money would have a negative activation for Filter 9. Seeing a stronger association between Filter 9 and 'no risk,' we can extrapolate that users who are not at risk are less likely to be preoccupied with their financial troubles on r/SW.

While not all subsets are clear, we can observe some patterns. For instance, Filter 2 has the highest positive weights for 'hear,' 'negative_affect,' 'death,' 'percept,' and 'see.' We could hypothesize that a user activating this filter is preoccupied with how they are perceived, and is also considering death (whether their own or that of a loved one). This filter may indicate both a feeling of being observed, perhaps stigmatized, and an experience of suicidal ideation, as discussed by Oexle et al. (2017).

5.4 Findings

Overall, this work demonstrates the power of combining human feature-engineering with deep learning in data-constrained situations. The Linguistic Inquiry and Word Count dictionary, in conjunction with a convolutional network, leads to a holistic consideration of each post and each user, all while reducing the overall number of parameters the network needs to learn. Within the constraints of a relatively small dataset, we find that our best model incorporates engineered features and all available data to outperform a 'baseline' re-implementation of Shing et al. (2018).

5.5 Acknowledgements

This material is based upon work supported by the National Science Foundation Graduate Research Fellowship Program under Grant No. DGE1252522. Any opinions, findings, and conclusions or recommendations expressed in this material are those of the author(s) and do not necessarily reflect the views of the National Science Foundation. Dr. Krishnamurti's time was supported by an Institutional K-award (NIH KL2 TR001856).

References

Scott R Braithwaite, Christophe Giraud-Carrier, Josh West, Michael D Barnes, and Carl Lee Hanson. 2016. Validating machine learning algorithms for Twitter data against established measures of suicidality. *JMIR mental health*, 3(2).

François Chollet et al. 2015. Keras. `https://github.com/fchollet/keras`.

Glen Coppersmith, Mark Dredze, and Craig Harman. 2014. Quantifying mental health signals in Twitter. In *Proceedings of the workshop on computational linguistics and clinical psychology: From linguistic signal to clinical reality*, pages 51–60.

Glen Coppersmith, Ryan Leary, Eric Whyne, and Tony Wood. 2015. Quantifying suicidal ideation via language usage on social media. In *Joint Statistics Meetings Proceedings, Statistical Computing Section, JSM*.

Munmun De Choudhury, Emre Kiciman, Mark Dredze, Glen Coppersmith, and Mrinal Kumar. 2016. Discovering shifts to suicidal ideation from mental health content in social media. In *Proceedings of the 2016 CHI conference on human factors in computing systems*, pages 2098–2110. ACM.

Yoon Kim. 2014. Convolutional neural networks for sentence classification. *arXiv preprint arXiv:1408.5882*.

Mrinal Kumar, Mark Dredze, Glen Coppersmith, and Munmun De Choudhury. 2015. Detecting changes in suicide content manifested in social media following celebrity suicides. In *Proceedings of the 26th ACM conference on Hypertext & Social Media*, pages 85–94. ACM.

Erin J Lightman, Philip M McCarthy, David F Dufty, and Danielle S McNamara. 2007. Using computational text analysis tools to compare the lyrics of suicidal and non-suicidal songwriters. In *Proceedings of the Annual Meeting of the Cognitive Science Society*, volume 29.

Catherine M. McHugh, Amy Corderoy, Christopher James Ryan, Ian B. Hickie, and Matthew Michael Large. 2019. Association between suicidal ideation and suicide: meta-analyses of odds ratios, sensitivity, specificity and positive predictive value. *BJPsych Open*, 5(2):e1.

Graham Neubig, Daniel Clothiaux, Vaibhav, Zhengzhong Liu, Danish Pruthi, and Zhiting Hu. 2019. Code samples from Neural Networks for NLP. `https://github.com/neubig/nn4nlp-code`.

N Oexle, V Ajdacic-Gross, R Kilian, M Müller, S Rodgers, Z Xu, W Rössler, and N Rüsch. 2017. Mental illness stigma, secrecy and suicidal ideation. *Epidemiology and psychiatric sciences*, 26(1):53–60.

Minsu Park, Chiyoung Cha, and Meeyoung Cha. 2012. Depressive moods of users portrayed in Twitter. In *Proceedings of the ACM SIGKDD Workshop on healthcare informatics (HI-KDD)*, volume 2012, pages 1–8. ACM New York, NY.

Jeffrey Pennington, Richard Socher, and Christopher Manning. 2014. GloVe: Global vectors for word representation. In *Proceedings of the 2014 conference on empirical methods in natural language processing (EMNLP)*, pages 1532–1543.

Jo Robinson, Georgina Cox, Eleanor Bailey, Sarah Hetrick, Maria Rodrigues, Steve Fisher, and Helen Herrman. 2016. Social media and suicide prevention: a systematic review. *Early intervention in psychiatry*, 10(2):103–121.

Han-Chin Shing, Suraj Nair, Ayah Zirikly, Meir Friedenberg, Hal Daumé III, and Philip Resnik. 2018. Expert, crowdsourced, and machine assessment of suicide risk via online postings. In *Proceedings of the Fifth Workshop on Computational Linguistics and Clinical Psychology: From Keyboard to Clinic*, pages 25–36.

Jost Tobias Springenberg, Alexey Dosovitskiy, Thomas Brox, and Martin Riedmiller. 2014. Striving for simplicity: The all convolutional net. *arXiv preprint arXiv:1412.6806*.

Yla R Tausczik and James W Pennebaker. 2010. The psychological meaning of words: LIWC and computerized text analysis methods. *Journal of language and social psychology*, 29(1):24–54.

M Johnson Vioulès, Bilel Moulahi, Jérôme Azé, and Sandra Bringay. 2018. Detection of suicide-related posts in Twitter data streams. *IBM Journal of Research and Development*, 62(1):7–1.

WHO. 2017. Policy options on mental health: a WHO-Gulbenkian mental health platform collaboration. Technical report, World Health Organization.

Ayah Zirikly, Philip Resnik, Özlem Uzuner, and Kristy Hollingshead. 2019. CLPsych 2019 shared task: Predicting the degree of suicide risk in Reddit posts. In *Proceedings of the Sixth Workshop on Computational Linguistics and Clinical Psychology: From Keyboard to Clinic*.

Dictionaries and Decision Trees for the 2019 CLPsych Shared Task

Micah Iserman, Taleen Nalabandian, and **Molly E. Ireland**
Department of Psychological Sciences, Texas Tech University, Lubbock, Texas
`first.last@ttu.edu`

Abstract

In this summary, we discuss our approach to the CLPsych Shared Task and its initial results. For our predictions in each task, we used a recursive partitioning algorithm (decision trees) to select from our set of features, which were primarily dictionary scores and counts of individual words. We focused primarily on Task A, which aimed to predict suicide risk, as rated by a team of expert clinicians (Shing et al., 2018), based on language used in SuicideWatch posts on Reddit. Category-level findings highlight the potential importance of social and moral language categories. Word-level correlates of risk levels underline the value of fine-grained data-driven approaches, revealing both theory-consistent and potentially novel correlates of suicide risk that may motivate future research.

1 Introduction

The shared task for this year's CLPsych workshop focused on predicting Reddit users' risk for suicide (none, low, moderate, and severe, as coded by clinical psychologists with suicide expertise) based on language used in their posts (Shing et al. 2018; for a review, see Zirikly et al. 2019). Reddit is a social media website that hosts over 138,000 active forums (or subreddits; as of 2017[1]) in which users can post on any topics of interest.

Social media sites like Reddit, Facebook, and Twitter have increasingly become an important source of data for researchers. Studies have demonstrated how language use in social media posts reflects various psychological processes, ranging from personality (Youyou et al., 2017) to mental health (e.g., postpartum depression; De Choudhury et al., 2014). For instance, Eichstaedt et al. (2018) were able to accurately distinguish depressed patients from non-depressed con-

trols based on Facebook statuses posted before the date of their diagnosis.

Certain language categories have been implicated as markers of mental health conditions (such as anxiety; Dirkse et al., 2015). Relevant to this shared task, suicidal ideation tends to be positively correlated with rates of first-person singular pronoun use (Stirman and Pennebaker, 2001) and negative emotion word use (e.g., anger, sadness; Coppersmith et al., 2016). Self-focused and negative language appear to be associated with psychological distress in general, relating to a variety of mental health issues, such as psychosis (Fineberg et al., 2016), neuroticism (Tackman et al., 2018), and depression (Rude et al., 2004). Notably, self-focused language correlates with psychological distress across a variety of contexts (such as across public Facebook posts; De Choudhury et al., 2014), whereas the use of negative emotional language tends to be limited to more private or intimate contexts (such as in conversations with romantic partners; Baddeley et al., 2012).

Based on previous research, we went into this year's shared task with a particular interest in first-person singular pronouns and overtly negative content words. Although our models cast a wide net, making use of all available lexicons, we expected categories relating to negative affect, self-focus, and social distance to be most predictive of suicide risk, as rated by expert coders.

2 Method

Preprocessing. We first removed any entries not from users in the task A or B sets, or with only "nan" as the post body. This left 11,856 posts from 329 users, which we cleaned automatically in order to (a) standardize encoding, such as for quotation or apostrophe marks; (b) remove some code elements, such as HTML tags or characters; (c)

---

[1] `https://www.redditinc.com/`

Proceedings of the Sixth Workshop on Computational Linguistics and Clinical Psychology, pages 188–194
Minneapolis, Minnesota, June 6, 2019. ©2019 Association for Computational Linguistics

remove some formatting that could make identifying word or sentence boundaries more difficult, such as periods within word; (d) standardize some common typing-related practices, such as repeating characters within some words for emphasis (e.g., "reeeeeaalllly"); and (e) replace some standard formatted elements with tags, such as URLs, references to subreddits, and simple emojis.

After cleaning and tokenizing texts, we applied a spelling correction processes in two phases: First, we applied a more generic version of the process (to be described), and checked its output for (a) miscorrections (such as specialized terminology like "reddit", "macbook", and "moba"), which we added to the list defining correctly spelled words, and (b) frequent misspellings not caught by the process, which we added to a map between correctly spelled words and their misspelled instances. This caught some of the most frequent miscorrections and missed misspellings, but was limited by available time. We applied the process again with these refinements and allowed it to correct the misspellings it identified.

The spelling correction process used the hunspell package (Ooms, 2018) and its US English dictionary to mark words as misspelled (on its own at first, then manually supplemented; only considering words over 3 characters long). The process then measured edit distance (optimal string alignment, calculated with the stringdist package; van der Loo, 2014) between each marked and unmarked (correctly spelled) word found in the text. If a misspelled word was within 2 edit distance of one and only one correctly spelled word, it was considered a matched to that word. If a word was within 1 edit distance of multiple words, these were considered potential matches, and the qgram and soundex distance were calculated between them and the original misspelling—a combination of these new distances and the frequency of the potential matches determined which of these would claim the misspelling (as shown in equation 1, where a is the misspelling, and b is each word in the set of words within 1 edit distance; document frequency is the number of posts in which the word appears).

$$\arg\min_{b\in matches} \frac{qgram(a,b) + soundex(a,b)}{document\,frequency(b)} \quad (1)$$

If a misspelled word did not meet the edit distance criteria, corrections suggested by hunspell

AFINN	Nielsen (2011)
Hu & Liu	Hu and Liu (2004)
General Inquirer	Stone and Hunt (1963)
labMT	Dodds et al. (2011)
LIWC	Pennebaker et al. (2015)
Lusi	Ireland and Iserman (2018)
Moral Foundations	Frimer et al. (2018)
Netspeak	Ireland and Iserman (2019)
NRC	Mohammad (2017)
Senticnet	Cambria et al. (2010)
SentimentDictionaries	Pröllochs et al. (2018)
SentiWordNet	Baccianella et al. (2010)
Slangsd	Wu et al. (2016)
Vader	Hutto and Gilbert (2014)
Whissell	Whissell (1989)
Age and Gender	Sap et al. (2014)
PERMA	Schwartz et al. (2016)

Table 1: Dictionaries/Lexicons.

were considered: If any of these were more frequent than the misspelling, the most frequent of them was considered its correction. Otherwise, if any suggested corrections contained spaces (i.e., the misspelling was suggested to be a combination of words), and if the individual suggested words were all found in the texts, the most frequent combination was taken to be its correction.

Most of the genuine spelling errors appeared to be typing related (e.g., *ddin't*, *favirite*), with other common errors seeming to be formatting related (such as words being combined, or parts of words being appended to others). Other corrections effectively standardized across certain word variants (e.g., forms of *highschool* to *high-school*, words with commonly omitted apostrophes to have apostrophes, or British to English spellings) or casual language (e.g., *wana*, *coulda*).

Features. Table 1 lists the dictionaries we used to score the texts. Those with multiple words or parts of words in single entries had each term searched for exactly in the raw text. Otherwise, terms were searched for in the tokens extracted from all texts, allowing for partial matches when words were marked at the beginning or end with an asterisk (as in the case of dictionaries intended for Linguistic Inquiry and Word Count; LIWC; Pennebaker et al., 2015). We also used LIWC to process its internal 2015 dictionary, prior to which we trimmed 3 or more sequential PERSON tags to 1, as some posts with many tags (such as posts containing code examples) caused entries to overflow.

Many individual categories were nearly identical, so we removed those correlating over .9 with any other category (done iteratively, such that only one of each similar category was retained, preferring to retain LIWC categories). In addition to these pre-built dictionaries, we considered each manually replaced tag (such as those for proper names, subreddits, and emojis) to be its own category, counting their instances up and including them as features. The final set of features included dictionary categories and counts of each token, as well as Language Style Matching (Ireland and Pennebaker, 2010) between (a) each post and the posting user's average language style across all of their posts, and (b) each post and the average language style of the subreddit in which it was posted.

Model. We ended up using a simple recursive partitioning model (as calculated by the rpart package; Therneau and Atkinson, 2018), with all features predicting the ratings for each task (with tasks simply defining the particular posts to be included). For final predictions, we trained each model on the full task specific training data (though the submitted task C model was accidentally trained on the task B data), then aggregated within user, assigning each user the rating that had the largest average probability across their posts (as depicted in Figure 2).

We also briefly considered other models (with a small set of features, selected by their correlations with any rating or the continuous rating scale), such as linear regressions predicting a numeric version of the ratings (with their predictions being binned), separate logistic regressions predicting each category, and multinomial logistic regressions (both with the subset of features, and an elastic net regularized version with all features), as well as a random forest model, but these all either performed worse than our final model in our own testing splits, or seemed to overly capitalize on priors (tending to predict only the most common ratings, even more than our final models). Of course, there are many strategies that might be explored to address the uneven distribution of ratings, but our first step in this brief analysis was to compare the performance of a few different models. We also considered mixed-effects models estimating a per-user intercept adjustment, but these did not work well, at least for task A, since most users had only 1 r/SuicideWatch post.

Task	a	b	c	d	mean	rank
A	.667	.200	.140	.600	.402	6th
B	.000	.000	.000	.591	.148	11th
C		.000	.000	.353	.118	8th

Table 2: F1 scores for each rating level in each task. Rankings out of 12, 11, and 8 for each task respectively.

3 Results and Discussion

As the results on the official test sample depict (Table 2), our models tended to only predict extreme ratings, capitalizing on the prior ratings distributions. Because the model could perform well in each task by identifying features that marked a- or d-rated users (with d being the most common rating; as ratings applied per user, across posts), trees in tasks B and C in particular tended to be very simple. This tendency was exacerbated by the fact that some users had multiple posts, which meant any idiosyncrasies in word use or topics of discussion among prolific posters could be used as a cue for their entire rating level.

In terms of differences in higher-level language dimensions, posts in r/SuicideWatch were more likely to be coded as high risk (category d) if they had higher Clout scores (used *I* more, and *we* and *you* less), talked about family (e.g., *dad*, *grandma*) at relatively low rates, and used less positive affective language (as indexed by sentimentr). With respect to moral language, higher-risk posts referred more often to care (e.g., *help*, *pity*; Moral Foundations Dictionary, Frimer et al. 2018) as well as both vice and virtue, as measured by the General Inquirer lexicons (e.g., *ability*, *burn*). In terms of sentence structure and punctuation, higher risk posts used more periods, fewer parentheses, and more hyperbolic or extreme statements (e.g., *quite*, *extreme*; overstatement, General Inquirer), and fewer third-person singular pronouns (e.g., *him*, *she*; LIWC shehe), relative to lower-risk posts.

At the word level, lower-risk posts (ratings a and b) seem to be more social, including more communicative words (like *called*, *said*, and *told*) and words connoting warmth (such as *comfortable*), more *we*, and specific family references (such as *brother*, *cousin*, and *mom*). Higher-risk posts (ratings c and d) seem to reflect more certainty, finality, or black-and-white thinking (*every*, *anymore*, *anything*, *end*), more focus on physical harm (*knife*, *hurts*) and life or death (*alive*, *die*). Higher risk posts also included a number of negations (*don't, can't, no*; see Weintraub 1989). Swearing (e.g., *fucking*) was indicative of the highest risk level as well, perhaps reflecting intense negative affect or disregard for social norms. Perhaps the most notable and theory-consistent word-level correlates of the highest risk level were self-focused pronouns, including *I, me,* and *myself.* Self-focused pronouns are commonly associated with depression (Rude et al., 2004), suicidality (Stirman and Pennebaker, 2001), or, more broadly, vulnerability to stress (Tackman et al., 2018). See Figure 1 for additional word-level correlates of risk-level ratings.

Some of the linguistic correlates of risk categorization are consistent with our prediction that posts would be viewed as indicating higher suicide risk to the degree that they used more negative and socially distant language. The interpersonal theory of suicide (Van Orden et al., 2010) is a leading psychological model of suicide risk. The theory proposes that people are more likely to attempt or die by suicide to the degree that they feel a thwarted desire to belong, believe they are a burden on their loved ones, and have acquired the capability to die (or no longer fear death). Talking infrequently about family and using fewer third-person singular references that might refer to other people in their lives could reflect social isolation.

Although not predicted *a priori*, the moral language correlates seem to be relatively face valid. People using care-related words from the revised Moral Foundations Dictionary (Frimer et al., 2018) may have simply been requesting help more explicitly than people who did not use words such as *help, mercy,* or *comfort* (Graham et al., 2009; Sagi and Dehghani, 2014). The General Inquirer *vice* and *virtue* categories (Stone and Hunt, 1963) are less intuitive, but discussing basic moral questions of good and evil may reflect the thwarted belonging dimension of the interpersonal theory of suicide (e.g., discussing wanting to be good but disappointing loved ones; Van Orden et al. 2010).

The punctuation categories are less straightforward to interpret. Using more periods and fewer parentheses seems to indicate simpler writing. Others have observed that writing about serious trauma is often better quality than writing about more mundane or lighter-hearted topics, partly due to its less convoluted sentence structures and more straightforward style (Pennebaker, 1997). Perhaps that is some of what experts were decoding in the severe-risk posts: Posts using simpler punctuation may have indicated a more urgent or certain desire to die, and thus were coded as high risk.

4 Conclusion

It is important to remember that the expert coders in Shing et al. (2018) had no more information than we do about these users. We do not know whether the people whose r/SuicideWatch posts comprised this sample have died by suicide since posting, either immediately following an expressed intention to die or later on, related to long-term complications of problems mentioned in their posts. Thus, there are bound to be some false positives in every risk category.

In lieu of additional information, it may be most productive to view these expert ratings as accurate. It could be the case that the main value of tasks like this—where teams aim to find specific linguistic features that correlate with holistic risk annotations—is to find variables that expert clinicians have procedural but not declarative access to in memory or everyday experiences with clients (Schneider et al., 1990). Clinical psychologists often note that they intuit someone's diagnosis or risk at a glance, without being able to easily verbalize what it is about that client that places them in a certain diagnostic category (Hamm, 1988). To the degree that those intuitions are accurate, it would benefit both computational linguists (to bolster the accuracy of predictive models) and clinicians (to improve treatment and diagnosis) if we could determine what behavioral variables are influencing those perceptions—perhaps particularly in the context of noisy, relatively low-fidelity samples of behavior, such as posts in mental heath forums on Reddit.

Figure 1: Word cloud based on posts in r/SuicideWatch, aggregated within user. Words are colored by the rating they most correlate with (*a* = green, *b* = yellow, *c* = orange, *d* = red), sized by correlation size, and shaded by document frequency (lighter words being used by more users).

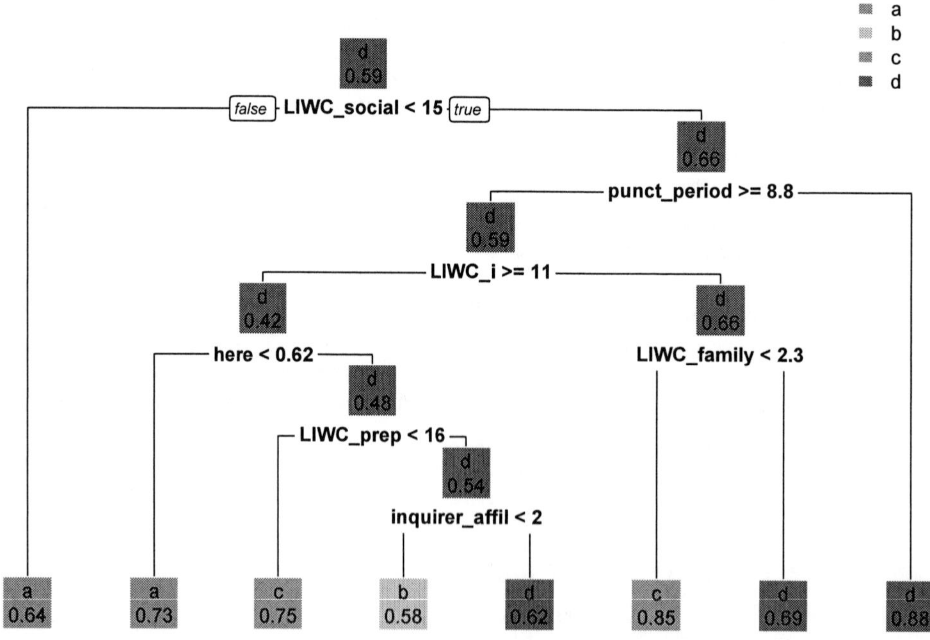

Figure 2: Decision tree fit to the full training set of posts in r/SuicideWatch (task A; in-sample, overall accuracy = 72%, macro F1 = .524). Variables are single words or dictionary categories, and values are percentages of total word count in each post. Each node is colored and labeled by the dominant rating, and displays the probability of that rating in the subset.

References

Stefano Baccianella, Andrea Esuli, and Fabrizio Sebastiani. 2010. Sentiwordnet 3.0: An enhanced lexical resource for sentiment analysis and opinion mining. In *Proceedings of the Seventh conference on International Language Resources and Evaluation (LREC'10)*, Valletta, Malta. European Languages Resources Association (ELRA).

Jenna L Baddeley, James W Pennebaker, and Christopher G Beevers. 2012. Everyday social behavior during a major depressive episode. *Social Psychological and Personality Science*, 4(4):445–452.

Erik Cambria, Robert Speer, Catherine Havasi, and Amir Hussain. 2010. Senticnet: A publicly available semantic resource for opinion mining. In *AAAI Fall Symposium: Commonsense Knowledge*, volume FS-10-02 of *AAAI Technical Report*. AAAI.

Glen Coppersmith, Kim Ngo, Ryan Leary, and Anthony Wood. 2016. Exploratory analysis of social media prior to a suicide attempt. In *Proceedings of the Third Workshop on Computational Lingusitics and Clinical Psychology*, pages 106–117.

Munmun De Choudhury, Scott Counts, Eric J Horvitz, and Aaron Hoff. 2014. Characterizing and predicting postpartum depression from shared facebook data. In *Proceedings of the 17th ACM conference on Computer supported cooperative work & social computing*, pages 626–638. ACM.

Dale Dirkse, Heather D Hadjistavropoulos, Hugo Hesser, and Azy Barak. 2015. Linguistic analysis of communication in therapist-assisted internet-delivered cognitive behavior therapy for generalized anxiety disorder. *Cognitive behaviour therapy*, 44(1):21–32.

Peter Sheridan Dodds, Kameron Decker Harris, Isabel M. Kloumann, Catherine A. Bliss, and Christopher M. Danforth. 2011. Temporal patterns of happiness and information in a global social network: Hedonometrics and twitter. *PLOS ONE*, 6(12):1–1.

Johannes C Eichstaedt, Robert J Smith, Raina M Merchant, Lyle H Ungar, Patrick Crutchley, Daniel Preoţiuc-Pietro, David A Asch, and H Andrew Schwartz. 2018. Facebook language predicts depression in medical records. *Proceedings of the National Academy of Sciences*, 115(44):11203–11208.

SK Fineberg, J Leavitt, S Deutsch-Link, S Dealy, CD Landry, K Pirruccio, S Shea, S Trent, G Cecchi, and PR Corlett. 2016. Self-reference in psychosis and depression: a language marker of illness. *Psychological medicine*, 46(12):2605–2615.

Jeremy Frimer, Jonathan Haidt, Jesse Graham, Morteza Dehgani, and Reihane Boghrati. 2018. *Moral Foundations Dictionary 2.0.*

Jesse Graham, Jonathan Haidt, and Brian A Nosek. 2009. Liberals and conservatives rely on different sets of moral foundations. *Journal of personality and social psychology*, 96(5):1029.

Robert M Hamm. 1988. Clinical intuition and clinical analysis: expertise and the cognitive continuum. *Professional judgment: A reader in clinical decision making*, pages 78–105.

Minqing Hu and Bing Liu. 2004. Mining and summarizing customer reviews. In *Proceedings of the Tenth ACM SIGKDD International Conference on Knowledge Discovery and Data Mining*, KDD '04, pages 168–177, New York, NY, USA. ACM.

C.J. Hutto and Eric Gilbert. 2014. Vader: A parsimonious rule-based model for sentiment analysis of social media text. In *Eighth International Conference on Weblogs and Social Media (ICWSM-14)*.

Molly E. Ireland and Micah Iserman. 2018. *LUSI Lab Development Dictionaries.*

Molly E. Ireland and Micah Iserman. 2019. *LUSI Lab Revised Netspeak Dictionary.*

Molly E Ireland and James W Pennebaker. 2010. Language style matching in writing: synchrony in essays, correspondence, and poetry. *Social Psychological and Personality Science*, 99(3):549–571.

Saif M. Mohammad. 2017. Word affect intensities. *CoRR*, abs/1704.08798.

Finn Årup Nielsen. 2011. A new ANEW: evaluation of a word list for sentiment analysis in microblogs. *CoRR*, abs/1103.2903.

Jeroen Ooms. 2018. *hunspell: High-Performance Stemmer, Tokenizer, and Spell Checker.* R package version 3.0.

James W Pennebaker. 1997. *Opening up: The healing power of expressing emotions.* Guilford Press.

James W. Pennebaker, Ryan L. Boyd, Kayla Jordan, and Kate Blackburn. 2015. The development and psychometric properties of liwc2015. *UT Faculty/Researcher Works*.

Nicolas Pröllochs, Stefan Feuerriegel, and Dirk Neumann. 2018. Statistical inferences for polarity identification in natural language. *PLOS ONE*, 13(12):1–21.

Stephanie Rude, Eva-Maria Gortner, and James Pennebaker. 2004. Language use of depressed and depression-vulnerable college students. *Cognition & Emotion*, 18(8):1121–1133.

Eyal Sagi and Morteza Dehghani. 2014. Measuring moral rhetoric in text. *Social science computer review*, 32(2):132–144.

Maarten Sap, Greg Park, Johannes C Eichstaedt, Margaret L Kern, David J Stillwell, Michal Kosinski, Lyle H Ungar, and H Andrew Schwartz. 2014. Developing age and gender predictive lexica over social media. In *Proceedings of the 2014 Conference on Empirical Methods in Natural Language Processing (EMNLP)*.

Wolfgang Schneider, Joachim Körkel, and Franz E. Weinert. 1990. Expert knowledge, general abilities, and text processing. In Wolfgang Schneider and Franz E. Weinert, editors, *Interactions Among Aptitudes, Strategies, and Knowledge in Cognitive Performance*, pages 235–251. Springer New York, New York, NY.

H Andrew Schwartz, Maarten Sap, Margaret L Kern, Johannes C Eichstaedt, Adam Kapelner, Megha Agrawal, Eduardo Blanco, Lukasz Dziurzynski, Gregory Park, David Stillwell, Michal Kosinski, Martin E P Seligman, and Lyle H Ungar. 2016. Predicting individual well-being through the language of social media. *Pacific Symposium on Biocomputing. Pacific Symposium on Biocomputing*, 21:516—527.

Han-Chin Shing, Suraj Nair, Ayah Zirikly, Meir Friedenberg, Hal Daumé III, and Philip Resnik. 2018. Expert, crowdsourced, and machine assessment of suicide risk via online postings. In *Proceedings of the Fifth Workshop on Computational Linguistics and Clinical Psychology: From Keyboard to Clinic*, pages 25–36.

Shannon Wiltsey Stirman and James W Pennebaker. 2001. Word use in the poetry of suicidal and non-suicidal poets. *Psychosomatic medicine*, 63(4):517–522.

Philip J. Stone and Earl B. Hunt. 1963. A computer approach to content analysis: Studies using the general inquirer system. In *Proceedings of the May 21-23, 1963, Spring Joint Computer Conference*, AFIPS '63 (Spring), pages 241–256, New York, NY, USA. ACM.

Allison M Tackman, David A Sbarra, Angela L Carey, M Brent Donnellan, Andrea B Horn, Nicholas S Holtzman, To'Meisha S Edwards, James W Pennebaker, and Matthias R Mehl. 2018. Depression, negative emotionality, and self-referential language: A multi-lab, multi-measure, and multi-language-task research synthesis. *Journal of personality and social psychology*.

Terry Therneau and Beth Atkinson. 2018. *rpart: Recursive Partitioning and Regression Trees*. R package version 4.1-13.

Mark P.J. van der Loo. 2014. The stringdist package for approximate string matching. *The R Journal*, 6:111–122. Version 0.9.4.7.

Kimberly A Van Orden, Tracy K Witte, Kelly C Cukrowicz, Scott R Braithwaite, Edward A Selby, and Thomas E Joiner Jr. 2010. The interpersonal theory of suicide. *Psychological review*, 117(2):575.

Walter Weintraub. 1989. *Verbal behavior in everyday life.* Springer Publishing Co.

Cynthia M. Whissell. 1989. Chapter 5 - the dictionary of affect in language. In Robert Plutchik and Henry Kellerman, editors, *The Measurement of Emotions*, pages 113 – 131. Academic Press.

Liang Wu, Fred Morstatter, and Huan Liu. 2016. Slangsd: Building and using a sentiment dictionary of slang words for short-text sentiment classification. *CoRR*, abs/1608.05129.

Wu Youyou, David Stillwell, H Andrew Schwartz, and Michal Kosinski. 2017. Birds of a feather do flock together: Behavior-based personality-assessment method reveals personality similarity among couples and friends. *Psychological science*, 28(3):276–284.

Ayah Zirikly, Philip Resnik, Özlem Uzuner, and Kristy Hollingshead. 2019. CLPsych 2019 shared task: Predicting the degree of suicide risk in Reddit posts. In *Proceedings of the Sixth Workshop on Computational Linguistics and Clinical Psychology: From Keyboard to Clinic.*

Author Index

Adhya, Soumya, 172
Aldayel, Abeer, 152
Alexandersson, Jan, 103
Allen, Kristen, 182
Ambalavanan, Ashwin Karthik, 172
Amini, Hessam, 34
Amir, Silvio, 114
Ayers, John W., 114

Bagroy, Shrey, 182
Bar, Kfir, 84
Baram, Heli, 84
Bekoulis, Giannis, 158
Belitz, Daniel, 177
Bernstein, Jared C., 137
Biemann, Chris, 121
Biernesser, Candice, 162
Bitew, Semere Kiros, 158
Bogoychev, Nikolay, 152
Brent, David, 162
Bröcker, Anna-Lena, 126
Brodkin, Edward, 45

Cerezo Menéndez, Rebeca, 148
Chandler, Chelsea, 137
Chen, Lushi, 152
Cheng, Jian, 137
Chernova, Natalia, 177
Cohen, Alex S., 137
Crestani, Fabio, 167

Davis, Alex, 182
Deleu, Johannes, 158
Demasi, Orianna, 1
Demeester, Thomas, 158
Dershowitz, Nachum, 84
Devarakonda, Murthy, 172
Develder, Chris, 158
Dey, Prajjalita, 177
Dredze, Mark, 114

Eckerström, Marie, 103
Elvevag, Brita, 137

Foltz, Peter W., 137

Fraser, Kathleen C., 55
Freire, Ana, 167
Funcke, Jakob, 126

Gayo-Avello, Daniel, 148
Giorgi, Sal, 39
Gong, Tao, 152
González Hevia, Alejandro, 148
Guntuku, Sharath Chandra, 39

Haegert, Erik, 126
Hauser, Michael, 45
Hearst, Marti A., 1
Hollingshead, Kristy, 24
Holmlund, Terje B., 137
Hull, Derrick, 12

Idnani, Akash, 39
Ireland, Molly, 62, 188
Iserman, Micah, 188
Itzikowitz, Samuel, 84

Jagtap, Pranjali Dileep, 172
Johannßen, Dirk, 121
Just, Sandra, 126

Kokkinakis, Dimitrios, 103
Konig, Alexandra, 55
Kořánová, Nora, 126
Kosseim, Leila, 34
Krishnamurti, Tamar, 182

Lee, Fei-Tzin, 12
Levine, Jacob, 12
Levitan, Sarah Ita, 74
Limbachiya, Parth, 39
Lindsay, Hali, 55, 103
Linz, Nicklas, 55, 103
Lundholm Fors, Kristina, 103

Matero, Matthew, 39
McKeown, Kathy, 12
Mohammadi, Elham, 34
Montag, Christiane, 126
Morales, Michelle, 177

Nalabandian, Taleen, 62, 188
Nenchev, Ivan, 126

Parish-Morris, Julia, 45, 94

Quan, Wei, 162

Ramírez-Cifuentes, Diana, 167
Ray, Bonnie, 12
Recht, Benjamin, 1
Resnik, Philip, 24
Rissola, Esteban, 167
Rosenfeld, Elizabeth P., 137
Ruiz, Victor, 162
Ryan, Neal, 162

Sariyanidi, Evangelos, 45
Scheffer, David, 121
Schultz, Robert, 45
Schwartz, H. Andrew, 39
Serper, Mark, 74
Shi, Lingyun, 162
Son, Youngseo, 39
Stede, Manfred, 126
Sterckx, Lucas, 158

Theisen, Thomas, 177
Tsui, Rich, 162
Tunc, Birkan, 45

Uzuner, Ozlem, 24

Vadim Harel, Eiran, 84
Vu, Huy, 39

Zamani, Mohammad, 39
Zampella, Casey, 45
Zaporojets, Klim, 158
Zilberstein, Vered, 84
Zirikly, Ayah, 24
Ziv, Ido, 84
Zomick, Jonathan, 74